Nomadic Narratives

The Thar, which is today divided by an international boundary, has historically been a frontier region connecting Punjab, Multan, Sindh, Gujarat and Rajasthan. *Nomadic Narratives* looks at the desert as a historical region shaped through the mobility of its inhabitants, who were warriors, pastoralists, traders, ascetics and bards, often in overlapping capacities, exchanging mobile wealth and equally mobile narratives.

It challenges the frames of Mughal–Rajput relationships generally employed to explore the histories of the Thar, arguing that Rajputana remains an inadequate category to explore polities located in this frontier region, where along with Rajputs, a range of groups like Charans, Bhils, Meenas, Soomras, and Pathans controlled circulation, and with whom the Rajput states had to constantly negotiate.

The narratives that emerged from Rajput courts, and later from the British administrator-historians, obfuscated the intertwined histories of Rajputs and other groups, giving primacy to the former and ascribing marginality and criminality to the latter. It is only in the oral narratives of these marginalized and criminalized groups that references to shared histories and indeterminate mixed identities are preserved.

Sifting through a wide range of Rajasthani written and oral narratives, travelogues of British administrators, and vernacular as well as English records, *Nomadic Narratives* explores long-term relationships between mobility, martiality, memory and identity in the desert expanses of the Thar.

This book will be useful for scholars, researchers and students of pre-colonial and colonial histories of western India, for those interested in processes of identity formation, as well as for general readers who are interested in knowing an alternative history of Rajasthan.

Tanuja Kothiyal teaches History at Ambedkar University, Delhi, and is at present a Fellow at Nehru Memorial Museum and Library, New Delhi. Her research interests include studying networks of circulation, people, resources and ideas in medieval and early modern western Rajasthan. She is also interested in exploring oral narrative traditions in western India, as ways through which alternate/counter narratives were produced and circulated.

Nomadic Narratives

A History of Mobility and Identity in
the Great Indian Desert

Tanuja Kothiyal

CAMBRIDGE
UNIVERSITY PRESS

4843/24, 2nd Floor, Ansari Road, Daryaganj, Delhi 110002, India

Cambridge University Press is part of the University of Cambridge.

It furthers the University's mission by disseminating knowledge in the pursuit of
education, learning and research at the highest international levels of excellence.

www.cambridge.org
Information on this title: www.cambridge.org/9781107080317

© Tanuja Kothiyal 2016

First published 2016

Printed in India by Thomson Press India Ltd., New Delhi 110001

A catalogue record for this publication is available from the British Library

Library of Congress Cataloging-in-Publication Data
Kothiyal, Tanuja.
Nomadic narratives : a history of mobility and identity in the Great Indian Desert /
Tanuja Kothiyal.
pages cm
Summary: "Discusses the emergence of socio-historical identities in the Thar Desert with
the mobility of its
inhabitants"-- Provided by publisher.
Includes bibliographical references and index.
ISBN 978-1-107-08031-7 (hardback)
1. Nomads--Thar Desert (India and Pakistan)--History. 2. Migration, Internal--Thar
Desert (India and Pakistan)--
History. 3. Group identity--Thar Desert (India and Pakistan)--History. 4. Human
geography--Thar Desert (India and
Pakistan)--History. 5. Human ecology--Thar Desert (India and Pakistan)--History. 6.
Thar Desert (India and Pakistan)-
-Social conditions. 7. Thar Desert (India and Pakistan)--Geography. 8. Thar Desert (India
and Pakistan)--
Environmental conditions. I. Title.
DS485.T476K68 2015
305.9'0691809544--dc23
2015014741

ISBN 978-1-107-08031-7 Hardback

Dedicated to the memory of my mother
Kaushalya Kothiyal

The purpose of wandering
Is to obtain a vision,
Otherwise there's bread aplenty
and salt anywhere.[1]

[1] Deccani Sufi couplet as quoted by Simon Digby, *Sufis and Soldiers in Aurangzeb's Deccan; Malfuzat -I-Naqshbandiya,* Delhi, 2001, 137.

Contents

List of Tables

List of Abbreviations

AAR	*Annals and Antiquities of Rajast'han*
FP	Foreign Political
IESHR	*Indian Economic and Social History Review*
JESHO	*Journal of the Social and Economic History of the Orient*
JPASB	*Journal and Proceedings of Asiatic Society of Bengal*
JRAS	*Journal of Royal Asiatic Society*
JWH	*Journal of World History*
JWSR	*Journal of World-Systems Research*
Khyat	*Munhata Nainsi ri Khyat*
MAR	Marwar Administration Report
MAS	*Modern Asian Studies*
MHJ	*The Medieval History Journal*
NAI	National Archives on India
NSIR	Northern India Salt Revenue
RSAB	Rajasthan State Archives, Bikaner.
SIH	*Studies in History*
SP Bahi	Sanad Parvana Bahis (Jodhpur)
Vigat	*Marwar ra Paraganan ri Vigat*
ZB	Zagat Bahi (Bikaner)

Acknowledgements

The writing of this book has been a very long journey, quite like the ones it talks about. In this journey to the frontiers of Thar, I have been accompanied by several fellow travellers, some remembered, others forgotten, thus accumulating debts, most of which I cannot even attempt to repay. Nevertheless, I take this opportunity to express my gratitude to some of the people whose generosity has helped me to complete this work.

I owe my greatest debt to Prof. Dilbagh Singh, my MPhil and PhD supervisor at the Centre for Historical Studies, JNU, who has been a grounding influence in my engagement with mobility. In course of my doctoral work, which forms the basis of this book, Prof. Singh, with his profound knowledge of histories of the Thar helped me question my own presumptions as I approached my sources. Without his faith in the questions I was asking as well as my ability to find answers to them, this work would never have been possible. Prof. Rajat Datta, who supervised me in some of the seminar papers I wrote, shaped my early ideas on regions, markets and trade. Late Nandita Prasad Sahai, who read my thesis as part of submission proceedings encouraged me to convert the thesis into the book. It is my lasting regret that she is no longer there to see it in print. Samira Sheikh and Aparna Kapadia's work on Gujarat support my arguments of continuities, geographical and historical, between what we today know as Rajasthan and Gujarat. Ramya Sreenivasan's work as well as personal interactions with her, have helped me look beyond my initial, very sketchy understanding of Rajput history. Salil Misra introduced me to Ernest Gellner's work which continues to influence my understanding of frontiers. Sanjay Sharma and Dhirendra Dangwal helped me improve upon my very inadequate understanding of modern Indian history. They along with Chirashree Dasgupta read parts of the manuscript, commented upon, edited and suggested changes that make this book readable. Discussions with Mahesh Rangarajan helped me engage with rather difficult questions regarding histories of frontiers, ecological as well as political.

I am grateful to Ambedkar University, Delhi where I teach, for providing an environment where I have been able to discuss a number of ideas I engage with as part of my research, in my classes. Discussions with colleagues and students have enriched me in innumerable ways, sometimes forcing me to reconsider what I was writing. I also thank NMML, New Delhi, for providing space to discuss some parts of my work in its seminar series, as well as for a fellowship to carry this work beyond its present form. Before joining AUD, I spent twelve long years teaching at the Government Girls College, Jhalawar, Rajasthan. These years were extremely instructive as they made me come to terms with the difference between my university-trained ways of history and 'local' histories. I have attempted to travel, rather unsuccessfully, this distance between my own training and 'local' ways of being historical.

Research work for this book was carried out at the Rajasthan State Archives at Bikaner, Archives Branch Office at Jodhpur, Rajasthan Oriental Research Institute, Jodhpur, National Archives of India, New Delhi, and the ICHR, NMML, JNU Central and DSA libraries. I am grateful to the staff of these institutions for the kind help extended at various times.

I extend my thanks to Qudsiya Ahmed at Cambridge University Press, who has displayed enormous patience towards my ever-extending deadlines. My gratitude also to my anonymous referees whose incisive reading and suggestions helped me clarify a number of issues and organize the book better. Suvadip Bhattacharjee at CUP helped me out with numerous details of publishing that I was quite unfamiliar with.

The actual writing of the book has been a difficult endeavor made possible by goodwill and support extended by friends and family. It is not possible for me to thank them enough for their sheer faith in me, which made it possible for me to overcome my own fears and actually sit down to write. Susan George has had more faith in me, than I myself ever did. She has been there every time I needed support. Aparna Kapadia gave wings to a small dream in a corner of my heart and pushed me beyond the limits of possibility. Samira Sheikh encouraged me when this book had not even been imagined and came forward with much needed support many times by way of supplying resources not readily available in India. Salil Misra, Chirashree Dasgupta, Sumangala

Damodaran, Dhirendra Dangwal, Sanjay Sharma, Surajit Mazumdar, Geetha Venkataraman, Ned Bertz, Aniket Alam, Kaushik Dasgupta, Aparna Balachandran, Farhana Ibrahim, Malavika Kasturi and CSR Shankar have heard me out through my unending 'authoring' woes and have humoured me at each stage.

Nivedita has shared my dreams from our childhood. I know that her sentiments mirror mine at this moment. My mother-in-law Uma Devi, as well as sister-in-law Jaya Tiwari, have supported me by extending care to my daughter, when work has kept me away. I also thank my brother-in-law Sanjay Tiwari for moral support extended on innumerable occasions. My daughter Arunima's ability to take pride in her unconventional mother has always been a source of assurance. She will finally be able to see this book in a shape that might make the hours I spent crouched on a laptop, while she waited to share her world with me, worth it. She along with Raghav, Pranav and Kabir has been a welcome distraction in an often arduous endeavor. Keshav has always believed that it is possible not to give up dreams, whatever be the circumstances, and has motivated me in more ways than I can thank him for.

Finally, my parents' unquestioned support for all I sought to do with my life has always been my strength. I have drawn enormous courage from my father Jagdish Prasad Kothiyal's ability to deal with vicissitudes of life, particularly after my mother's demise. My mother Kaushalya Kothiyal always wanted her daughters to be courageous as she was in her life. It is to her memory that I dedicate this book.

While I extend my gratitude to all who have helped in making this work possible, for any error in the text I hold myself responsible.

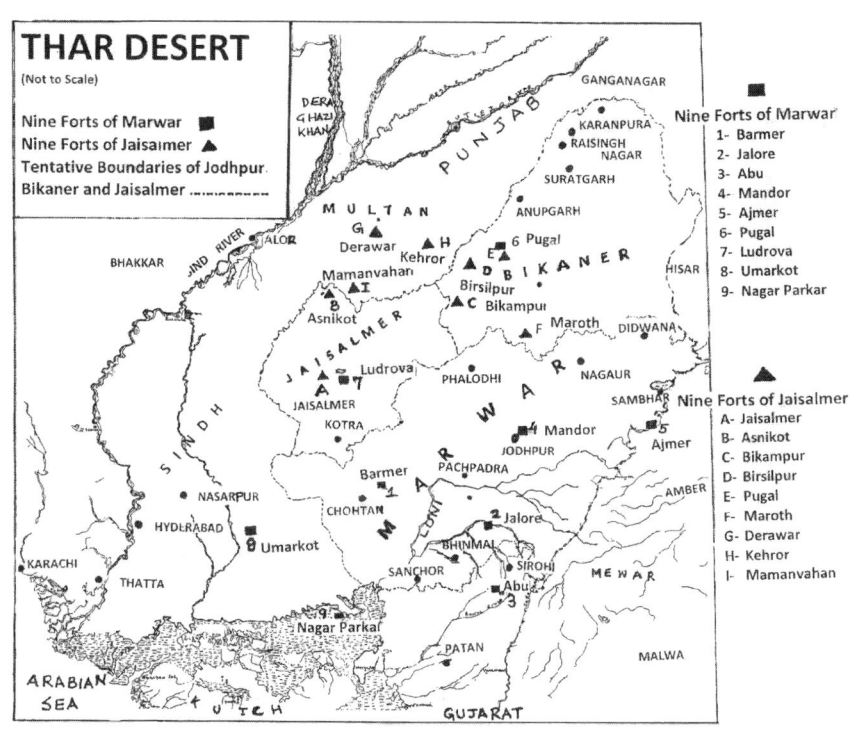

THAR DESERT
(Not to Scale)

Nine Forts of Marwar ■
Nine Forts of Jaisalmer ▲
Tentative Boundaries of Jodhpur.
Bikaner and Jaisalmer ....

Nine Forts of Marwar ■
1- Barmer
2- Jalore
3- Abu
4- Mandor
5- Ajmer
6- Pugal
7- Ludrova
8- Umarkot
9- Nagar Parkar

Nine Forts of Jaisalmer ▲
A- Jaisalmer
B- Asnikot
C- Bikampur
D- Birsilpur
E- Pugal
F- Maroth
G- Derawar
H- Kehror
I- Mamanvahan

Glossary

Agar	Salt mine or pan.
Akalwant	Someone with divine knowledge
Akharsidh	Brave and successful in battle. Also one able to control battlefield through miraculous powers
Arhat	Persian Wheel
Bachri	Heifer (Cow)
Baldiya/Banjara	Community involved in trade and carriage of salt and grain
Bambhi	Member of a community engaged in skinning of dead cattle
Bapoti	Inheritance from father
Begar	Forced labour
Behtiwan	Transit tax
Bhai-bandh	Kinsman
Bhomiya	Hereditary right holders entitled to collect share in the produce.
Bhunga	Grazing tax in Bikaner
Bhuraj	Tax on the bhurat grass
Bolawo	Local protection tax levied by Thikanadars
Bund	Dam
Chakar	Serviceman
Chanch	Shallow well
Chappaniya Kal	Famine of VS 1856/1900AD
Chattra	Royal Umbrella
Chira	Revenue circle in Bikaner

Chirayat	Revenue assignee in a Chira
Dalali	Tax paid by middlemen in the markets
Dan	A commercial tax
Des	Clan land
Dhad	Raids usually understood to be cattle raids
Dhani	Overlord
Dhavala	A woolen garment worn by Jat women
Dugani	Local currency in Western Rajasthan
Ganveti	Resident Cultivator
Ganayat	Relatives by marriage
Ghasmari	Grazing tax on animals grazing on grass.
Ghiyayi	Tax on sale of ghee
Ghora Kambal	Tax on wool production
Goria	Skin of cows and buffaloes
Ijara	Lease against cash
Jhota	Heifer (Buffalo)
Jhumpi	Temporary shelters constructed by pastoralists in forests and grasslands
Jod-dar	Official designated to care for the jod
Jod	Grassland
Kamdar	Official at village level
Kankar kunta	Estimation made on the standing crop
Katariya	Shearer
Kayali	A Commercial tax
Khadin	A series of dammed channels
Khatik	Member of butcher community also engaged in shearing.
Korad	Tax on grass of dry moth or til
Kosito	Shallow well
Kunta	Estimation arrived at by guessing at threshing floor
Lata	A system of collecting all the grain and measuring it with kalsis after which the darbar share was taken on spot
Looe	Coarse blalnket
Mapa	A commercial tax
Nagara	Drums
Naree	Skin of sheep and goats
Navikhati	Unbroken, unconquered land
Nesar (nikal)	Exit Tax
Oran	Sacred grove

Pancharai	Grazing tax on animals that ate leaves
Partal	Fallow land
Pattadars/Pattayat	Revenue assignees of Pattas (Rajputs)
Pesar	Entry tax
Phadiya	Local currency in Marwar
Qabulat	Fine or tribute
Rahadari	Transit tax
Raigar	Community involved in tanning of hides
Rajvi	Pattadars who did not pay rekh
Rawanna	Travel permit for commodities mentioning amount, type, destination and route
Rekh	Income of the Patta
Rel/ Bahla	Seasonal overflow of the rivers
Sagani	Interpreter of omens
Sasan	Revenue free grants to Charanas, Brahmins, Jogis, Sevags etc.
Sayer Jihati	Record of commercial taxes.
Sehat	Tax on Sewan grass
Serino	Tax on wool
Siladibab	Tax on leather workers
Singhoti	Grazing tax on sheep
Solkalar	Saline lands
Subhraj	Ceremonial lore of praise
Tanda	Caravan of the Banjaras
Thakurai	Overlordship
Tikayat	Anointed chief
Vikhau	Periods of displacement faced by Rajputs in the medieval period.

Note on Transliteration, Translation and Dates

Assuming that readers familiar with Rajasthani languages will not require diacritical marks, and the unfamiliar will find them cumbersome, I have omitted all diacritical marks. I have retained original spellings and diacritical marks while citing the works of other scholars, which accounts for variations in spellings of place names in the text. However, in my own writing, I have used contemporary English spellings for Indian language words, names and places. Most of the translations from Rajasthani works are mine, unless indicated otherwise. All dates mentioned in the book are in Common Era, unless stated otherwise.

Contemporary Place Names and their Nineteenth Century Spellings

Contemporary Name	Various Spellings in the Text
Abu	Aboo
Ajmer	Ajmere
Bikaner	Bikanir
Harauti	Harowtee
Jaipur	Jeypore
Jaisalmer	Jeysulmere, Jaisalmir
Jodhpur	Joudhpore, Joudhpur, Jodhpoor
Luni	Loonee
Mewar	Meywar
Nagaur	Nagore
Pachpadra	Puchbudra, Pachbhadra, Puchpadra
Phalodi	Phallodee, Filodi
Pokhran	Pokharan, Pohakaran
Rajasthan	Rajast'han
Sambhar	Sambur
Sindh	Sind, Scinde

Introduction

What will you do with a mare? You should just live off your land. But now
that you have the mare, it looks like you will raid and plunder.[1]

The Thar Desert in South Asia is at present divided by an international
boundary between India and Pakistan. However, the Thar has historically
existed as a frontier connecting regions like Punjab, Multan, Sindh, Gujarat and
Rajasthan with each other. The Thar desert can be defined as a region through
the mobility of its inhabitants, who were warriors, pastoralists, traders, ascetics
and bards, often in overlapping capacities, exchanging mobile wealth and
equally mobile narratives.[2] The historical understanding of the Thar Desert,
like that of other such spaces, is couched in familiar frames of barrenness and
waste. Yet, a closer look at the Thar Desert reveals a rich history of movements
of a large number of itinerant groups, of settlements and depopulations, as well
as of a cultural milieu where memories of movements have been immortalized
in the rich folkloric traditions of the region.

---

[1] Badri Prasad Sakaria, (Ed.) *Nainsi Ri Khyat*, Vol III, Rajasthan Oriental Research
Institute, Jodhpur, 1962, (Reprint 1993), 63, *Vat Pabu ji ri*. In the epic of Pabuji when
he acquires a mare coveted by all other Rajputs around him, his sister-in-law chides
him fearing that now he would engage in raiding and pillage.

[2] I carry the argument forward from Jos Gommans' idea of frontiers in South Asia. Jos
Gommans, 'The Silent Frontier of South Asia, c AD 1100–1800', *JWH*, Vol 9, No 1,
1998, 1–23.

A wide corpus of historical research on the Thar focuses on 'Rajputana' as the physical and intellectual area of study.[3] The former refers to definitive political spaces divided by fixed boundaries and ruled by Rajput clans. The latter alludes to Rajputs as the pre-dominant reference to its socio-historical identity as well as, to a misplaced emphasis on land, agrarianism and territoriality as the basis of social and political relations in the region. Both identifications are misplaced and highly problematic. The arid desert of the Thar has historically existed as a frontier region that could be defined better through the mobility of its peripatetic residents than through the political boundaries that divided it. The recurrent patterns of circulation of people, resources and lore united the Thar as a region encompassing several political states with shared histories of mobility. From the sixteenth century onwards, the overarching endogamous category of 'Rajput' increasingly focused on land and territoriality as means for extending control. Bardic traditions patronized by Rajputs reiterated protection of land and forts, particularly against Muslim invaders, as being central to 'Rajput ethos'.[4] Poetic and prose compositions like *Raso, Vachanika, Khyat, Vat* etc focused on bravery as well as generosity as attributes of Rajputs. In the writing of later histories of Rajasthan, these were read and interpreted as accounts of heroic Rajput struggles against expanding Muslim polities. The idea of a Rajput struggle against Muslim invaders became the axis around which histories of Rajasthan like G H Ojha's volumes of *Rajputane ka Itihas* and Dasratha Sharma's *Rajasthan through the Ages* were written. As Ramya Sreenivsan argues, selective appropriations from earlier traditions like *Padmavat* became instrumental in "reformulation of new national identities along tacitly communal lines".[5]

By the nineteenth century, the Thar was primarily identified as 'Rajputana', an assortment of princely states ruled by the Rajputs. James Tod viewed Rajput

---

[3] I would like to mention G H Ojha's *Rajputane Ka Itihas* (1927), J S Gehlot, *History of Rajputana* (1937), V N Reu, *Glories of Marwar and the Glorious Rathors* (1943), J N Asopa, *Origin of Rajputs*, Dasratha Sharma, *Rajasthan through the Ages* (1966) and Shiv Dutt Dan Barhat, *Jodhpur Rajya ka Itihas, 1753–1800* (1991) as representatives of this idea over a long period of time.

[4] Texts like *Gora Badal Padmini Chaupai, Kanhadde Prabandh, Hammir Mahakavya, Achaldas Khichi ri Vachanika* can be seen as examples of texts that focus on narratives of protection of forts against Muslim invasions.

[5] Ramya Sreenivasan, *The Many lives of a Rajput Queen: Heroic Pasts in India, 1500–1900*, Permanent Black, Ranikhet, 2007, 14.

ethos as central to the polity and culture of the Rajput kingdoms and thereby of Rajputana in the nineteenth century, literally the land of the Rajputs. He saw 'Rajast'han' as "the collective and classical denomination of that portion of India which is 'the abode of (Rajpoot) princes'. In the common dialect it is termed as Rajwarra, but by the more refined Raéthana, corrupted to Rajpootana, the common designation amongst the British to denote the Rajpoot principalities".[6] Jason Freitag points out that between 1872 and 1998, Tod's *Annals* have been translated into five major Indian languages that is Hindi, Gujarati, Urdu, Marathi and Bengali, signalling an Indian appropriation of British valourization of Rajputs.[7] First to use the word 'Rajast'han', Tod became instrumental in defining both Rajputs and their land.

On March 25, 1948, after bitter negotiations between the Rajput states, the formation of the state of Rajasthan was announced, though it would take another eight years for the state to take its present form.[8] After the end of princely rule, there appears to be an increasing awareness of common heritage based on Rajasthan's broader political, historical and cultural traditions among the residents of former princely states.[9] Today this identification is further bolstered by the projection of Rajputs as the proud protectors of this land of shifting sand dunes, *dharati dhoran ri*.[10] The phrase *dharati dhoran ri* rather than allude to dry barren stretches actually refers to an ingrained sense of pride in identification with a land that bred bravery and chivalry in circumstances of adversity.

---

[6] James Tod, *Annals and Antiquities of Rajast'han or the Central and Western Rajpoot States of India*, (Ed.) Douglas Sladen, Rupa and Co., Delhi, 1997, (First Pub. London, 1929), Vol 2, 1. Henceforth, *AAR*.

[7] Jason Freitag, *Serving Empire Serving Nation: James Tod and the Rajputs of Rajasthan*, Brill, Leiden, 2009, 174–179.

[8] The first union of Rajasthan contained nine princely states of Banswara, Bundi, Dungarpur, Jhalawar, Kishangarh, Kota, Pratapgarh, Shahpura and Tonk. The larger states of Jaipur, Mewar, Marwar, Bikaner and Jaisalmer joined over the next two years after considerable persuasion.

[9] Deryck O Lodrick, 'Rajasthan as a Region: Myth or Reality' in *The Idea of Rajasthan: Explorations in Regional Identity*, Vol I, Karine Schomer, Joan L Erdman, Deryck O Lodrick and Llyod I Rudolph, (Eds.) Manohar, 2001, 1–35.

[10] Joan L Erdman, 'Becoming Rajasthani: Pluralism and the Production of Dharti Dhoran Ri' in *The Idea of Rajasthan*, I, 45–79. Erdman argues that cultural idioms like *Dharti Dhoran Ri* were used to encapsulate the vitally and conceptually important features of Rajasthan including forts, chronicles of bravery, Padmini, Chetak and Maharana Pratap.

However, while the awareness of Rajasthan as a region draws heavily on its heroic 'Rajput' past, erstwhile rulers of the princely states, the Rajputs, have increasingly taken a backseat in the contemporary political milieu. With demographic strength and control over land ensuring political control, a number of non-Rajput groups, including lower caste groups have edged Rajputs out of the political arena. As Rajputs increasingly turn to hospitality sector, turning their palaces and *havelis* into heritage hotels, it is other castes like Jats and Gujars that reconstruct heroic pasts for themselves, while increasingly challenging the tribe/caste status ascribed to them.[11] These challenges emanate from an increasing awareness of a past, contested as well as shared with Rajputs. In a state that is suffused with history, with its forts, palaces, temples, water bodies, inscriptions and manuscripts, this increasing awareness relies on oral narratives, the only references to history that these communities have. It is this historical and historiographical lacuna that motivates this work.

This book explores the relationships between mobility, martiality, memory and identity in the frontiers of the Thar Desert, arguing that emergence of social identities was closely entwined with mobility on circulatory networks in the Thar. Exploring multiple narratives of itinerants in the desert, I trace a long history of relationships between martiality and mobility in the Thar. I map out networks of mobility and circulation in the Thar region from the sixteenth to the nineteenth centuries to demonstrate how these networks were sites of struggle for authority and control. This long period of four hundred years becomes significant given the political shifts that occurred in this region. The early sixteenth century witnessed the rise of Rajput polity, as it contested with several powers along with the emerging Mughals to claim a central position. By late sixteenth century, a major part of the Thar had been incorporated into the Mughal empire and the early nineteenth century witnessed the establishment of indirect rule in Rajput states of the Thar region.

In the following discussion I will take up four distinct strands that connect in the arid frontiers of Thar and allow us to imagine identities differently from what historiography of Thar has permitted us to.

---

[11] In 2007 Gujars of Rajasthan launched a movement demanding to be placed in the Schedule for Tribes rather than in the Schedule for Other Backward Castes that they were in. While the stir was largely motivated by the politics of reservations, it nevertheless raised important questions about the manner in which tribe/caste identities had been ascribed in the state.

The Frontiers of Thar

Deserts, like oceans, are frontier zones providing passage to people, commodities and ideas essential for the existence of states. However, much of our understanding of socio-political processes like state formation and emergence of social hierarchies has emerged from core agrarian regions. Deserts and other non-agrarian spaces like oceans, mountains and forests have often been used to engage with ideas like statelessness, the absence of agrarian surplus appearing to signify the absence of state in the most conventional sense. Deserts have been understood as dangerous territories harbouring inhabitants hostile to states. Considering that large parts of the earth are either hot or cold deserts, the representation of such spaces poses a methodological question about the way in which these have been understood.

In recent years understanding of political and ecological frontiers and their relationship with state formation has received considerable attention, particularly in Central and West Asia and North Africa. Sinologist Owen Lattimore suggests frontier represents a paradox between the need for exercise of centralized control as well as the expression of centrifugal forces of separation.[12] Frontiers, while present opportunities of expansion for core agrarian polities, also transform the core by introducing 'unsettled' elements into it. This has been true of frontiers in South Asia, that Jos Gommans views as a "wide, open ended zone which not only favoured circulation of people, animals, goods and ideas but also agricultural expansion".[13] The openness of these frontiers generated social, cultural and political flux, transforming societies with emergence of newer forms of organization and control, with agrarian expansion following tribal conquest in an Ibn Khaldunian cycle. Ernest Gellner suggests, "peripheral areas harbour cohesive participatory, segmentary communities, endowed with great military potential. Thus, they constitute a kind of political womb, a source of new rulers".[14] But what was the nature of these 'new' polities? Were the forms of authority and control exercised by the emerging polities any different from the older?

---

[12] Owen Lattimore, *Studies in Frontier History: Collected Papers*, OUP, 1962, www.archives.org/details/ studiesinfrontie017780mbp.

[13] Jos Gommans, 'The Eurasian Frontier after the First Millennium AD : Reflections along the fringe of Time and Space 1', *MHJ*, 1998; 1, 125–143, 142.

[14] Ernest Gellner, 'War and Violence', in *Anthropology and Politics: Revolutions in the Sacred Grove*, Oxford, 1995, 160–179, 164.

A frontier polity that has been widely studied and has important implications for studying a non-agrarian frontier like the Thar, is the Mongol empire. The Mongol Empire stretching from Central Asia to Poland was a land empire larger than any other in history, and one created by a pastoral nomadic group. However, according to Thomas Barfield, the Mongols were 'tamed' by the very societies they conquered. The accounts of visitors to the courts of the Khans underline the enormous wealth and sophistication that they had accumulated over time and the ferocious means that they employed to maintain their authority intact. Barfield argues that both 'stable sedentary' and 'nomadic political' structures were necessary preconditions for the existence of both, the sedentary Chinese states and the Mongol empires in Central Asia. In Chinese Turkestan, extensive Mongol hordes controlled important trade routes like the silk route while the Manchu state controlled the commodities.[15] In Anatoly Khazanov's framework 'nomadic feudalism' seems to be an appropriate milestone in the journey of nomadic societies towards state formation whereby nomadic societies successfully managed to synthesize tenets of kinship with retainership, thereby reorganizing community forces into a bureaucracy. In the process, newer forms of social organization emerged. Once a pastoral nomadic society evolved beyond clan and lineage-based organization, it was forced to develop elaborate hierarchical structures.[16] Nikolay Kradin explains that the politico-cultural integration of frontier societies can be understood at three levels. The first, of segmentary clan and tribal formations, second, of tribe and chiefdom and the third, of nomadic empires and quasi-nomadic polities of smaller sizes. He explains 'nomadic empires' were "actually 'peripheries' in themselves, organized on the military-hierarchical principle, occupying a quite large space and exploiting the nearby territories, as a rule, by external forms of exploitation (robbery, war and indemnity, extortion of "presents", non-equivalent trade and tribute)".[17] Kradin also claims that the nomadic societies were doomed to remain peripheral, as to become a 'centre' it was necessary to cease to be a nomad. Citing the Mongol example he reiterates the ancient Chinese wisdom, "although you inherited the Chinese Empire on horseback, you cannot rule it from that position".[18] However, in a study of state in nomadic

---

[15] Thomas J Barfield, *The Perilous Frontier - Nomadic Empires and China*, Basil Blackwell, Cambridge, Massachusetts, 1989, vii-ix.

[16] Anatoly Khazanov, *Nomads and the Outside World*, tr. Julia Crookenden, University of Wisconsin Press, Madison, 1984, 29.

[17] Nikolay N. Kradin, 'Nomadism, Evolution And World-Systems: Pastoral Societies in Theories Of Historical Development', *JWSR*, Viii, 1ii, Fall 2002, 368–388, 373–4.

[18] Ibid., 380.

inner Asia, David Sneath argues that the "history of region shows no clear dichotomy between highly centralized, stratified 'state' society and egalitarian, kin based 'tribal' society, but rather displays principles of descent deployed as technologies of power in a range of more or less centralized polities, ruling subjects engaged in various kinds of productive practices- pastoral, artisanal, and even agricultural".[19]

All these positions have some relevance to the study of frontier societies in South Asia as well. As Jos Gommans demonstrates, a large part of the Indian subcontinent stretching from Sindh and Rajasthan towards Deccan was arid and semi arid.[20] The Thar Desert, while historically has had a high component of nomadic and semi-nomadic pastoralist population however differs significantly from both Central Asia and North Africa in the scale of pastoralism practised. While studying frontier regions in India, Andre Wink argues that the lack of sufficient pasture prevented the Indian plains from being occupied and nomadized by Mongols.[21] The development of Indo-Islamic polity took the route of purposeful agrarian-fiscal-military formulation. However, the Thar was a frontier crossed by a large number of "vigorous inhabitants.... in their often overlapping capacities of nomads, warriors and ascetics".[22] As an arid frontier while the Thar Desert connected sedentary centres, its fringes also provided space for settlement as well as mobility. According to Gommans around the first millennium AD, the arid frontiers of South Asia witnessed the emergence of nomadic and semi-nomadic warrior-pastoralist groups exercising control over mobile wealth.[23] The emergence of these warriors, identified under the umbrella category of Rajputs, unsettled older power sharing arrangements, marginalising groups like Bhils, Mers, Minas, Gujars and Jats.

Rajputs in the Frontiers

The question of origin of Rajputs has received considerable attention in the historiography of Rajput kingdoms of the Thar. Summarising some of these

---

[19] David Sneath, *The Headless State: Aristocratic Orders, Kinship Society, and Misrepresentations of Nomadic Inner Asia,* Columbia University Press, New York, 2007, 4.

[20] Jos Gommans, *Mughal Warfare: Indian Frontiers and High Roads to Empire, 1500–1700,* Routledge, London, 2002, 11.

[21] Andre Wink, *Al Hind The Making of the Indo Islamic World: The Slave Kings and the Islamic Conquest,* (Vol II) Brill, Leiden, 1997, 3.

[22] Gommans, 'The Eurasian Frontier', 131.

[23] Ibid., 134.

positions helps in contextualising Rajput identity in the Thar. Genealogical traditions emerging from Rajput courts as well as the historiography of the Rajputs, link their origins to celestial sources of light, the sun, the moon and the fire, older *kshatriya* clans as well as Vedic Aryans, and Central Asian Scythians.[24] However, as recent work suggests, emergence of Rajputs as landed aristocrats involved the transformation of tribal and nomadic groups into royal lineages. B D Chattopadhyaya's extensive work on emergence of Rajputs in early medieval Rajasthan points towards tribal and nomadic origins of Rajputs. In the initial stages of these transformative processes he views 'Rajput' as a category that was assimilative in nature and could be seen as a recognizable channel of transition from tribal to state polity.[25] Richard Fox in his early study of Rajput kinship structures in northern India identifies 'Rajput' as a 'class' or 'status' category rather than a 'caste'.[26] Norman Zeigler points out that, complex codes of service and kinship structures, as well as marital alliances helped in the formulation of Rajput identity in the Mughal period.[27] D H A Kolff views 'Rajput' as a social category subscribed to by lineages of varied social origins through the practice of *naukari* or military entrepreneurship. He suggests that, however by late sixteenth century, a 'Great Rajput Tradition' representing a genealogical orthodoxy emerged, which focused on ideas of blood purity and linkages with older *kshatriya* traditions. Nevertheless, the peregrinations of adventurous young men with claims to Rajputhood, in the

---

[24] Detailed genealogies form part of Rajput historical narratives like *Munhata Nainsi ri Khyat*, which traces the origins of various Rajput clans to celestial sources as well as older Kshatriya clans like Gahadwalas. Among colonial and modern histories of Rajputs, while James Tod forwards the Scythian origin theory, Dasaratha Sharma, G N Ojha and J N Asopa's early works focus on celestial origins and older Kshatriya lineages. Dasaratha Sharma, *Early Chauhan Dynasties*, (Second Revised Edition), New Delhi, 1975. J N Asopa, *Origin of the Rajputs*, Bharatiya Vidyapith, New Delhi, 1976. James Tod, *AAR*, I, 49–50, 450, 471.

[25] B D Chattopadhyaya 'Origin of the Rajputs: The Political, Economic and Social Processes in Early Medieval India' in *The Making of Early Medieval India*, Second Edition OUP, Delhi, 2012, 59–92. A comprehensive review of these processes for Mewar can be seen in Nandini Sinha Kapur, *State Formation in Rajasthan: Mewar during the Seventh-Fifteenth Centuries*, Manohar Publishers, Delhi, 2002.

[26] R G Fox, *King, Clan, Raja and Rule: State-Hinterland Relations in Preindustrial India*, University of California Press, 1971, 16–23.

[27] Norman Zeigler, *'Action Power and Service in Rajasthani Culture: A Social History of the Rajputs of Middle Period Rajasthan'*, Unpublished PhD Dissertation, University of Chicago, Illinois, 1973.

form of errant migrant soldiers remained in practice even till as late as the nineteenth century.[28] Stewart Gordon views the development of 'Rajput' martial ideology as developing through service to the Mughal Empire, though "hypergamous pattern of Rajput marriage tacitly acknowledged that it was somewhat open caste category".[29]

However, with the coming of the Mughal empire the open category of 'Rajput' was replaced by a 'genealogical orthodoxy' with Mewar as its seat.[30] The resultant overarching endogamous category of 'Rajput' is increasingly understood to have focused on land and territoriality as means for extending control through kin networks. An important part of these traditions were the genealogies, which stressed on direct descent and affinal relationships, contemporaneous to the genealogies that were becoming a part of Mughal identity. Besides, chronicles like the Padmini narratives explored by Ramya Sreenivasan, were instrumental in constituting and preserving glorious pasts of Rajputs, even though as she demonstrates that these narratives also underline the complexities of the Rajput responses to Turkish and Mughal advances.[31] In fact, varied vernacular forms like *kavya* or *masnavi* were used to foreground claims to Rajputhood in the interregnum between the Sultanate and the Mughal Empire.[32] Cynthia Talbot's examination of the Kyamkhani lineage and *Kyamkhan raso* composed in mid seventeenth century illustrates

[28] D H A Kolff, 'The Rajput of pre-Mughal North India', in *Naukar, Rajput and Sepoy: The Ethnohistory of the Military Labour Market in Hindustan, 1450–1850*, Cambridge University Press, Cambridge, 1990, 2007, 73.

[29] Stewart Gordon, *The Marathas: 1600–1800*, The New Cambridge History of India II.4, Cambridge University Press, 1993, 16.

[30] Kolff, *Naukar, Rajput and Sepoy*, 73.

[31] Sreenivasan, *The Many Lives*, 202. In recent years Sreenivasan's engagement with Rajasthani heroic verse chronicles like *Kanhadde Prabandh* and *Kyamkhan Raso* underlines the complexity of Hindu-Muslim interactions in the Rajput context. She argues that sharper intra-Rajput hierarchies were being articulated through heroic poetry that has prima-facie been read only through the trope of resistance. See, Ramya Sreenivasan, 'The 'Marriage' of 'Hindu' and 'Turak': Medieval Rajput Histories of Jalor', *MHJ*, 7:1, 2004, 87–108. Also, Ramya Sreenivasan, 'Faith and Allegiance in Mughal Era', in *Religious Interactions in Mughal India*, (Eds.) Vasudha Dalmia and Munis Faruqui, OUP, New Delhi, 2014, 157–191.

[32] Ramya Sreenivasan, 'Warrior Tales in Hinterland Courts in North India, *c* 1370–1550' in *After Timur Left: Culture and Circulation in Fifteenth Century North India*, (Eds.) Francesca Orsini and Samira Sheikh, OUP, New Delhi, 2014, 111–130. For an illustration of the life of *Purbiya* warlord Silhadi and contestations for Rajputhood, see Kolff, *Naukar, Rajput and Sepoy*, 71–110.

the ways in which seeking martial pasts and warrior lineages caused no contradiction in belonging both, to the communities of Rajput warriors as well as Muslim gentry, as warriorhood appears to be the common ideal for both identities.[33] In Rajput court chronicles, 'honour' was consciously constituted as a Rajput virtue, which then became the keystone of British construction of Rajput identity in the nineteenth century, as well as the trope through which nationalist imaginations of Rajput response to Muslim invasions were articulated.[34] However, Talbot's recent examination of the Sanskrit poem *Surajanacharita*, about the Rajput warrior Surjana Hada who surrendered the fort of Ranathambhor to Akbar in 1569, shows how the idea of dishonour as associated with defeat, was articulated in multiple ways in the Rajput as well as the Rajput-Mughal worlds.[35] In the same period, a number of elite Rajput lineages displayed increasing devotion towards *vaishanavism*, leading to the establishment of *Vallabha* deities in Rajput capitals. This allowed Rajputs to reinterpret the Rajput concept of *chakri* through the Vaishnava idea of *seva*.[36] Interestingly, while the idea of an honourable 'Rajput' death increasingly disappeared in later Rajput court chronicles, it continued to find space in oral narratives circulating in the Thar.[37] Dying the Rajput way, either in battles or while protecting cattle, continued to remain central to oral narratives like that of Pabuji and Tejaji foregrounding claims to Rajput status if not caste, on behalf of a number of non-Rajput groups.

Another turn in reconstruction of Rajput identity came about in the nineteenth century, when colonial administrators re-imagined the Rajput

---

[33] Cynthia Talbot, 'Becoming Turk the Rajput Way: Conversion and Identity in an Indian Warrior Narrative', *MAS*, 43, no. 01, 2009, 211–243.

[34] Sreenivasan, *The Many Lives*, 117–200. Also, see Ramya Sreenivasan, 'Honoring the Family: Narratives and Politics of Kinship in Pre-colonial Rajasthan' in *Unfamiliar Relations: Family and History in South Asia*, (Ed.) Indrani Chatterjee, Permanent Black, Delhi, 2004, 46–73.

[35] Cynthia Talbot, 'Justifying Defeat : A Rajput Perspective on the Age of Akbar,' *JESHO*, 55, 2012, 329–68.

[36] Norbert Peabody's work on the relationship between the Kota Raos and Vallabha *sampradaya* demonstrates how in the eighteenth century the Rajput notion of *chakri* was decentered in Kota, in particular by Zalim Singh. Norbert Peabody, 'From 'royal service' to 'maternal devotion' during the Jhala Regency: Local politics at the end of the old regime' in *Hindu kingship and polity in precolonial India*, CUP, Cambridge, 2003, 112–147.

[37] Janet Kamphorst, *In Praise of Death: History and Poetry in Medieval Marwar*, Leiden University Press, 2008.

in several ways, ranging from 'tribe', 'caste', 'race' to 'nation', but certainly as a category that was set apart due to its lineage as well as conduct. Thomas R Metcalf refers to the 'medievalist vision' that sought to create orders of knighthood in India.[38] Rajputs were best suited to be modelled as Anglo-Saxon knights, with similar relationship with land, particularly in the context of Muslim invasions. In the processes of land settlements, compiling castes and tribes compendia, as well as of writing 'Rajput' histories, genealogies were hunted for by British administrators. These genealogies became the basis of a new genealogical orthodoxy, which distinguished between the genuine and the 'spurious' Rajput. Further, as Malavika Kasturi suggests, Rajput identity continued to be shaped by colonial political economy, as Rajputs engaged in a struggle for land resources and power with agrarian castes in northern India. In this struggle banditry, feuding and rebellion became ways of redefining the self for the Rajputs.[39]

In 'Rajputana' while on the one hand the 'Rajput' was represented through categories like the noble chivalrous warrior and the feudal lord, his other persona was the *barwuttea* or the outlaw, the marauding Rajput highwayman. In my understanding this distinction is an amorphous one. It actually refers to the distinction between the Rajput located in the confines of agrarian landed aristocratic polity and one operating on the frontiers of this polity. This distinction is sharpened with the inclusion of 'Rajputana' in the Mughal Empire and further with the implementation of British indirect rule in the region. However even till the nineteenth century, these categories remained interchangeable. I see these two representations of the Rajput as reflective respectively of explicit sovereignty of the core areas, and "ambiguous, plural and shifting" sovereignty of the frontiers.[40] It was also increasingly understood through the binaries of 'civilized' settler and the 'barbaric' nomad. It is important to note here that the category of 'barbaric' nomad included both the upper caste Rajput *barwuttea* as well as a range of mobile groups that moved on the frontiers of the Rajput states.

---

[38] Thomas R Metcalf, *Ideologies of the Raj*, Indian Edition, Foundation Books, New Delhi, 1998, 2005, 77.

[39] Malavika Kasturi, *Embattled Identities: Rajput Lineages and the Colonial State in Nineteenth Century North India*, OUP, New Delhi, 2002.

[40] James Scott, *The Art of Not being Governed: An Anarchist History of Upland South East Asia*, Yale University Press, New Haven, 2009, 50–61.

Indirect Rule and the Frontiers

By the early nineteenth century as Rajputs states came under British indirect rule, distinctions between the 'civilized' settler and the 'barbaric' nomad were reflected in two kinds of ideas. The first was the idea of 'improvement', which was aimed at settling and rendering useful, the wastes that frontier zones like forests, deserts and mountains were seen as. The second was extension of control over groups that wielded authority in such frontiers, through instruments like the Criminal Tribes Act of 1871. As Neeladri Bhattacharya points out, the "uncultivated country side was not only barren and desolate but also dreary and ugly. Tamed, ordered, inhabited and productive landscape was beautiful. The 'well fenced field' was a sign of industry and order, as also a picture pleasing to the eye. The clearance of 'wastes' and colonization of land were therefore processes which transformed dreary landscapes into beautiful ones, activities by which 'wild' nature was tamed and ordered".[41] 'Wild' nature was equated with 'wild' inhabitants, with wildness signifying what came before and outside civilization.[42] The danger to state and civilisation lay both in untamed landscapes as well as untamed inhabitants. Dry and desolate spaces like Thar represented an unproductive space inhabited by dangerous outlaws. The colonial state was also intrigued by the way there appeared to be no fixed identities that could be attributed to the mobile groups. Therefore, it became important for the British to carry out extensive surveys and censuses that helped them locate communities, particularly mobile ones, in definitive frames. The colonial state sought to increase the "legibility" of the society by employing "reliable means of enumerating and locating its population, gauging its wealth and mapping its lands".[43] These censuses contained not only the numerical estimates of population but also descriptions of tribes and castes in the region, "larded with normative judgements about their qualities".[44] Bernard Cohn argues that the information regarding caste, customs, religion and language was practically garnered on the basis of enumerator's own knowledge

---

[41] Neeladri Bhattacharya, 'Pastoralists in a Colonial World' in *Nature, Culture, Imperialism: Essays on the Environmental History of South Asia*, (Eds.) D Arnold and R Guha, OUP, Delhi, 1995, 49–85, 73.

[42] Ajay Skaria, 'Being jangli: The politics of wildness', *SIH*, 1998; 14; 193–215, 196.

[43] James C Scott, *Seeing Like a State: How Certain Schemes to Improve Human Condition Have Failed*, Yale University Press, Yale, 1998, 46.

[44] Bernard Cohn, 'The Census and Objectification in South Asia', in *An Anthropologist among Historians and Other Essays*, OUP, New Delhi, 1987, 232.

of the communities, and therefore turned native perspectives into colonial information. In the latter half of the nineteenth century, the British, "anxious to rule India without disrupting its social institutions, and driven by an ever more compelling commitment to 'scientific understanding', set out to reduce to a comprehensible order what they saw as the baffling variety of India's myriad peoples".[45] This not only codified the society in a particular manner but also led to the wider circulation of ideas about the 'vagrant', 'lazy', 'irresponsible', 'lawless' nomad as against the 'sedentary', 'industrious', 'responsible' and 'law abiding' peasant.

Neeladri Bhattacharya argues that, "nomads, pedlars and pastoralists faced a more univocal opposition under the colonial regime as the state attempted to discipline and settle them, as the institutions of disciplinary power crystallized over them. The conflicting images, with all their ambiguities and possible variations in meaning, fused into one stereotypical image of nomad as vagrant. Watched, hounded, harassed and frequently persecuted by the police, nomads henceforth lived a life of eternal persecution".[46] The increasing control exercised by the British on mobile communities has been underlined by Mahesh Rangarajan who points out that, "stock keepers and animal trappers, shifting cultivators and itinerant communities were never subject to the kind of control that was to be attempted by the British. A functional explanation would be misleading but the British had a more antagonistic attitude to mobile groups".[47] Meena Radhkrishna's work *Dishonoured by History*, on 'criminal' tribes in the nineteenth century also emphasizes that it was particularly the mobile communities that were targeted by the Criminal Tribes Act of 1871.[48] Radhika Singha suggests that it was the fear of wandering service groups that were out of the ambit of control exercised by village elites and the access that they had to rural societies, that motivated their notification as criminal tribes.[49] Sanjay Nigam points out that "the experiences in Punjab and the North Western Provinces suggested that if predatory tribes were to be controlled, their mobility would have to be restricted and a systematic record

[45] Metcalf, *Ideologies of the Raj*, 116.

[46] Bhattacharya, 'Pastoralists in a Colonial World', 83.

[47] Mahesh Rangarajan, *Fencing the Forest: Conservation and Ecological Change in India's Central Provinces, 1860–1914*, OUP, Delhi, 1996, 16.

[48] Meena Radhakrishna, *Dishonoured by History: Criminal Tribes and British Colonial Policy*, Orient Longman, 2001.

[49] Radhika Singha, 'Settle, mobilize, verify: Identification practices in colonial India', *SIH*, 16, 2, 2000, 151–198, 159.

of their history and criminal activities be developed".[50] In my understanding, 'vagrancy' did not merely pose a challenge to the idea of 'improvement' and the settled order of life, it was dangerous because what was understood as criminal was closely related to control in frontiers like mountains, forests and deserts, which remained unassimilated in pre-colonial regimes for all practical purposes. Cattle raids, pillage and plunder were all methods employed by groups who laid and exercised claims on the frontiers. In such frontiers there had always existed multiple controls by groups familiar with the complex geographies of the region. What were considered to be 'predatory' tribes were all situated on frontiers of princely states and of the expanding colonial state. In the Thar, the extent of actual control exercised by the bardic community of Charans on routes of travel is yet to be understood. The coincidence between the fact that Charans had played important role as negotiators in periods of Rajput ascendance and their criminalisation and marginalisation in the 19th century is too uncanny to ignore.

The colonial view of the raids by the Bhils, Mers, Minas and such communities led to the emergence of notion of habitual criminality. Ajay Skaria argues that raids or *dhads* can be best understood as particular claims to power rather than being driven by subsistence needs alone.[51] This has also been pointed out by Shail Mayaram in the context of Meos, who views banditry as a reflection of fierce resistance offered by Meos to the sedentarisation and subordination fostered by a series of state formations.[52] The stringent control was evident in the dim view taken by the British regarding acts of 'crime' of cattle theft. David Gilmartin explains it having originated from the way they related milch and draught cattle with agrarian economy in the context of colonial Punjab. The increasing focus on private landed property, that was seen as a productive asset in the colonial economy, also necessitated the need to protect cattle that were seen as imperative for agrarian economy. Besides, the settlement of such disputes by communities themselves was seen as the overriding of the 'rule of law' and 'order' that was being promoted by the British.[53] In attempts to pacify such tribes, their inclusion in armed forces was

---

[50] Sanjay Nigam, "Disciplining and Policing the 'criminals by birth', Part 1: The making of a colonial stereotype - The criminal tribes and castes of North India", *IESHR*, 27, 2, (1990), 131–164, 138.

[51] Ibid., 200.

[52] Shail Mayaram, 'Kings versus Bandits: Anti-Colonialism in a Bandit Narrative' in *JRAS*, Third Series, Vol. 13, No. 3 (Nov, 2003), 315–338.

[53] David Gilmartin, 'Cattle, crime and colonialism: Property as negotiation in north India', *IESHR*, 40 (1) 2003, 33–56.

undertaken, as is clear from the example of Mewar, Malwa and Khandesh Bhil corps as well as Mhairwara Local Battalions in the 1830s.[54] The inclusion of these groups in the category of martial groups was a reflection of tacit acceptance of their martiality and the fact that inclusion of such groups was essential in settling the frontiers of the empire. Interestingly, another British enterprise that was at the receiving end of resistance from locals in the frontier was mapping. Mathew Edney records instances where survey parties were faced with mobs of unruly villagers commanded by local rulers who refused to allow mapping flags to be fixed or even pulled them down forcibly.[55] Mapping was closely related to fixing boundaries between states, in areas that were contested and also inhabited by recalcitrant groups. These were frontiers where the control exercised by Rajputs states was minimal and one that often had to be negotiated with multiple claimants to power, often the *barwuttea*. The British attempts at boundary settlements were actually aimed at converting the frontier into boundary, and thus rendering the multiple controls into singular that could be regulated through the Political Agents and Residents in the Rajput capitals.

The Frontiers of Memory

As frontiers were home to multiple political possibilities, they also fostered multiple historical imaginations. While 'authoritative' accounts of Rajput history emerged from Rajput courts and later from British historian-administrators like James Tod, multiple narratives hinting at ambivalence of identities circulated in the frontiers of Thar. These were largely oral narratives that relied on memory, both individual and historical. In the Thar, the basic frame through which such accounts are recollected and recited is the *bat* or *vat*, emanating from *varta*, which could mean both a tale and an epic, available both in oral and written forms from the late seventeenth century onwards. The *vat*, containing a number of elements like genealogies, poetic verses and explanations were technically to be recited but were certainly also composed in written forms by the community of Charans. With the decline of Rajput courts while Charanic *vats* stood marginalized, a number of oral narratives pertaining to non-Rajput groups remained in circulation. These narratives while ostensibly dedicated to Rajput folk deities like Pabuji, Harbhuji, Ramdeoji, or to heroic tales of non-Rajput heroes like Tejaji or Devnarayan, actually commemorate very complex

---

[54] Mark Brown, *Penal Power and Colonial Rule,* Routledge, 2014, 72–75.

[55] Mathew Edney, *Mapping an Empire: The Geographical Reconstruction of British India, 1765–1843,* Chicago University Press, 1999, 329–331.

and nuanced accounts of inter-community relationships, structures of power and control, and claims over heroic pasts in the frontiers.

In recent years memory has been viewed as a channel to approach collective pasts, particularly among groups marginalised by dominant practices of history writing. Memory becomes a device to recollect and recite origin myths, tales of valour, honour and sacrifice that form the basic frame of construction of community identity. It has been proposed by Maurice Halbawchs and Pierre Nora that the divergences between memory and history can be traced to modernity, with history being regarded as the rational scientific method to access a past from which communities had dissociated.[56] Memory on the other hand was associated with a lived past, continuing into the present, which was divested of its legitimacy by history. Nora differentiates between spontaneous and artificial memories, the latter being created through commemorative practices, through what Eric Hobsbawm and Terence Ranger describe as 'invented traditions'.[57] A similar dissociation has been assumed between orality and literacy, for example, by Walter Ong, who styles, "the orality of a culture totally untouched by any knowledge of writing or print, 'primary orality'".[58] These binaries are untenable as recovery of a pure oral, untouched by the literal is as difficult as recovering memory untouched by history. Memory in itself is a deeply conscious exercise, acts of remembering and forgetting often deliberate and conscious choices. Memory is often also a construct, catering to the needs of a community. As Ramya Sreenivasan points out, "memory is not history's Other but is itself a deeply historical practice".[59]

Over the past few years several historians and anthropologists have engaged with oral narratives to recover alternate or counter narratives to dominant historical traditions. Shahid Amin's exploration of the narratives of Ghazi Miyan, underlines the processes involved in "refashioning of sagas of 'religious' conflict in order to create communities in the past and in the present".[60] The idea

---

[56] Maurice Halbawchs, *On Collective Memory,* (Ed. and Tr.) Lewis Coser, The University of Chicago Press, Chicago, 1992. Pierre Nora, *Realms of Memory: Rethinking the French Past,* 3 Vols, Columbia University Press, New York, 1996.

[57] Eric Hobsbawm and Terence Ranger, *The Invention of Tradition,* Cambridge University Press, 1983, 4.

[58] Walter Ong, *Orality and Literacy: The Technologizing of the Word,* Routledge, 2002.

[59] Sreenivasan, *The Many Narratives,* 6.

[60] Shahid Amin, 'On retelling the Muslim Conquest of North India' in *History and the Present,* (Eds.) Partha Chatterjee and Anjan Ghosh, Permanent Black, Delhi, 2002, 24–44, 31.

of the Muslim warrior saint as the cow protector falls in the tradition of several Rajput *junjhars* who give up their lives in the act of cow protection.[61] These traditions appear to belong to regions with histories of mobility, particularly of young men seeking *naukari* in distant lands. Shail Mayaram's use of Meo oral traditions to recount Meo resistance is aware of the fact that through their inconsistencies and contradictions, the Meo oral traditions articulate a definite point of view. Rather than being caught in proving or disproving the facticity of these accounts, Mayaram examines the "interweave of the imaginary and the real that oral traditions represent, capturing perspectives on the contested past as well as shifts in group perspectives".[62] Brenda Beck demonstrates the construction of Kovantur identity in Tamil Nadu through the narrative of The Three Twins, situating the epic in deeply conflicted zones between agrarianism and forest based livelihoods. The recent textualisation of Annanmar epic as well as its popular cinematic versions while essentialising the epic tradition also introduce newer elements into it, reflective of the ways in which community identities have been shaped by contemporary events.[63] Similarly, in her study of the *Dhola* epic, Susan Wadley points out the deeply historical roots of the *Dhola* tradition in the conflict of identity between Jats and Rajputs in the Braj region.[64] Aditya Malik's exploration of the Devnarayan tradition warns against treating oral narratives as sources from which 'tangible' histories could be winnowed.[65] In her work on the Pabuji narrative, Janet Kamphorst traces a range of transferences between written and oral narratives of Pabuji tradition indicating the complex ways in which Rajput, Bhil and Charan identities were entwined.[66] All these studies however firmly place the oral narratives in the communities for whom they hold a meaning. All these arguments regarding oral narratives are located in a debate between Stuart Blackburn and Alf Hiltebeitel. While Blackburn considers the vernacular epic traditions to be

---

[61] The idea of headless warrior itself appears to be drawn from the shaktic tradition of *Chhinmasta*, the beheaded goddess.

[62] Shail Mayaram, *Resisting Regimes: Myth, Memory and the shaping of a Muslim Identity*, OUP, 1997, 47.

[63] Brenda Beck, *The Three Twins: The telling of a South Indian Folk Epic*, Indiana University Press, 1982.

[64] Susan Wadley, *Raja Nal and the Goddess: The North Indian Epic of Dhola in Performance*, Indiana University Press, Bloomington, 2004.

[65] Aditya Malik, *Nectar gaze and Poison Breath: An Analysis and Translation of the Oral Narrative Tradition of Devnarāyan*, OUP, New York, 2005.

[66] Kamphorst, *In Praise of Death*.

in opposition to Sanskrit epic traditions, Hiltebeital views the vernacular traditions as re-emplotments of the Sanskrit epic traditions, evolving through various regional frames and seamlessly merging into each other.[67]

For this work, examining oral traditions became essential given the absolute paucity of 'historical' material about a number of mobile communities. Engaging with oral narratives raised several questions about historicity. However while at no point attempting to locate a historical Pabuji or Devnarayan, I have tried to see the larger historical context in which communities refashioned, re-emplotted and remembered these epics. Narratives like Pabuji, Dhola and Devnarayan circulated in frontier regions, where low caste bardic groups narrated multiple versions catering to varied social groups. These narratives themselves were the frontiers between history and memory.

This book is an attempt at understanding the relationships between changing perceptions of mobility and transitions in the nature of polity between periods of Mughal and British indirect rule in the Thar. My engagement with polity through mobility explores the relationship between mobility, control and power in the emerging contours of Rajput polity from the sixteenth century onwards. As sedentarizing Rajput states employed multiple tools of legitimization, aimed at both their kinsmen as well as their indirect rulers, they still had to continue to deal with mobile groups that wielded immense control on the frontiers of the Thar. The 'ordering' of the Thar in the nineteenth century resulted not only in physical demarcation of boundaries but also into reconfiguration of networks of circulation as well as of the groups circulating on these networks. With references to shared histories of itinerancy having been increasingly obfuscated in the new emerging Rajput histories, the oral narratives of the region offer the only possibilities of understanding links between mobility, martiality and power in the Thar.

A Note on Sources

The time frame of this work necessitated use of several different kinds of sources, which together were able to afford clarity on several issues discussed. Beginning from Nainsi's works composed in late seventeenth century to vernacular English Records and Memoirs of the nineteenth century, each

---

[67] Stuart Blackburn, Peter J Klaus, Joyce B Flueckiger, Susan S Wadley (Eds.), *Oral Epics in India,* University of California Press, Berkley, 1989. Also Alf Hiltebeitel, *Draupadi among Rajputs, Muslims and Dalits: Rethinking India's Oral and Classical Epics*, Oxford University Press, New Delhi, 2001.

source revealed a different perspective. Yet, every kind of source also presented a limitation by its own nature. Therefore, while using these sources to garner information, each source has had to be critically evaluated.

Nainsi's *Khyat* and *Marwar ra Paraganan ri Vigat*[68] are important sources for gaining an insight into the history of the Thar. While the former deals exclusively with the history of Rajput clans, the latter has been composed in a gazetteer like fashion to give an account of the paraganas of Marwar during the reign of Jaswant Singh. *Khyat*, which has been understood to mean an account of glory (*khyati/akhyan*) emerged from the tradition of composition of panegyric works of poetry and prose in the Rajput courts. These were primarily oral accounts that were being textualised by the seventeenth century. *Khyat* and *vat/bat* were recited in Rajput gatherings and were meant to convey the magnificence of both the Rajputs and the composers that were the Charans. These accounts contained histories of battles, sacrifice, valor, and chivalry, values that came to be associated with Rajputs and were often exhortation to the Rajputs.

Nainsi's *Khyat* also contains the genealogies of various Rajput groups, as they were composed around the seventeenth century. It has been pointed out that the process of creating close associations between new Rajput groups and older Kshatriya clans involved converting king lists into ascendant genealogies, in order to create an impression of a long unbroken line of descent. The time of composition of these genealogies around the period when Akbar as the Mughal emperor was himself attempting to forge a similar association, is reflective of the fact that Rajput polity by the seventeenth century was to a certain extent oriented towards the Mughal Empire. The attribution of values like bravery and chivalry was meant to create a particular kind of image with respect to the Mughal court. However, the fact that Nainsi composed two separate accounts, one framed in the accepted historical traditions of the Rajputs and another of the Mughal court suggests that these accounts had separate circulations.

The tradition of *khyats* and *bats* came under severe criticism in the nineteenth century when Dr. L P Tessitori rejected them as hyperboles.[69] James Tod and Tessitori both discussed mutual relations of Rajputs and Charans and labeled

---

[68] Narayan Singh Bhati (Ed.), *Munhata Nainsi ri Likhi Marwar Ra Paraganan Ri Vigat*, 3 Vols, Rajasthan Oriental Research Institute, Jodhpur, 1968–74. Henceforth Nainsi, *Vigat*.

[69] L P Tessitori, 'A Progress Report on the Preliminary work done during the year 1916 in connection with the Proposed Bardic and Historical Survey of Rajputana', *JPASB* (New Series) Vol. XIII, No: 4, 1917, 228.

these traditions as "empty praise against solid pudding".[70] The Charans were accused of manipulating facts and blowing them out of proportions in order to heap praises upon the Rajput patrons. In the nineteenth century Charans were also charged with using the fear of *bhumd* or ridicule in order to extort gifts and grants from the Rajputs, particularly on the occasions of marriages. This evaluation of the *khyat* and *bat* tradition was primarily the outcome of changing relationships of Rajputs, Charans and the British, a context in which Charans became dispensable for Rajputs. Besides, *khyats* and *bats* were accounts that were recited according to the occasions. The process of narration and hearing caused the *bat* to be transformed in each recitation. The act of writing made them definitive and later these were evaluated in the context of norms of history writing as they were emerging in the nineteenth century.

Munhata Nainsi was however not a Charan but an Oswal *mutsaddi* in the court of Jaswant Singh. He remained the Diwan of Marwar till he lost favour with Jaswant Singh following which he was imprisoned. Nainsi's *Khyat*, nevertheless depends on the Charan accounts in order to put together a comprehensive history of the Rajput clans. The *Khyat* contains a collection of *bats* as well as *kavitts, dohas, vamsvallis* and *pidhivallis*, gathered by Nainsi through his official means. Nainsi, at places acknowledges the Charans who composed the *bats* and at others it is quite evident that the authorship is lost. Nainsi does not claim to author the accounts and at places even refers to the anonymous source as '..*aa bat suni hai*' (this has been heard). This clearly refers to *bats* having been a part of a tradition that was to be narrated and heard. Nainsi's *Khyat* while referring to Rajputs as descendants of older Kshatriya clans also details a long history of their movements indicating how Rajput movements and settlements were perceived and remembered.

Marwar ra Paraganan ri Vigat, on the other hand is a *paragana* vise description of the land and people and can be considered more of a gazetteer. The *Vigat* also provides a chronological account of the more recent past of the Rathors and their relations with the Mughal Empire. Nainsi describes battles, marital alliances, cadet lines, court rituals, *sasan* grants, *pattas* apart from the descriptions of events during the reign of Jaswant Singh, his patron. Therefore, we are able to get an idea of the geographical lay of the *paraganas* of Marwar state in the late seventeenth century including production patterns, description of towns and villages, castes and communities. The *Vigat* however deals only with the *khalsa* villages and the *raiyyat* in these villages. As a result,

---

[70] Tod, *AAR*, I, xv.

we do not get an account of the *patta* villages and communities that did not figure in the purview of the state taxation mechanisms.

The official records of the states however provide us with details of communities that came in contact with state mechanisms. The *Sanad Parwana Bahis* of the eighteenth century Jodhpur state for instance, are detailed records of petitions and orders. The people of the state initially resorted to mechanisms like jati panchayats. However if their grievances could not be redressed at the local level they addressed these to the state officials. These grievances were then resolved at the paragana headquarters and suitable orders passed on to the *patayat* or *daroga*. It needs to be pointed out however, that the petitions were often versions created by the state officials themselves as the petitioners were more likely to present their complaints orally. Therefore, *Sanad Parwana Bahis* present the official case histories of grievances. Yet, a perusal of *Sanad Parwana Bahis* gives the idea of issues that were being raised by people irrespective of their social status. The *Sanad Parwana Bahis* range from VS 1821–1995 (1764 to 1938), a total of hundred and two volumes. Each Bahi provides us with paragana wise details of the *arzees* and the *sanads* thus issued beginning from *Chaitra sudi* 1. Each document bears the date and place of writing. About fifty *Bahis* were consulted in course of research for this work. The references cited refer to the date and place of the individual petition as well as the number of the *Bahi*. Apart from *Sanad Parwana Bahis*, *Haqeeqat Bahis* of the Jodhpur state have also been consulted. Since the book also deals with the commercial networks, a number of *Sayer Bahis* were consulted to get an idea of the movement of commodities through Marwar. *Sayer Bahis* provide details of taxes, routes and the merchants that moved on these routes. In context of Bikaner state a number of *Zagat Bahis* of mandis were consulted. These *Bahis* contain detailed references to the communities engaged in commercial exchange. However, references to the communities that remained outside the purview of both rural and urban societies and maintained their own redressal mechanisms like the Kanjars, Nats and Kalbelias etc are very rare.

The references to these communities were difficult to find till I referred to the ethnographic works of the nineteenth and early twentieth centuries. The compendia of tribes and castes particularly those of R E Enthoven, William Crooke, H A Rose and R V Russell, provide wide-ranging information about the history and status of various communities in the late nineteenth century. However, what these compendia offer is the administrators' and enumerators' perceptions of these communities, which were highly antagonistic. In order to classify India and Indian people in a variety of classificatory systems, local histories were collected with the help of informants who often were literate groups Brahmins, Bhats, Charans etc. Special attention was paid to a study of physical appearances and social customs in order to compare communities with the classification systems as perceived by the British administrators. The turning of communities into ethnological curiosities was also accompanied with a description of peculiar characteristics of the communities that particularly in case of the mobile communities underlined their differences from the settled communities. The comparisons were often negative and described most mobile communities as being engaged in criminal activities. These compendia established and reinforced the specific images of mobile communities that continued to be reiterated by subsequent works. The early census reports of Marwar and Bikaner, like the detailed *Mardumshumari Raj Marwar* collated by Munshi Hardayal were also instrumental in realigning of the rural and urban societies in newer frameworks.

This imagery was often restated in the memoirs and travelogues of written by itinerant travellers and British officials in the nineteenth century. A dominant feature of these works was the expression of desolation in the Thar. Most European travellers perceived the desert as hostile to settlement and travel. The travel accounts of Col. James Tod, Capt. Boileau, Archibald Adams, and Col. Holbein Hendley all underline the hazards of travel in the hostile desert. This notion of hostility as ingrained in the psyche of the British travellers and administrators became the basis of the ideology of settlement that sought to make the desert hospitable for human settlement. The notion of land lying waste was repeatedly underlined by Capt. P W Powlett and P J Fagan, who respectively wrote the *Gazetteer of the Bikaner State* and *Settlement Report of Bikanir.* Hostile landscapes were also seen as harbouring hostile inhabitants moving with an ease that the administrators could not garner. The experienced hostility of landscapes and its itinerants became a reason for 'criminalization' of mobile communities, as evident in policies of the Rajput states in the late nineteenth century.

A work that has majorly impacted subsequent histories of the Thar has been Col James Tod's *Annals and Antiquities of Rajast'han.* As a Political Agent, Tod negotiated treaties between Rajput states and the British. In this endeavor he travelled through several princely states and began the ordering of the past and the present of these states in his work. Tod attempted to create a comparison between the Rajput martial systems and the European feudal orders. Rajputs were to be resurrected as Anglo-Saxon knights with comparable values of honour, chivalry, truth and valor. So impressive were Tod's arguments that

Rajputs attempted to model themselves to this imagery. In Tod's romantic imagination remnants of ancient civilizations were spread all over Rajasthan and needed to be rescued from the oblivion. It was Tod's imagery that fostered the use of the word 'Rajast'han', a clear identification of the region with Rajputs, despite the fact that Rajputs were a demographic minority. For the next hundred years or so Rajputs became the dominant context of history writing in Rajasthan. Yet, Tod's works lay bare for us the context in which the mutual relations of the Rajput principalities and the British operated. Tod's geographical imagination also creates the image of the desert as no other work does. With his travels we are able to see what it meant to traverse through the Thar in the nineteenth century.

Dr. L P Tessitori's work *Bardic and Historical Survey of Rajputana* is an important work that details the process of identification, acquisition and editing of the texts that were to become the basis of history writing on Rajasthan. The importance of Tessitori's work lies in his critical understanding of the process of composition and preservation of historical texts. Tessitori's views on the role and the position of the Bhats and the Charans are also indicative of the changing perception of these communities and their 'histories' in the early twentieth century. Tessitori, as a grammarian and an antiquarian, was interested in both the manner in which texts that he was unearthing and collecting could be used as sources for history writing as well as the changing contours of the texts themselves. He critically examined the poetic and the prose literature and correlated the changes in the form, structure and language of these texts with the structures of patronage as well as to the socio-cultural transformations like the rise of lower caste composers who, according to Tessitori, used Braj Bhasha instead of older Dingal to compose in the eighteenth and the nineteenth centuries. Tessitori's astute understanding of processes of composition of genealogies and the significance of lacunae in genealogies contributes significantly to the understanding of processes of emergence of Rajputs.

By the middle of the nineteenth century, the British methods of governance necessitated the compilation of detailed reports on administration, customs, boundary surveys, forests, PWD, Railways, salt, famine and famine relief etc. Separate departments were formed that were accountable to the Musahib Ala, who was responsible for overall administration. Reports like those by Loch and Hewson on Customs, Famine Relief Reports and Criminal Administration Reports by Sukhdeo Prasad, Manuals of the Northern India Salt Revenue, Boundary Settlements among others have been used to get an idea of the functioning of state mechanisms in the nineteenth century. Routine correspondence under the Mehkama Khas records of Jodhpur, Bikaner and Jaisalmer has also been consulted. The Foreign and Home Department Records at the National Archives of India have also been used to understand the administrative processes of the nineteenth century.

Plan of the Book

The first chapter, "Geographical Imagination and Narratives of a Region", explores how the term 'Thar' referred not merely to a given region, but rather to a wider space where people, commodities and lore constantly circulated on overlapping networks that were political, commercial, pastoral and cultural at the same time. Extending beyond political boundaries of what came to be understood as 'Rajputana', the Thar was united through shared history of itinerancy as well as shared experiences like famines. Ravelling through a number of narratives from the seventeenth century to the nineteenth century I show how the region evoked varying geographical imaginations, yet itinerancy always featured prominently. I contrast geographical and political imaginations of two chroniclers of the Thar, Munhata Nainsi in the seventeenth century and James Tod in the nineteenth. I explore how the idea of Thar shifted from the open frontier traversed by itinerant Rajputs that Nainsi views it as, to the region circumscribed by the boundaries that James Tod drew on his map.

The second chapter "Mobility, Polity and Territory", examines the process of emergence of Rajput states and thus of shift from a polity of mobility to a polity of landedness. The chapter explores the shifts in constructions of Rajput identities from mobile exogamous pastoral warrior groups to endogamous landed clan aristocracies in vernacular literature like the *Khyats* as well as colonial accounts, like that of James Tod. Thereby, it traces the emergence of the predominant identification of the Thar as synonymous with Rajputana, the land of Rajputs and subsequently, Rajasthan. It also explores the emergence distinctions between the 'noble' Rajput and the *barwuttea* and argues that these distinctions should be understood in the context of the core and the frontier. Rajput-Mughal marital alliances legitimised through extension of Rajput kinship norms to Mughals helped in emergence of the 'Mughal' Rajput who represented the brave, loyal chivalrous Rajput. While the emergence of Rajput as a 'caste' category closed the ranks of Rajputhood for various groups with martial aspirations, Rajput 'status' remained open and was claimed by several genealogically obscure groups like Marathas, Jats, Sikhs, Satnamis among others. The dominance of Rajput historical narratives, which themselves were a part of the process of 'the making of the Rajput', led to the erasure of narratives of groups like Bhils, Mers, Gujjars and Jats who at various times had exercised control over large parts of the Thar desert.

The third chapter "Itinerants of the Thar" documents the circulatory networks in the Thar desert as well as a variety of itinerant communities travelling on them in this period. It traces the itinerancy of Rajputs, commercial groups like Marwaris, Lowanas, Banjaras, pastoralists like Raikas and Gujars, artisanal groups like Sansis, Kalbelias and Gaduliya Luhars and bardic groups like Charans, Bhats, Manganiyars, Sevags, Kamads etc. It demonstrates that social identities attributed to these groups were formulated with reference to each other and to degrees of mobility of these communities. It also argues that for a long time, mobility and sedentarism were not fixed attributes in this region. The fragile ecological base of the Thar called for a prudent balance of resource use, which could only have been possible through constant circulation of resources and persons.The chapter also demonstrates that the assertions of a Rajput past on part of mobile groups like Raikas, Gujars, Bhils, Gaduliya Luhars etc do not merely result from a aspirational claim over past, but from a history of mixed, ambivalent and indeterminate social identities.

The fourth chapter "Expanding State Contracting Space", examines the extent and the mechanisms of control exercised by state on the circulatory networks and thus on mobile communities in the ninetenth century Thar. The chapter explores why mobility came to be viewed with suspicion in the nineteenth century, when extensive efforts were made to 'settle' the region and its inhabitants by expanding cultivation and irrigation. Not only did this result in decline in availability of pasture to pastoral herds, it also laid grounds for subsequent conflicts between cultivators and herders, by privileging the former. The nineteenth century princely states of Jodhpur, Bikaner and Jaisalmer also imposed extensive restrictions on the movements of mobile groups like Baories, Sansis, Banjaras, Charans and Bhats, and established departments like *Mehkama Jurayampesha wa Baoriyaan,* to monitor and settle such groups. Such measures reiterated the state's view that ownership of agrarian land ingrained orderly subjecthood while mobility was equated with vagrancy and criminality. The chapter also examines the reordering of circulatory spaces through construction of roads and railways. These were built primarily to regulate the traffic in salt, a commodity monopolised by the British in the late nineteenth century, as well as to extend greater control in frontier regions like Mallani and Sindh. However, the new networks replaced an older circulatory regime, which was composed of a wide range of traders and commodities. The chapter includes a case study of salt trade in the region and demonstrates how monopoly over salt trade, decline of old market networks and "criminalisation" of a trading community like Banjaras were related processes.

The fifth chapter "Narratives of Mobility and Mobility of Narratives", explores some oral narratives of the region and shows that travelling with these narratives can enable us to see how social identities were shaped in the region. Examining oral narratives like those of Pabuji, Tejaji and Devnarayan, the chapter explores the relationships between mobility and memory. The chapter argues that the recitation and propagation of these oral narratives could well be understood as acts of preservation and reiteration of a claim over past by communities like Bhils, Raikas, Jats and Gujars. Exploring the circulation of these oral traditions, temporally and spatially, not only helps me understand these as counter narratives to dominant written Rajput histories, but also in the process provides me with new ways of examining the formulation of complex social identities.

1

Geographical Imagination and Narratives of a Region

Marwar is a corruption of *Maroo-wár*, classically *Maroost'hali* or *Maroost'han*, 'region of death.' It is also called *Maroo-dèsa*, whence the unintelligible *Mardés* of the early Mahomedan writers. The bards frequently style it *Mord'hur*, which is synonymous with *Maroo-désa* and when it suits their rhym *Maroo*.[1]

The Thar Desert, classically identified as *Marusthali* or the land of death, signifies the entire arid and semi arid stretch of land enclosed by the Sutlej basin on the north, Aravali on the east and south east, the Rann of Kutch and plains of Kathiawar on the south and the Indus basin on the west. It is the world's seventh largest desert and if plotted on a map, is part of a chain of hot and cold deserts between North Africa and Central Asia, what Jos Gommans, refers to as 'Saharasia'.[2] The temperature in the Thar during summer days may reach up to 50°C and fall to freezing point at night in winter. The average rainfall is about 254 mm and high rate of evaporation causes the water to disappear quickly.[3] Sources of water are few seasonal rivers, lakes and reservoirs around which the Thari people make their settlements. The origin of the word Maru lies in the Sanskrit word *mri* meaning death, which evokes the vision of sandy plains devoid of water. When James Tod, the nineteenth century annalist of Rajputana wrote his *Annals*, the word Maru had long come

---

[1] Tod, *AAR*, II, 1.

[2] Gommans, *Mughal Warfare*, 26.

[3] *Rajasthan Development Report*, Planning Commission GOI, 2003.

to represent Marwar, the political domain of the Rathors. However, for Tod, its "ancient and appropriate appellation comprehended the entire 'desert' from Sutlej to Ocean".[4]

While the Thar is largely understood as a geographical region, a dry arid zone divided by an international boundary between India and Pakistan, its history has been explored separately through the political units that it was divided into. The Thar Desert contains in itself the older regional divisions of Marwar, Jaisalmer, Bikaner and parts of Kutch, Multan and Sindh, which further consist of smaller ethno-regions. The southern and northeastern portion of the Thar is a vast sandy tract comprising of parts of Marwar and Amber known as Bagar.[5] Shekhawati, Bidawati, Pugal, Asigarh, and Beniwal and Bhatner are the northern parts of the desert. The northwestern part of the desert is called Chitrang. The land around Umarkot is addressed as Dhat, which neighbours regions like Thalaicha, Tharel, Tirrud, Khawar, Khairalu, Khadal and Sam. The older appellation for the area around Barmer is Mallani, or the land of Mallinath. South of it are Sirohi or Deorawati, Endowati, Mahewa, Sewanchi and Bhakar. The fertile region adjoining the foothills of Aravali that meets the basin of the Luni River is known as Godwar, Merwara and Nyar. Apart from sand dunes the other dominant feature of the Thar, from Pokhran to Jaisalmer and Barmer, are the high rocky plains blocked by isolated hills. The gravely kankarised plains of Kolayat, Chandan and Phalodi as well as sandy plains with shallow soils lie almost in the middle of the Thar. Pokhran, Bap, Phalodi, Pachpadra, Lunkaransar, Chappar, Kuchaman, Didwana are the major saline depressions. [6] A part of this desert was also addressed as 'Jangal', 'Jangal Des' or 'Jangalu' referring to the arid wilderness of the region. Contrasted with the neighbouring fertile plains of the Punjab rivers and the Indus, rich loamy lands of Gujarat and the thickly forested plateau of Malwa, its aridity renders a singularity to its character, that of a desert.

Munhata Nainsi's seventeenth century chronicle *Marwar ra Paraganan ri Vigat* refers to an ancient expanse of the desert marked by the nine forts of Marwar or *Naukoti Marwar* belonging to the early Paramaras, namely, Barmer,

---

[4] Tod, *AAR*, II, 1.

[5] P W Powlett, *Gazetteer of the Bikaner State*, 1874. Reprinted at the Government Press Bikaner, 1932, Bikaner, 91–92. Henceforth, *Gazetteer.*

[6] A Adams, *The Western Rajputana States: A Medico Topographical and General Account of Marwar, Sirohi and Jaisalmer*, 1889, Vintage Books, Reprinted, 1990, 43. Henceforth, *The Western Rajputana States.*

Abu, Parkar, Pugal, Jalor, Umarkot, Ludrovo, Ajmer and Mandor.[7] If plotted on a map this expanse ranges from Ajmer in the east to Parkar, Sindh in the west; and from Pugal and Ludrovo (near Bikaner and Jaisalmer respectively) in the north and northwest to Abu in the south. Along with the nine forts of Bhatis, that is, Jaisalmer, Pugal, Bikampur, Barsalpur, Mamanvahan, Marot, Derawar, Aasnikot and Kehror, these forts appear to circumscribe the entire Thar Desert.[8] Interestingly a popular couplet referring to the scourge of famine refers to more or less a similar space, "My feet are in Pugal, torso in Umarkot, and stomach in Bikaner...I wander frequently into Jodhpur but I reside permanently in Jaisalmer". This couplet is perhaps the most apt representation of the geographical-ecological continuum of the desert, where the scarcity is an ever- present feature.

The traditions of Thar view the desertification of the region as part of an older process resulting from both natural and supernatural causes. The region abounds in mythologies of desertification that refer to long abandoned ways of life, of rivers and oceans. Myths attribute the present state of land to sins of rulers and banes of deities. Narayan Singh Bhati in his commentary on the 17th Century chronicle of Munhata Nainsi refers to an old myth that says, "at one time a distributary of Sutlej passed through Sindh, because of which the region was fertile. Umar Soomra evoked the ire of Goddess Sangviyan, the presiding deity of the Bhatis, by expressing a desire to marry her, who then changed the course of the river and made it to flow through Multan. Ever since there is a famous saying that, the water went to Multan".[9] Nainsi also attributes the desolation to a bane, which resulted in the land becoming "destitute of knowledge, dignity and water".[10] A H E Boileau surveying the desert in 1835–36, came across a myth according to which Bikampur, "once

---

[7] Nainsi, *Vigat*, I, 1.

[8] Nainsi, *Khyat*, II, 11.

[9] "Paani Multan gayo" Narayan singh Bhati's commentary in Munhata Nainsi's *Vigat*, III, 88. Histories of Rajput States of Jodhpur, Bikaner and Jaisalmer refer to important roles played by virgin goddesses of Charan origin. *Tawarikh Jaisalmer* refers to a similar incident where by the Charani goddess Avar and her six sisters caused the river to be dammed turning water towards Multan. Likhmi Chand, *Tawarikh Jaisalmer: Jaisalmer Rajya ka Itihas*, (First Published 1891), Rajasthani Granthagar, Jodhpur, 1999, 226. Harald Tambs-Lyche also refers to several such myths in his work *Power, Profit Poetry: Traditional Society in Kathiawar*, Manohar, New Delhi, 1997.

[10] The people of Pokhran were cursed, "*Vedheen huvo, kriya bhrust huvo, nirjal des baso, man heeen huvo*". Nainsi, *Vigat*, II, 289.

stood on the bank of a river which was drank dry by a divinity taking up the
water in the hollow of his hand".[11] By the nineteenth century it was observed
that the availability of river water was reduced to seasonal overflows during
monsoons. Capt. Powlett writes in the *Gazetteer of the Bikaner State* that,

> In the rainy season, a nallah sometimes flows from Shekhawati over the eastern
> border, but is soon lost in sands. The Kagar also called the 'Sotra or Hakra' once
> flowed through the northern part of the Bikaner territory... but it is now dry.[12]

James Tod observed that, the "traditions assert that these regions were not
always arid or desolate". Their 'deterioration' was a result of the "drying of
the Hakra river which came from Punjab and flowing through the heart of
the country emptied itself into the Indus between Rory Bekhar and Ootch".[13]
Geological researches in the Thar Desert attribute these changes to an
old history of geological shifts resulting in the altered flow of rivers and
their drying up, gradual silting leading to formation of sandy plains and
salty marshes among others. Several Neolithic, Chalcolithic and Harappan
sites dating to 6000–2000 BC have been found in this area. These include
Harappan settlements like Kali Bangan, Banawali and Baror that developed in
the dry bed of Ghagghar in northwestern Bikaner.[14] In the mature Harappan
phase, there appears to have been a large-scale migration of nomadic and food
gathering communities to urban centres and to the hinterlands surrounding
them. Nevertheless, these Harappan urban centres were abandoned within the
next 500 years. Mohammad Rafique Mughal attributes this abandonment to
hydrological deterioration, for instance the westward shift of the Indus river
system.[15] This led to the engulfing of entire settlements by the expanding
desert, as stated by Mughal in case of Cholistan, the Pakistani side of the Thar.
This expansion of the Thar Desert has been dated to around 2000 - 1500
BC, around which time Ghaggar appears to have ceased to be a major river
and terminated in the desert. It has been observed through remote sensing

---

[11] A.H.E Boileau, *Narrative of a Tour Through Raiwara Embracing the Princely States
of Jaisalmer, Jodhpur and Bikaner in 1836*, Baptist Mission Press, Calcutta, 1837, 94.
Henceforth, *Narrative*.

[12] Powlett, *Gazetteer*, 92.

[13] Tod, *AAR*, II, 144.

[14] B. Allchin, and F R Allchin, *The Rise of Civilization in India and Pakistan*, Cambridge,
1982, 54–56, 157–159.

[15] M. Rafique Mughal, *Ancient Cholistan: Archeology and Architecture*, Ferozesons, Lahore,
1997, 29–48.

techniques that late quaternary climatic changes and neo-tectonics have played a significant role in modifying the drainage courses in this part and a large number of paleo-channels exist in the region. Some studies have accepted the existence of paleo-channels coinciding with the bed of present day Ghaggar, and believe that the Sutlej along with the Yamuna once flowed into the present Ghaggar riverbed. It is also speculated that the Sutlej was the main tributary of the Ghaggar and that subsequently the tectonic movements might have forced the Sutlej westwards, and the Yamuna eastwards leading to the drying up of the Ghaggar.[16] Mughal concludes that these tectonic shifts might have occurred either in late Harappan period or in the PGW period or both as attested by the presence of both Harappan and PGW sites on one of the paleo-channels known locally as Hakra.[17] The present disorganized state of many streams in the Thar Desert can thus been attributed to tectonic shifts and change in climate.[18] It is not surprising that mythologies and folk wisdom also refer to similar changes.

However, though the mythology appears to refer to a well watered and populated land ruined either by geological changes or banes, the recent remembered or written accounts allude to a recognizable picture of a desert with sandy plains, rocky formations and scarce water resources. In the well known seventeenth century poetic work, *Dhola Maru ra Duha*, Dhola's second wife Malwan points out,

> Maru is colourless…where water is to be found so deep in the wells that people have to start fetching water at midnight, where there are only shepherds, where there is either a famine or an invasion of locusts, where grasses like *kareel* and *untkataraa* are regarded as trees, the only shadow to be found is that of *phog* and *aak* and people have to eat seeds of *bhurat* to assuage their hunger, where there are only rough blankets to wear and people roam around all the time.[19]

In early nineteenth century Tod points out, "from Bhalotra on the Looni, throughout the whole of Dhāt and Oomrasoomra, the western point of

---

[16] A. B. Roy and S. R. Jakhar, 'Late Quarternery drainage and disorganization and migration and extinction of Vedic Saraswati' in *Current Science*, Vol 81, no 9, November 2001, 1194.

[17] Mughal, *Ancient Cholistan*, 29–48.

[18] Ghosh and Jakhar, 'Late Quarternery drainage', 1195.

[19] Narottam Das Swami (Ed.), *Dhola Maru ra Duha*, Rajasthani Granthagar, Jodhpur, 1995, 185–186.

Jessulmér, and a broad strip between the southern limits of Daodpotra and Bikanir, there is real solitude and desolation".[20] In this region, "the villages are far apart, and though grass and jungle bushes often abound, the aspect of the country is dreary and desolate, except, as often happens in the cold weather, when a mirage places a lake in the horizon".[21] The western part of Jaisalmer was considered to be, "a vast expanse of sand, thinly populated, that could hardly present a more desolate appearance; the villages few and far apart consist of some circular huts round a brackish well of water".[22] Through his travels Tod discovered that,

> From the north bank of Luni to the south, and the Shekhavati frontier to the east, the sandy region commences. Bikaner, Jodpoor, Jessulmer, are all sandy plains. All this portion of the territory is incumbent on a sandstone formation; soundings of all new wells made from Jodpoor to Ajmer, yielded the same result; sand, concrete siliceous deposits, and chalk.[23]

As Capt. Powlett saw,

> The north-west and part of the north are in the Great Indian Desert. From the city of Bikaner, south-west to the Jaisalmer border the country is hard, but, throughout the greater part of the territory the plain is undulating or interspersed with shifting sand-hills, slopes of which, lightly furrowed as they are from the action of the wind, suggest the sea-shore.[24]

In approximately the same period, Archibald Adams noted in Jaisalmer that,

> The western portion of the country is the most desolate tract with large extents of shifting sands termed dhurian. They are very difficult to cross.... the path shifting daily. The inhabitants say they are travelling northwards…and swallow up and occupy a large portion of the country, depriving the inhabitants of their pasture grounds.[25]

This desolation was referred to in more ways than one by travellers and chroniclers. For, Nainsi, *soono khedo* was the term most expressive of barren

---

[20] Tod, *AAR*, I, 15.

[21] Powlett, *Gazetteer*, 92.

[22] Adams, *The Western Rajputana States*, 22.

[23] Tod, AAR, I, 14.

[24] Powlett, Gazetteer, 92.

[25] Adams, *The Western Rajputana States*, 22.

and destitute lands, abandoned by residents.[26] For Tod, the best expression of the desert lay in the terms, '*t'hul*' and '*rooé*, "the first means an arid and barren desert; the other is equally expressive of the desert but implies the presence of natural vegetation".[27] As Adams saw, the country north of Jodhpur was, "one vast sandy plain or *thal*, broken by sand hills or *tebas* which commencing in Marwar stretch north into Bikaner and west and south into Jaisalmer and Sindh".[28] Tod delineated major *thals* as being; the "*t'hul* of Luni, embracing the tracts on both banks of the river, *Gogadeo ca t'hul*, immediately north of Eendovati"; the *t'hul* of *Tirruroé*, lying between that of Gogadeo and the present frontier of Jessulmer; *t'hul of Khawur*, lying between that of Jessulmer and Barmair and abutting at Girap into the desert of Dhat; and *Mallinat'h ca t'hul* or *Barmair*".[29] Nainsi also points to the *thals* as being associated with the communities that occupy them.

> Mangaliya Bhatis live in the *Mangaliya ka thal*, 20 kos west of Jaisalmer. It is such a desert where sand dunes keep changing because of the wind. Only the people who know the land can travel here. Unknown horsemen are lost in the sand with their horses, if they stray they can die.[30]

The natural vegetation of the Thar is thorny and consists of grasses and shrubs suitable for grazing of camels, sheep, goats and cattle at different times and levels. It is also home to several species that act as sand stabilizers and check expanding desertification. For instance *Phog (Calligonium polygonoides), Kheemp (Lepatadenia)* and *Thor (Euphorbia caducifolia)* are found to be dominant in the rocky gravelly areas of Barmer and Jaisalmer with negligible soil cover. *Khejri (Prosopis)* and *Babul (Acacia)* trees and shrubs are dominant between Jodhpur and Shekhwati. According to Tod, these shrubs provided the shifting dunes with a degree of permanence in winter and almost gave them an "appearance that amounted to verdure".[31] However, the vegetation cover is scanty as noticed

---

[26] Nainsi, *Vigat*, II, 1.

[27] Tod, *AAR*, II, 238.

[28] Adams, *The Western Rajputana States*, 4.

[29] Tod, *AAR*, II, 241–43.

[30] Nainsi, *Khyat*, II, 31.

[31] Tod, *AAR*, I, 235. In recent decades a non-indigenous variety of Babul has been planted all over the desert giving an appearance of a green cover even in summer, but in reality these shrubs do not serve the purpose of either providing fodder for the animals or holding the sand together as did the older varieties.

in the late nineteenth century by the settlement officer of Bikaner, Lt. Fagan, while conducting his surveys. He writes,

> Bikanir can boast of very scant vegetation. The commonest tree is the *jand* or *khejre*; in the more level and firmer soils the *beri* is sometimes seen, and the *kikar* is occasionally met with in the north of Ghagghar; the *babul*, a stunted *kikar*, and the *phog* are common on the sand hills, especially the latter. The *jal, kair* and *ak* are not also infrequently met with.[32]

However, it is the *orans,* or the sacred groves often preserved in the memory of local deities, that have traditionally been excellent repositories of plant bio-diversity. *Orans* were located on the fringes of the hamlets and have since time immemorial been the sites of ritual and ceremonial practices. In the *oran*, grazing and lopping of firewood was strictly prohibited and allowed only in times of famine and scarcity. This led to the perpetuation of local species that best survived the harsh weather conditions and helped in maintaining the water table as well as the botanical balance in the area. Besides, the grazing requirements were met through wide stretches of uncultivated land. Nainsi mentions the dry pastures spread over Marwar as well as the ones located along seasonal rivers and streams, which were inundated in the rainy season.[33] He refers to the *jod* of Pokhran as, "a big *jod* 3 kos from Pokhran on the road to Jaisalmer, where abundant *sewan* and *burganthiya* grass grows. This *jod* is 5 *kos* in area and is fed by a *bahla*".[34] Sojhat and Jaitaran are mentioned as wet grasslands watered by *bahla* or *rel*.[35] Land was sometimes specifically allotted for *jods*, as is apparent from the example of the 22 ploughs of land granted by Maharaja Sur Singh in Siwana in 1605.[36] As in case of *orans*, there existed norms for the method and period of use of the fodder from the village *jods*. The flouting of these rules could attract punishment from the village community bodies.[37]

---

[32] Fagan, *Settlement Report of Bikanir*, 3.

[33] Bhadani, *Peasants, Artisans and Entrepreneurs: Economy of Marwar in the Seventeenth Century*, Rawat Publications, Jaipur and New Delhi, 1999, 26.

[34] Nainsi, *Vigat*, II, 315.

[35] Nainsi mentions the *jod* of Village Mahew in Sojhat as one a *jod* that could produce 200 cartloads of grass and was situated near a pond, *Vigat*, I, 457. He also mentions two *jods* of Udesi Kuvo in Jaitaran that were inundated by *rel* and could produce 100 cartloads of grass, Vol I, 527. Another large pasture is mentioned between Sojhat and Jodhpur located along the river Sukri that produced 2000 cartloads of grass, *Vigat*, II, 394.

[36] Nainsi, *Vigat*, II, 444.

[37] Ann Grodzins Gold and Bhoju Ram Gujar, *In Times of Trees and Sorrow: Nature, Power and Memory in Rajasthan*, Duke University Press, Durham, 2002, 241–277.

Besides, a lot of uncultivated land was used primarily as pasture. In 1874, Capt. Powlett observed, "all Bikaner may be said to be pasture ground. To the north about Hanumangarh (the old Bhatner) the grazing is famous, but after fair rains good grazing is everywhere to be found. The banks of Chappar lake vie with the pastures of Bhatner".[38] *Karad, Mogra, Murut, Sapri, Bhongri, Saredi, Burado, Ratheio, Khad* and *Dhakda* were grasses that could either be as consumed green or as dried fodder by various types of herds.[39] Powlett notes the several varieties of grasses in Thar in the nineteenth century as,

> Bikaner abounds in best cattle grasses ... *'ganthil'* a fine grass so called from its top knot (*ganth*), ranks first, both for the production of wool and ghi ...*'sewan'* a rather tall grass is very good for sheep. *'Kiu'* the chief grass on the banks of Chappar lake is particularly good for cattle; *daman* too is very common in the north; other good grasses are *'karr'* more common in Marwar, *'narara'* and *'ganthia'*. *Bhurat* is most abundant in the southern part of the state and ranks after those just mentioned.[40]

Even though with the expansion of the Rajput states, from the mid fifteenth century onwards, large expanses of land were brought under cultivation, the net area sown in the arid districts of Bikaner, Jaisalmer and Jodhpur remained low.[41] While parts of south-eastern Marwar like the Luni basin and Godwar, northern parts of Bikaner, Rajgarh, Nohar and Rinni and were double cropped and partially artificially irrigated, the rest of the Thar was largely dependent on the rains for the single *kharif* crop. Often these too were coarse grains and pulses like *Bajra, Moth* and *Gwar*. The cultivation of these crops too was a difficult proposition and depended either on the meager rains or the conservation techniques followed by the farmers. In the sandy areas towards the west, the sturdiest crop was *Bajra* that grew better in these regions. In fact, sand allowed minimum surface drainage hence even scanty rainfall could actually lead to bumper crop of *Bajra*.[42]

Apart from rain fed cultivation a rich tradition of management of surface drainage for cultivation of *rabi* crops, as well as that of construction of water bodies, existed in the Thar. The main sources of irrigation were either small

---

[38] Powlett, *Gazetteer*, 95.

[39] Purnendu Kavoori, *Pastoralism in Expansion: The Transhuming Herders of Western Rajasthan*, OUP, New Delhi, 1999, 30.

[40] Ibid.

[41] *Rajasthan Development Report*, Planning Commission GOI, 2003, 65 and 283.

[42] Adams, *The Western Rajputana States*, 62.

seasonal rivulets in the Aravali basin or wells with Persian wheels.[43] When the
rainwater overflowed from the rivulets or depressions into neighbouring fields,
crops like wheat could be cultivated. These overflows called *rela* or *bahala* were,
"rapid mountain torrents, carrying in their descent a vast volume of alluvial
deposit, to enrich the siliceous soil below". [44] In the drier and kankerised
areas around Jaisalmer and Barmer, surface drainage was managed through
chains of *khadins* that tapped and dammed the rainwater torrents, by creating
low bunds to flood the fields enough to grow wheat, gram, vegetables and
fodder.[45] Besides, water bodies like tanks, stepwells etc. endowed by kings,
chieftains and merchants were foci of agrarian expansion and settlement
in this region as is suggested by a number of epigraphs scattered all around
the Thar.[46] In arid areas, settlement patterns most logically followed the
course of water. Based on his study of inscriptions relating to construction
of wells, tanks, stepwells and Persian wheels in early medieval Rajasthan,
B D Chattopadhyaya points out that the expansion of agriculture as well as
patterns of crop production were closely related to the construction of water
bodies.[47] Often legends of magicalism or spiritualism were also associated
with them, not only to facilitate better management and conservation, but
also to underline the relationship between availability of water and political

---

[43] Nainsi, *Vigat* I, 231, 234, 237 (village Badali in Jodhpur had 6 *arhats*), 242, 252 (village
    Jhuddli had 10 *arhats*).

[44] Tod, *AAR*, I, 9.

[45] Mayank Kumar, 'Situating the Environment: Settlement, Irrigation and Agriculture
    in Pre-colonial Rajasthan', *SIH*; 24; 2008, 211–233, 221.

[46] B D Chattopadhyaya, 'Villages, Wells and Rulers in South Eastern Marwar: Aspects
    of Rural Settlements and Rural Society in the Kingdom of Nadol, Cahmanas' in
    Aspects of Rural Society and Rural Settlements in Early Medieval India, Calcutta, 1990,
    70–92. Chattopadhyaya lists the proliferation of rural settlements particularly in the
    Jodhpur, Pali-Bali region based on inscriptions regarding tanks, wells, reservoirs as well
    as *arghattas* and *dhikus*. Such references continue in Nainsi's accounts not only in the
    Pali- Godwar region, but also in the sandier expanses of Jaisalmer. The construction
    of the Ghadsisar lake can be treated as an example. Also, Devraj Bhati is attributed
    with the digging of wells and one lake in Derawar as well as three tanks in Ludrovo.
    Nainsi, *Khyat*, II, 22 and 28.

[47] In the first half of the twelfth century, cesses in the form of wheat, barley and sesame
    were extracted from fields irrigated by Persian wheels in Kekind, Lalrai and Sevadi in
    southeastern Marwar. 'Irrigation in Early Medieval Rajasthan' in B D Chattopadhyaya,
    The Making of Early Medieval India, 38–59, 40–50.

control.[48] Urban and rural settlements usually developed in places that were, "almost always in depression, and often where 'kankar' or hard ground is on, or not far from, the surface, so that the drainage of the neighbourhood is caught either in covered pits, called 'kunds' or simple excavations. 'Sar' which is the final syllable of so many Bikaner villages, implies a tank or depression".[49] In the entire Thar Desert a number of settlements bear testimony to this fact as is seen in their names suffixed with *sar, tal, talai, beri*, all referring to water bodies.[50]

Despite these efforts, failure of rains and thus of crops was a recurrent phenomenon. Thomas Hendley in his Rajputana Medical Gazetteer noted,

> Rainfall in these districts, being irregular and variable, especially in Bikaner, and parts of Marwar, and also Shekhawati, there are frequent periods of scarcity if not regular famines, both as regards grain and fodder, which act prejudicially on the health of the people and their cattle, the latter being even more important matter, for if cattle die from want of nourishment, the fields for a long time are thrown out of cultivation.[51]

Famines were like unwanted but expected visitors, signs of whose arrival were keenly watched for. They resulted not only in death and devastation in large measures, but also in large-scale desertion of villages. Adams reports that failure of crops and grass was "so frequent in the western desert, that the people leave their homes with their herds to find pasture before the animals become

---

[48] Ann Grodzins Gold also refers to tales of magicalism relating to water bodies in shrines, where certain kinds of uses of water are prohibited. Ann Grodzins Gold, and Bhoju Ram Gujar, 'Of Gods, Trees and Boundaries: Divine Conservation in Rajasthan', *Asian Folklore Studies*, Vol. 48, No. 2, 1989, 211–229.

[49] Powlett, *Gazetteer*, 93.

[50] Among the kinds of water bodies mentioned, there existed further differentiations. For example, *Kosito* was a shallow well while *kuvo* or *beri* could be deeper. *Jhalaro* was a step well while *kohar* or *talao* meant a tank or pond. *Samands* were lakes natural or man-made. The brackish water wells were called *kharo* and *kharchio* was the special kind of wheat cultivated from brackish water. *Nada, nadi, talai, chanch, kund, tebho, drah, vai, bawri, bahlo* were other kinds of water bodies used for drinking as well as irrigation purposes. *Tin* was the time settled for drawing water from a well. A *betino* water body was one from which water could be drawn twice.

[51] Col. Thomas Holbein Hendley, *General Medical History of Rajputana*, Indian Medical Service, Calcutta, 1900, 4.

too impoverished to make long journeys into Kotah, Malwa, and Sindh". [52]
Traditional wisdom expected 3 years of extreme famine, 7 of famine, 63 average
years and 23 good years in a span of hundred years.[53] In local perceptions the
famines were placed in hierarchies according to severity. When scanty rains
led to shortage of water it was named *akaal*; when there was shortage of water
and grain, it was called *dukaal*; when water, grain and fodder fell short, it was
tinkaal and the shortage of water, food, fodder and fuel was named *mahaakal*.
The folk wisdom of the Thar contains several warnings that indicate the arrival
of a dry spell, and thus of yet another migration. In an exchange a wife from
Gujarat tells her husband,

> The *Revati* is bright and the *Aswini* is dull on the *amavasya* of the dark half
> of the moon, ...and it was a Saturday on the tenth of *Jeth* of the dark half, and
> there has been no lightning in the month of *Bhadva*... therefore, you my dear
> go to Malwa...and I will go to Gujarat.[54]

The earliest recorded evidence of famine is in 1309–13, in the reign of Rao
Raipal.[55] The chronicler Badauni records another great famine in 1570.[56] An
annexure to Nainsi's *Vigat* notes in a commentary on paragana Sanchor,

> The year 1754 VS (1697) was year of *mahakaal*, there were no crops anywhere.
> All the villages of the paragana were deserted and all men went away to
> Gujarat. When it rained diseases killed people and cattle.[57]

In Bikaner, it is reported in 1763, that the Jats of village Ranasar relocated
to another village because of severe drought.[58] Similarly, the entire hamlet of
Lunkaransar was deserted in 1783 and the peasants had to be resettled with
assurances.[59] About twenty major famines were recorded between the 17th

---

[52] Adams, *The Western Rajputana States*, 153.

[53] S P Malhotra, 'Man and the Desert' in *Desert, Drought and Development*, (Eds.) Rakesh
Hooja and Rajendra Hooja, Jaipur, 1999, 17.

[54] K P Sahel, 'Rajasthan ki Varsha Sambandhi Kahavatein', *Maru Bharati*, Vol IV, 17–18.

[55] O P Kachchwaha, *Famines in Rajasthan, (1900–1947 AD)*, Hindi Sahitya Mandir,
Jodhpur, 1985, 24.

[56] W H Moreland, *India at the Death of Akbar: An Economic Study*, Delhi, 56, 1962.

[57] Nainsi, *Vigat*, II, 368.

[58] *Jeth Sudi 8, Kagad Bahi*, VS 1820/1763 CE.

[59] *Katik Vadi 3, Kagad Bahi*, VS1840/1783 CE.

and the 19th centuries, which led to distress migration from the Thar.[60] The distress of the population during the 1848–49 famines is described as follows, "the year 1905–06 VS (1848–49) compelled people to migrate…all was lost… cultivators went to Malwa…good rains came when all were gone".[61]

Of the suffering caused by the 1868 famine Adams writes,

> Mighty streams of human beings and cattle poured from the southeastern portions of the country. The poor people who arrived in Malwa late had to pass on, and were reduced to last extremity of distress. The second great stream of immigrants passed via Palanpur into Gujerat, too were doomed to bitter disappointment. The great floods which devastated Gujerat in August left nothing but a desert for the Marwar herds. They were driven farther, dying of starvation at each stage, till a few only reached the jungles east of Baroda, here, difference of climate and forage killed off large numbers.[62]

Of the same famine Powlett reports,

> Nowhere was the great famine of 1868–69 felt more fearfully than in Bikaner. Early in October 1868, the starving had begun to flock the towns…Bikaner lost permanently more than third of its population…cattle was reduced to less than one-twentieth.[63]

Despite famines being a regular feature, the nature of land did not allow people to store grain for lean years and left migration as the only recourse. Powlett noticed that, "though famines are frequent and general famine occurs at least once in ten years, the mass of people possess no stores of grain wherewith to meet them and when they occur they either emigrate or depend on charity or grass seeds."[64] Adams notes that the Grasias, Minas and Bhils supplemented their "precarious meals by adding thereto a large proportion of bark of the *khejra* tree, or the barbed seed of *gokroo*, ground to a fine flour. The carcasses of dead

---

[60] Kachchwaha, *Famines*, 24–27. Major famines being in, 1616, 1661, 1711–1717, 1742–1747, 1756, 1783, 1796, 1812–1813, 1833, 1848–1849, 1861, 1868–1869, 1877–1878, 1891–1892, 1896–1897 and 1899–1900.

[61] H B Sarda, *Ajmer: Historical and Descriptive*, Fine Art Printing Press, Ajmer, 1941, 315.

[62] Adams, *The Western Rajputana States*, 144.

[63] Powlett, *Gazetteer*, 107–108.

[64] Ibid.,104.

cattle, which perished in numbers, were eagerly devoured by these people".[65] The recurrent famines led to large-scale desertion of villages by farmers and cattle that fled to Sindh, Malwa, Gujarat and Punjab in large numbers, some perishing on the way. During the 1891–92 famine 8 percent of the population of Marwar, numbering 199,600 migrated with 661,906 cattle. The distribution of migration was as follows; 203,789 animals were taken to Malwa, 1,64,964 to Hadauti, 121,702 to Sindh, 115,000 to Gujarat and about 56,461 to Delhi and Bhiwani.[66] Of these only 46% of the men and cattle that migrated made it back.[67]

The states were thus required to constantly undertake philanthropic measures in order to mitigate the impact of the famines. This included waiver of taxes, construction of water bodies as well as relief by way of distribution of grain.[68] In the year 1896–97, the VP of the Regency council wrote to the Political agent, Bikaner,

> Distress is in every way probable in the Bikaner state and has already begun to be felt by the people, but as famine relief works, have already been taken in hand, it can necessarily be asserted that the number of emigrants to Punjab would not be very large, though it can never be affirmed that there would be no emigration to the Punjab at all.[69]

Yet, in the *Mahakaal* of 1899–1900, called the *chappaniya kaal*, 12% population and 20% livestock migrated to neighbouring areas like Malwa, Gujarat, Central Provinces, Sindh and southern Punjab.[70]

---

[65] Adams, *The Western Rajputana States*, 145.

[66] Kavoori, *Pastoralism in Expansion*, 66.

[67] *Report on Relief Operations Undertaken in the native States of Marwar, Jaisalmer, Bikaner and Kishengarh during the scarcity of 1891–92*, 5, RSAB. Also, Munshi Hardayal, *Report Mardumshumari Raj Marwar 1891* (First Pub 1896), Jagdish Singh Gehlot Shodh Sansthan, Jodhpur, 1997, 10. Henceforth *Report Mardumshumari*.

[68] The state records of Jodhpur, Bikaner and Jaisalmer have routine references of tax waivers granted owing to scarcities. Apart from waivers states also offered relief to the affected population through measures like distribution of grain etc. An interesting reference to such philanthropic practice is found in Adams' description of the 1868 famine, in which he reports that the queen of Jodhpur distributed seven *maunds* of cooked food, besides double handfuls of grain, after nightfall, to classes who were too respectable to beg. Adams, *The Western Rajputana States*, 145.

[69] Letter No 1882, dated 19/10/1896 From VP Regency Council to Political Agent, Famine and Famine Relief work, 1896–97, RSAB.

[70] Kachchwaha, *Famines*, 180 and Kamala Maloo *The History of Famines in Rajputana (1858–1900)*, Himanshu Prakashan, Udaipur, 1987, 187.

Recurrent famines along with ambiguous land rights of the peasantry made desertion of villages seem to be an escape that the peasants often undertook. The peasantry was largely constituted of upper and middle castes like the Rajputs, Jats, Malis, Sirvis, Patels and Kolis. They were sedentary cultivators, but also often owned some animal stock. The term *kisan* or *karsa* is used to describe peasants in general though *bhomia, chaudhari, muqaddam* or *padariya* were the superior rights holders. The resident cultivators were called *ganveti,* while the migrant cultivators are referred to as *pahi, muqati* or *baharla gaon ra.*[71]The real control of both land and water resources lay with the Rajput *bhomiyas* responsible for the revenue assessment and collection who also exercised strict checks on peasant mobility. However, in the event of famines, the state often instructed the *patayats* and *bhomias* to forgo revenue collection and resettle deserted villages.[72] Yet, peasant mobility was difficult to restrict, as sedentary cultivation and mobile pastoralism were closely interlinked in this region. Just as herders returned to their homesteads at the end of a seasonal cycle, the peasants could also seasonally migrate with their small herds, assimilating with several circuits of mobility through the Thar.

Networks of Circulation in the Thar Desert

The understanding of the Thar Desert as a region involves engaging with a rather complex interplay of its political, ecological, social, and cultural dimensions. The Thar was also the site of vigorous networks of circulation whereby communities, commodities and traditions travelled between the arid expanses in the Thar Desert; the semi arid stretches of Aravali hills and the river basins of Indus and Sutlej. The mobility of peripatetics of several kinds like warriors, pastoralists, merchants, ascetics and bards the who circumambulated the Thar can be seen as a major factor in conceptualizing it as a mobile region. The circulatory regime of the Thar Desert was constituted of overlapping networks, whereby political, commercial, social and religious circulations coincided with each other. Recurrent movements between arid, semi-arid and wet zones led to continuous exchange of people and commodities on these networks. The travellers on these networks were 'occupational travellers', as their overlapping social and cultural identities were defined though their mobility on the circulatory networks.[73] The itinerant culture of the Thar

---

[71] Bhadani, *Peasants, Artisans and Entrepreneurs,* 114.

[72] Ibid.

[73] Jos Gommans identified such travellers as, "vigorous inhabitants of the arid frontiers in their often overlapping capacities of nomads, warriors and ascetics". Gommans, 'The Eurasian Frontier', 131.

has been defined through the *marag* or the road, which had a life of its own, dotted with check posts called *sayer thanas*, as well as resting places, *sarais*, temples, shrines, water posts and often just a bit of shade for the odd traveller. In a constantly itinerant culture, the routes held the poetic imagination of the bards like no other element, and constituted a vibrant migratory space. This migratory space was also the site for expression of political control over networks of mobility, which consisted of both settlements as well as the routes on which movements took place. In my understanding, settlements and routes existed in continuity with each other, but were increasingly seen as distinct and in binary opposition in the nineteenth century, with each representing a different kind of value. In the following paragraphs, I will try to sketch an image of the Thar as a dynamic migratory space, as its travellers and chroniclers between the seventeenth and the nineteenth centuries viewed it.

Migration and mobility have often been studied only in the context of distress migrations caused by famines and scarcity. It needs to be emphasized here that in case of occupational travellers like pastoralists, traders and bards, migration can be seen as a livelihood practice, as much a part of their existence as planting and reaping for a peasant might be. While socio-economic developments of the last two centuries have altered the migratory networks in most regions, some of these movements occur in accordance with age-old patterns. For example, Purnendu Kavoori in his work, *Pastoralism in Expansion: The Transhuming Herders of Western Rajasthan*, has traced the patterns of movements of pastoralists across the Thar, and comes to the conclusion that pastoralists have continued to follow very old patterns of movement. Pastoralists have traditionally moved in several short and long circuits of transhumance across the Thar, some migrating to pastures close to their homesteads, for short periods, others moving seasonally to far off places. In the Thar, Kavoori identifies three major patterns of pastoral migration. First, the northeastern pattern, where by herds belonging to Marwar move into present day Uttar Pradesh and Haryana. Second, the southeastern pattern, in which herds from Jalor, Sirohi, Nagaur, Pali and Ajmer districts migrate into Central India. Finally, the southern pattern, in which the herds from Jaisalmer, Jodhpur, Barmer and Jalor migrated into Gujarat as far as into Rajkot, Junagarh, Ahmedabad and Baroda. A reciprocal migration also occurs from Kathiawad into Western Rajasthan.[74] There existed two more routes before the partition, one into Sindh and the other into Multan, which are no longer operational.

---

[74] Kavoori, *Pastoralism in Expansion*, 27.

The routes of seasonal migration of pastoralists in the Thar have also been the pastoral trade routes catering to exchange in cattle as well as pastoral by-products like ghee, wool and hides. These trade routes consisted of both, permanent cattle marts as well as periodic cattle fairs, often held in honour of folk deities like Pabuji, Gogaji, Ramdeoji, Tejaji, Mallinathji, Harbhuji, Jambhaji etc. who are seen as protectors of cattle and cattle herding communities. By the end of nineteenth century, major fairs held in the Thar were Ramdeoji fair held twice a year in Pokhran, *Chaitri ka mela* held in Tilwara or Balotra, *Kesariya Kanwar ji ka mela* held in Jodhpur, *Tejaji* fair at Parbatsar, two fairs at Jaisalmer, apart from numerous local fairs held in villages. These fairs lay on the periodic routes of pastoral groups allowing them to buy and sell cattle and livestock. The fairs were also very closely linked with the agrarian calendar. Often held in the months of *Magh* (February/March) and *Bhadwa* (August/September), which were the months of leisure after the harvesting of the *rabi* and *kharif* crops respectively, these were sites of exchange for the peasant, pastoralist and the merchant.

Apart from being sites of commercial exchange, fairs were also festive occasions, a relief from monotony, and meeting ground for people from all walks of life. They were reflective of the ways in which legends, customs, traditions and religion formed the basis of exchange between various strata of society and the way they circulated all over the Thar. These fairs were occasions for the display of reverence and faith and the restating of lores of the deities. They were visited not only by buyers and sellers but also by bardic and minstrel groups like Bhats and Manganiyars. They became occasions for the recital of folklore and also ceremonial recitation of genealogies. Therefore they became appropriate grounds for exchange of news, which was circulated by these wandering bards. These fairs were reflective of the ways in which legends, customs, traditions, and religion formed the basis of exchange between various strata of society and thus of a Thari culture constantly on the move.

The network of periodic markets or fairs fed into a larger network constituted of old market towns in this region. The rise of market towns in this region was closely allied to emergence of Rajput polities, which allied with old mercantile lineages during what B.D Chattopadhyaya terms as the 'third phase of urbanism' in the early medieval period.[75] Towns like Ghatiyala, Hastikundika, Sevadi, Nadol, Kairatapura (Kiradu), Nadulagadika (Narlai),

---

[75] B D Chattopadhyaya, 'Urban centres in Early Medieval India: An overview' in *The Making of Early Medieval India*, 160.

Dholapasthana, Vahadmeru (Juna Vadmer) Mandavyapura (Mandor), Ratanpur, Jalor, Srimalla (Bhinmal), Palli (Pali), Sakambhari (Sambhar), Ahichtrapura (Nagaur), Phalavardhika (Phalodi) and Sanchor illustrate the close linkages between the centres of power identifiable with emerging ruling lineages like Nadol Cahamanas and Pratiharas, and centres of exchange in early medieval Rajasthan. The establishment of *hattas* and *mandapikas* in places inhabited by resident populations of *vaniks* and manufacturers points towards the importance of sedentary sites of exchange that also were sites for collection of cesses and taxes.[76] These urban centres where various social groups interacted on a regular basis, were also seats of old merchant lineages like Dharkatas, Pragvatas, Oswals and Srimals, whose networks expanded to integrate wider resource bases and centres of exchange. The towns were not merely significant centres of exchange but also important Jain and Vaishnav religious sites patronised alike by ruling groups as well as merchant lineages where a number of temples and *upashrayas* were constructed. These towns therefore lay on the intersections of circulatory networks of multiple kinds.

The emergence of Rajputs in the Thar also saw the establishment of seats of Rajput lineages in some of these towns with construction of fortresses. Mandavyapuradurga at Mandor, Srimaliyyakota at Bhinmal and Suvarnagiridurga at Jalor could be seen as examples. The emerging Rajput lineages like Guhilas, Chauhans, Solankis, Parmaras established their fortifications in or near some of the existing towns and these became the focal point of further expansion of Rajput lineages. On the other hand, the centres that no longer were identifiable with emerging ruling lineages or were abandoned by merchant families, steadily declined. Nadol is one such example, where the declining Cahamana influence led to the fading away of the importance of Nadol as a market town. Towns like Sevadi, Kiradu, Narlai also seem to recede in importance as both political and commercial centres. The town of Osian associated with Jain merchants also appears to have been deserted by the merchant community of Oswals leading to its decline. As a matter of fact, the relationship between ruling lineages, fortifications and markets seems apparent in the forts of Bhatis and Paramaras as listed by Nainsi. Some of the nine forts of Bhatis, like Jaisalmer, Bikampur, Marot and Derawar were also located in market towns. Similarly, some of the nine fort-towns of Parmaras that define the boundaries of Marwar that is Barmer, Abu, Parkar, Pugal, Jalor, Umarkot, Ludrovo, Ajmer and Mandor were also market towns.

---

[76] B D Chattopadhyaya, 'Markets and Merchants in Early Medieval Rajasthan' in *The Making of Early Medieval India*, 93–124.

The conflict between the emergence of new Rajput lineages and the expansionary moves of the emerging Sultanate in Delhi, was also most visible in fortress towns like Mandor, Jalor, Sambhar, Sanchor, Siwana and Nagaur. These towns lay on the route between Delhi, Punjab and Gujarat and were seen as strategic centres by the Sultans of Delhi. Throughout the thirteenth and fourteenth centuries, intense struggle to control these towns waged between various Sultans of Delhi and Rajput lineages like Chauhans. Jalor, located as it was between the regions of Gujarat and Marwar; and Nagaur located as the gateway to the desert routes emerged as important Sultanate strongholds as well as garrison towns in this period. In fact, all of the aforementioned towns continued to thrive as markets. On the other hand a number of older market towns like Nadol, Kiradu, Sevadi, Osian etc. declined, as the power centres shifted from these towns.

This remained a continuous process as is apparent from the Rathor move towards establishing their capital in the fort town of Mandor, which they received from the Eendas as dowry after establishing domination over Pali, already an important market town in the fourteenth century.[77] The Rathors eventually built the Mehrangarh fort in Jodhpur in 1459 and established their seat in the new city, which also became an important trading centre. The spread of Rathor subclans in old towns like Merta, Sojhat, Jaitaran, Siwana, Sanchor, Jalor and Bhinmal led to the reaffirmation of these towns as politico-administrative centres of dominance for Rathor subclans.[78] These towns already had resident populations of Jain and Vaishnav merchant groups, who were closely associated with administration as well commercial enterprise in these towns.[79] Under Mughal indirect rule, the market towns of Marwar and Bikaner continued to thrive commercially and developed systematic commercial institutions and practices with officials being appointed for the purpose of smooth conduct of trade in market towns. The construction of imperial highway between Agra and Ahmedabad, which was zealously maintained and guarded by the kings of Marwar, indicates the importance of networks of travel and exchange in the region. In each of the market towns *sayer thanas* were instituted with *sayer daroga* being appointed to levy import and export duties on commercial traffic.[80]

---

[77] Nainsi, *Vigat*, I, 9, 25.

[78] Ibid., 40–46

[79] *Report Mardumshumari*, 421–432. Oswal *Mutsaddis* for instance played important administrative roles as Diwans in Marwar.

[80] Bhadani, *Peasants, Artisans and Entrepreneurs*, 331.

The *Vigat* of Munhata Nainsi, which is a gazetteer and a substantial account of paragana towns under Raja Jaswant Singh, provides a detailed description of market towns in this region in the late seventeenth century. Nainsi does not appear to apply a consistent parameter for distinction between market towns, and employs two distinct categories of *sahar* and *qasba*. In case of some towns like Sojhat, towns appear to have rural features like the presence of pastures in vicinity. This, however, could be seen as a necessity in market towns given that trading caravans included large numbers of draught animals that would have required sufficient fodder and grazing grounds. Besides, given the importance of cattle trade in the region, some of the market towns would also have been sites for cattle trade. Nonetheless, a parameter for deciding upon a hierarchy of market towns could be to look at the proportion of merchant households formed, of the total number of households in the region. Based on Nainsi's *Vigat,* Bhadani has calculated the percentage of merchant and banker households in towns of Marwar to vary between 29.51% to 58.09% in the late seventeenth century. [81]

By the seventeenth century most towns appear to have developed genealogies that refer to old histories of settlement, of which some had already been abandoned. Nainsi collates some interesting histories and accounts of the towns. According to him, Mandor, "an old town mentioned in Padampuran was established by Mandodar Daitya for his daughter Mandodari".[82] But it does not appear to be an important market town to Nainsi, overshadowed as it was by the new capital Jodhpur. However, others like Sojhat, whose older name according to Nainsi was Sudhdanti Nagari, survived and was considered important enough to be addressed as a *sahar*. [83] It had a small fort, but it was dominated by Mahajans and artisanal castes. On the other hand Nainsi described Phalodi, which was earlier known as Vijaynagari and Phalavardhika, as a *qasba*, even though it too had a large population of Oswal and Maheshwari Mahajans as well as Pushkarna Brahmins who were also engaged in trade.[84] Phalodi already appears to have been an important Jain centre of pilgrimage in the seventeenth century. The *paragana* of Phalodi also had a number of salt

---

[81] Ibid., 345. According to Bhadani, Sanchor had 58.09% of Mahajan households, Pohkran 55.65%, Jaitaran 47.3%, Siwana 43.08%, Phalodi 36.83%, Sojhat 32.74% and Jalor 29.51%. The merchants belonged to Oswal, Maheshwari, Agrawal Bania subcastes as well as of Khatri caste in Phalodi.

[82] Nainsi, *Vigat*, I, 1.

[83] Ibid., 391–393.

[84] Nainsi, *Vigat*, II, 8–9.

mines that made it an important town on salt networks.[85] Similarly Pokhran, which had a fort built by Rao Malde, had a large population of Maheshwari Mahajans, who patronized several temples in the town and contributed towards the construction of a large number of water bodies.[86] Pokhran had a market where merchants from Gujarat came to trade and was an important stopover for caravans going to Jaisalmer and Sindh.[87]

Most of these towns were not merely in continuous existence but were seen as thriving by European travellers and administrators in the early nineteenth century. For instance, in the seventeenth Century, Jodhpur and Pali appear to have been the two large emporia towns of Marwar. The market of Jodhpur in 1664 had 815 shops.[88] The number seems to have increased to 4000 by 1836 as accounted by Alexander Boileau.[89] The walls of the city had seven gates each bearing the name of the place to which it led like Mertia, Sojatia, Jalori, Nagori and Sewanchi. In the main street, locally known as "Ooba bazaar... there were shops of perfumers, grocers, pansaris, cloth sellers, halwais, patwas, shroffs, kaseras, betel and fresh fruit sellers".[90] Jodhpur was a big wholesale market of cloth and thriving with dyers and printers.[91]

Pali was another important market town that received attention of many travellers of the nineteenth Century. Tod describes Pali as,

> The entrepot for eastern and western regions where the productions of India, Cashmere and China were interchanged for those of Europe, Africa, Persia and Arabia. Caravans (Katars) from the ports of Cutch and Guzzerat imported elephant's teeth, copper, dates, gumarabic, borax, coconuts, broad cloths, silks, sandal- wood camphor, dyes, drugs, oxide and sulphate of arsenic, spices, coffee etc. In exchange it exported chintzes, dried fruits jeeroh, asafoetida, from Mooltan, sugar, opium (Kotah and Malwa), silks and fine cloth, potash, shawls, dyed blankets, arms and salt of home manufacture.[92]

---

[85] Ibid., 34–36.

[86] Ibid., 311.

[87] Ibid., 323–326

[88] Nainsi, *Vigat* I, 186–188.

[89] Boileau, *Narrative*, 238.

[90] Adams, *The Western Rajputana States*, 79.

[91] Nainsi notes that in the town of Jodhpur there were shops of dyers and printers known as *chinpas*, Vigat I, 187. From *Sanad Parwana Bahi* No: 4, VS 1823/1766 AD *Kartik vadi* 7, we find that the Jodhpur court made its annual purchase of cloth from the Jodhpur bazaar for the *nava kapada ra kothar*, This included yards of plain and printed cotton, silks, *kinkhab, meesru, odhnis, dhotis* and turbans. Henceforth, *SP Bahi*.

[92] Tod, *AAR*, II, 127–128.

Such was its commercial organization that James Tod found that, Pali, like "Bhilwara, Jhalarapattan, Rinnie and other marts enjoys the right of electing its own magistrates, both for its municipal regulations and arbitration of all matters connected with commercial pursuits."[93] These were called *panchmahajans* who arbitrated in all commercial matters. Pali's importance can be gauged from the fact that, in the nineteenth century, Pali, "had its own currency which it retained undebased".[94] Nagaur was another town of varied commerce though it was renowned for livestock and wool trade. Being a connecting link between the Jodhpur and Bikaner states, in its markets commodities from both the areas were exchanged in the market of Nagaur. The adjacent town of Merta, by the nineteenth century was seen as a town that, covered a "large space of ground, enclosed with a strong wall and bastions" and was "said to contain twenty thousand houses".[95] Merta, according to Tod showed "a mixture of magnificence and poverty, a straw hut adjoins a superb house of free stones."[96] Bhinmal and Sanchor were two other towns on "high road to Cutch and Guzzerat", which gave them, "a commercial celebrity, with most wealthy Mahajans residing there".[97] The old fort and market town of Jalor that had been taken over by the Sultans of Delhi, had 1156 Mahajan households in 1813.[98]

The networks of periodic and permanent markets in the Thar were located within larger networks of travel and exchange that connected riverine plains of northern India to the western coast. Thar was strategically located as a corridor connecting trading networks of north and south with those of west and northwest. On the one hand, traders and commodities travelled from Doab to the ports of Gujarat and Sindh, through Thar, on the other Deccan was connected to far off markets in Punjab, Multan and Baluchistan. A large number of major and minor trunk routes passed through the Thar forming lines of long and short distance trade. Also, the distinctions in the geographical patterns within the region and around it led to the formation of intensive networks of local exchange, whereby the commodities from one area found their way into the market centers and were exchanged for commodities of other areas. These networks and routes had been old and had been used for

---

[93] Tod *AAR*, I, 553.
[94] Ibid.
[95] Tod, *AAR*, I, 583.
[96] Ibid.
[97] Tod, *AAR*, II, 241.
[98] Ibid.

purposes like pilgrimage and battles. The permanent and periodic markets formed nodes and internodes of these networks, where the physical act of exchange took place. The distance between points of exchange and transit and the movement of mercantile groups and commodities on these routes formed important aspects of the structure and organization of trading networks in the Thar Desert. This constant movement of men and commodities across the desert created a mercantile culture that was based on continuous exchange of commodities. The movements on these routes were not limited by political boundaries. Commercial transactions took place between far off places and were carried out with the help of sophisticated financial instruments like the *hundis*. Thus, in the seventeenth and the eighteenth centuries, the Thar emerged as an open, yet widely traversed coherent space.

Table 1.1. Trade Routes in the Thar[99]

S. no.	Towns connected	Route
1.	Gwalior to Jodhpur	Gwalior- Karauli -Jaipur-Sambhar -Parbatsar-Merta-Jodhpur
2.	Jodhpur to Ahmedabad/ Surat	Jodhpur-Jalor-Bhinmal -Palanpur- Ahmedabad-Surat
3.	Jodhpur to Kashmir	Jodhpur-Nagaur- Hardesar-Nohar -Sirsa -Bhatinda-Amritsar-Kashmir
4.	Bikaner to Deccan	Bikaner-Nagaur- Merta-Tatoti-Bundi-Kota-Jhalarapatan-Ujjain-Deccan
5.	Agra-Ahmedabad/Surat	Agra-Fatehpur Sikri-Hindaun-Lalsot- Mozamabad -Bander -Sindree-Ajmer- Merta- Jodhpur-Jalor-Bhinmal -Palanpur -Ahmedabad-Surat
6.	Multan to Bikaner	Multan-Dunyapur -Kahror -Miskota -Maroth -Sobasar -Bikaner
7.	Multan to Jodhpur	Multan-Bahawalpur-Pugal -Bikaner-Nagaur-Jodhpur
8.	Jodhpur to Delhi	Jodhpur- Merta-Ajmer-Mewat -Delhi
9.	Delhi to Ahmedabad	Delhi-Bhiwani -Rajgarh -Churu -Nagaur-Jodhpur-Jalor-Sirohi -Palanpur -Ahmedabad
10.	Bikaner to Jodhpur	Bikaner-Nagaur-Khinvsar -Jodhpur
11.	Rajgarh to Pali	Rajgarh -Navalgarh -Didwana -Nagaur-Pali
12.	Jodhpur to Jaipur	Jodhpur- Merta-Parbatsar -Sāmbhar –Jaipur
13.	Jaisalmer to Jodhpur	Jaisalmer-Pokhran -Lawa -Mandla -Jodhpur
14.	Jhunjhunu to Pali	Jahunjhunu-Fatehpur-Ladnu-Didwana - Nagaur-Pali
15.	Jodhpur to Kishengarh	Jodhpur- Merta-Kishengarh

Contd.

---

[99] B L Gupta, *Trade and Commerce in Rajasthan in the Eighteenth Century*, Jaipur, 1991.

Contd.

S. no.	Towns connected	Route
16.	Pachpadra to Kota and Baran	Pachpadra-Bilara-Kotra-Tatoti-Pander-Jahazpur-Thun-Bundi-Kota-Baran
17.	Bikaner to Thatta/ Hyderabad	Bikaner-Nagaur-Jodhpur-Khinvsar-Pachpadra-Balotra-Barmer-Chohtan-Umarkot-Thatta/ Hyderabad
18.	Jodhpur to Derawar	Jodhpur-Khinvsar -Nagaur-Bikaner-Birsilpur -Derawar
19.	Thatta to Bhatinda	Thatta-Umarkot-Chohtan-Barmer-Balotra-Pachpadra-Balotra-Jodhpur-Osian-Phalodi-Bikampur-Birsilpur-Bhatner-Bhatinda

The (Un)Making of the Thar in the Nineteenth Century

Interestingly at the same time, in the nineteenth century rather than being seen as a mobile region that provided free passage to a range of mobilities, the Thar emerged as a region difficult to travel through, particularly in the experiences of European travellers. Despite the Thar region having existed as an open mobile zone between Punjab, Multan, Sindh and Kutch, on which people and commodities circulated freely, it surprisingly appeared a 'closed' region to the British in the nineteenth century. Perhaps the reason was that the inaccessibility of certain frontiers to the ruling clans and their British administrators was seen as an impediment in administering those areas. The fact was that these areas were not inaccessible to the itinerants of the region, and authority on frontiers was negotiated through systems like *bolawo* or local networks of protection. However, a discourse of 'incomprehensibility' and 'hostility' evolved in the nineteenth century, according to which the Thar was increasingly seen as a difficult and hostile terrain, which required extensive reordering.[100] Despite the fact that Thar had for centuries served as a passage connecting settlements, as well as fostered cultures that were linked by networks of mobility, in the nineteenth century it was largely viewed as an arid, hostile region, described through adjectives like 'barren' and 'desolate'. On the one hand, the descriptions of the Thar emphasized its low agricultural

---

[100] The British Agents corresponded at length on the issue of 'Disturbed State of Marwar' and expressed concern over the lack of safety on the routes passing through the region. The frontier region of Mallani was seized owing to 'lawlessness' and depredations. I discuss these issues in detail in Chapter Four.

productivity, on the other they also stressed upon difficulties in traversing through the region. For instance, Boileau in his travels around Jaisalmer encountered a stony desert, barren, sterile and quite unprofitable.[101] In his narrative he writes of "the whole country west of Balmer and south of Girab" as a "desolate waste".[102] While commenting upon the nature of roads in the desert, Boileau observes that the internal communications between various cities in northern Thar were limited in consequence of extreme sandiness of the soil, while roads in Bahawulpur were merely camel paths.[103] Given that most travellers of early nineteenth century including Boileau himself, as well as Tod, Powlett and Adams attest to the presence of vibrant market towns bustling with trade, the contrast between these thriving market towns and unpassable routes is difficult to explain. A possible way of understanding it could be by exploring ways in which the region was experienced, changed over time. After all, how did European administrators coming from cold, wet and lush European countryside engage with hot, dry, arid desert landscapes? The perceived hostility of the desert landscape evoked conflicting emotions of awe and revulsion in these officers. The years spent in the desert were seen as years of "continuous tissue of toil and accident".[104]They saw themselves as brave warriors not only for having controlled the Rajputana, but also for surviving the harsh landscape, which they constantly struggled against and sought to tame. Besides, the Thar was unmapped, a 'vacant space' which had to be rendered comprehensible, which they did by travelling, mapping and measuring. By the time these officials were able to leave the desert, diseases contracted during numerous travels often plagued their bodies and spirits.[105] They attempted to find 'home' in these unfamiliar landscapes and yet yearned to return to their native land. [106] Thus, possibly the specter of desolation that

---

[101] He also underlines the desolation through this story that he recounts, "a wild doe had lost her fawn, and taxed a hynaea with having devoured it: the hynaea denied the charge with indignation by the most solemn oath: 'If I have eaten your fawn may I be condemned to dwell in the desert between Bikampoor and Poogul!'". Boileau, *Narrative*, 94.

[102] Ibid., 178–79.

[103] Ibid., 188–189.

[104] "The bow must be unbent, or it will snap", wrote Tod as he ventured towards the end of his jounalising. Tod, *AAR* II, 611.

[105] As Tod laments, "Geography has been destructive to those who have pursued it with ardour in the East". Tod, *AAR*, I, 4.

[106] Tod, *AAR*, II, 611. James Tod attempted to recreate some of his British environs by planting peaches and apples in the gardens in his residences at Udaipur and Merta.

the Thar presented to these travellers was because of the manner in which they experienced the Thar, which was different from the way it had been experienced by native travellers. For native travellers, routes and settlements, whether villages, towns or periodic fairs, were part of a continuum occupied by the same people. A binary of the settler and the traveller, revealed by safe thriving towns and dangerous desolate routes, appears to be emerging in the nineteenth century.

In this context it may also be worthwhile to see how perceptions of the region varied over time, particularly the kind of geographical imagination that the region evoked in various travellers. Did the desolate nature of the land hold different kinds of meanings for travellers of different sorts? What did the barrenness of the lands signify for the Rajput rulers whose bards composed panegyric accounts that held land as the core of Rajput kingship from the sixteenth century onwards? For the British travellers and administrators, was describing the nature and extent of desolation a way of overcoming their own sense of awe and revulsion as they encountered a landscape so radically different from the lands they came from? What then is the region that emerges as 'Thar' when we attempt to construct the idea of a region through these accounts? In the following sections, I present two accounts of administrator-travellers of two different centuries. Munhata Nainsi, the seventeenth century Diwan of Raja Jaswant Singh of Marwar and Col. James Tod the early nineteenth century Political Agent of Rajputana, wrote extensive accounts of the region. However, the Thar appears to be different regions in both accounts, not only because of the different periods in which the works were written, but also because of two different vantage points from which these two men view the region.

Munhata Nainsi's Marwar: Frontiers of Polity and Geography

Munhata Nainsi (1610–1670), an Oswal Jain Mutsaddi, who had been appointed the Diwan by Maharaja Jaswant Singh of Jodhpur came from a family of Rajput-converts to Jainism. Like their employers the Rathors of Jodhpur, the Oswals too had a martial past as well as, a long history of migration. Nainsi had participated in several military campaigns for Jaswant Singh before being appointed the Diwan from 1657 to 1666.[107] As an administrator not only had

---

[107] In his last years, Nainsi was accused of maladministration and asked to pay a *qabulat* of one lakh rupees. Nainsi, considering this to be a grave injustice refused to pay the *qabulat*. He and his brother Sundarsi were imprisoned in Aurangabad and were ordered to be taken to Jodhpur as prisoners, where their families had already been imprisoned.

Nainsi travelled far and wide but he also had access to administrative records as well as older historical narratives. As the Diwan of Jodhpur, Nainsi composed two major works, *Munhata Nainsi ri Khyat* and *Marwar ra Paraganan ri Vigat*. The former has come to be recognized as a comprehensive source to Rajput history, while the latter is a detailed seventeenth century gazetteer like account of Marwar. Nainsi relies on pre-existing accounts, information from officials as well as oral accounts in the composition of his narratives. He employs multiple kinds of narrative devices like *hakikat* (account), *y/adidasht* (memory/note), *vat* (tale/telling), *varta* (dialogue) apart from poetic compositions like *geet, kavitt, doha* and *sortha*.[108]

Nainsi's *Khyat* and *Vigat* complement each other. Nainsi's *Khyat*, in compiling a Rajput history, imagines a historical Rajput space, extending from Baluchistan to Kutch and from Ganga Valley to Indus valley. On the other hand, his *Vigat* details a contemporary space that Rathors were granted in the seventeenth century Mughal Empire. Nainsi, while being aware of politico-geographical divisions like Mewar, Amber, Marwar, Gujarat, Saurashtra, Kutch, is also aware that these divisions had been flexible through time. Thus, Nainsi's imagination of a geographical space effortlessly traverses multiple temporal and spatial levels, providing a glimpse into how imagination about spaces altered over time. In the first instance, Nainsi appears to be following no particular order in his *Khyat*. What Nainsi proceeds through, is the interrelationships of Rajput clans or individuals with each other as they move into each other's spaces to fight or make alliances through marriage. So what Nainsi appears to focus on are the networks that Rajputs create through their movements. The edition published by the Rajasthan Oriental Research Institute divides

---

Shaken by this prospect, Nainsi and Sundarsi committed suicide by stabbing themselves in 1670, near village Phoolmari. Nainsi's defiance is visible in the following couplet believed to have been composed by him,

> '*lakh lakharan neepje, bad peepal ri sakh*
> *natiyo munto Nainsi, tamba den talak*'

Badri Prasad Sakaria, 'Jodhpur ke Maharaja Jaswant Singh Pratham ke Diwan Prasiddh Khyat Lekhak Muhnot Nainsi' in *Khyat*, IV, 25–36.

[108] Like a historian trained in modern methods, Nainsi unhesitatingly acknowledges the bards whose compositions he uses. In places where he cannot quote an authoritative source he refers to the anonymous source by attributing it to hearsay (*aa baat suni hai*: I have heard/it is heard), thus referring to a vast oral and popular archive of historical imagination about the land and the people who inhabited it.

Nainsi's *Khyat* in four volumes, though it is not clear if Nainsi himself wrote the *Khyat* in this order. The first part follows the origins and histories of the Sisodiyas, a number of Paramara clans, including Chauhans, Solankis, Sodhas, and Sankhlas etc. It appears to move from Chittaur through Bundi, Sanchor, Jalor and Sirohi to Gujarat, Kutch and finally Umarkot. The second part focuses on Bhatis of Jaisalmer and their associated clans in Saurashtra, Kutch and Sindh. The third part is dedicated largely to the Rathors of Jodhpur and Bikaner. It is through these sub-regional/local networks that orientations of Rajput movements can be understood. While the Rajput world could be seen as a wide network of alliances, these alliances were made within a local domain. So, Rajput clans in Amber and Central India appear to belong to one network, while the Rajput clans of western Rajasthan, Saurashtra, Kutch and Sindh appear to belong to another, forming two separate east-west orientations.

Nainsi views Puranic knowledge related to origins of Rajputs as well as information about contemporaneous events, as located in a wider Rajput world. However, Nainsi's locating of the clan spaces in the Rajput world, provides a glimpse into the ways in which Nainsi viewed and heirarchised the Rajput world. It is important to note here that Nainsi's *Khyat* begins with descriptions of Rajput history with the Sisodiyas of Mewar, even though he is in the Rathor employ. Kolff has suggested that a new Great Rajput Tradition with Mewar as its seat emerged by seventeenth century, around the time when Nainsi compiled his *Khyat*. It was exclusive and forts like Chittaur and Ranthambhor were its foci. Tod's emphasis on the *Suryavanshi* Sisodiyas as the first of the 'thirty-six royal tribes' with no doubts with respect to their purity of descent indicates that by nineteenth century the central position occupied by Mewar was firmly entrenched.[109] In fact, in the negotiations for the state of Rajasthan, it was strongly suggested that the state be named Mewar, in keeping with the centrality of the princely state in history of the region.[110]

Yet, Mewar was only one of the many Rajput states that had emerged from the several migrations of itinerant warriors in this region. Its political space in the Aravali hills had been wrested from the previous occupants, the Bhils. It was bitterly contested by other Rajput clans, particularly, the Rathors who had achieved similar ascendancy over Bhils and Mers in this region. The houses of Mewar and Marwar had continued to engage in bitter boundary disputes, that could only be settled by regarding the Aravali mountains as a

---

[109] Tod, *AAR*, I, 173.

[110] M S Jain, *Concise History of Modern Rajasthan*, Vishwa Prakashan, New Delhi, 1993, 189–202.

natural boundary between Mewar and Marwar, explained by an epithet, *aonl aonl rana ra, banwal banwal rao ra.*[111] The division on the basis of geographies that distinguished the dry arid Thar from the mountainous Mewar, needs to be taken rather seriously, and forms an important marker of how Nainsi sought to separate and distinguish Thar while referring to a larger historical space of the Rajputs. The nineteenth century imagination of 'Rajputana' as a political space stands challenged by multiple visions of Rajput space visible in Nainsi's seventeenth century account, where space acquires a social meaning rather than merely a political one. It is important to note that Nainsi does not use the word Rajputana in either the *Khyat* or the *Vigat*. Interestingly, he uses the word *Hindusthan* as a political space centering around Jaisalmer. Nainsi separates *Hindusthan* from both Delhi and Ghazni, where the two other power centres or *chatras* seem to be located.[112] In fact, if there is a single clan to whose history, the greatest part of the Khyat is dedicated; it is the Bhatis of Jaisalmer and their associated clans.

Thus, while in the nineteenth century, Mewar owing to its stiff opposition to the Mughal advance came to be seen as the space where the Rajput world was centered, for Nainsi, it was Jaisalmer where the wider Rajput world lay. While Mewar defined the Rajput through its inward looking clan orthodoxy, it was Jaisalmer through its wide network of clan alliances that defined its expansiveness. In this wider world, opposition to Turks or Mughals was only one way out of many others to define the Rajput. Nainsi defines Bhati historical geography through the movement of Bhatis from Mathura, the place where all Bhatis are reported to have originated from Pradyumn, the son of Krishna, through Bhatner in Lakhi Jangal, to Kutch, Sindh, Gujarat, Jaisalmer, Multan and Punjab, where numerous allied clans were located in the seventeenth century.[113] Jaisalmer with a long line of Bhati rulers formed the

---

[111] Nainsi, *Khyat*, III, 89. The land where *Aonla* blooms belongs to the Ranas of Mewar, where as the land where *Babool* blooms belongs to the Raos of Marwar. This border dispute continued in the 19th century and is also referred to by James Tod. Tod, *AAR*, I, 547.

[112] '*Ek gaddhan mahe Dilli chhatra, ek Ghazni chhatra, Hindusthan ra gaddhan upper hai Jaisalmer chhatra. Tin karan Bhati chhatrala kahije*' (There is a *chhatra* in Delhi, one in Ghazni, but the *chhatra* over the forts of *Hindusthan* is located in Jaisalmer. That is why Bhatis are called *chhatrala*). Nainsi, *Khyat*, II, 15.

[113] 'The first was Raja Jadu, his descendants are Jaduvanshis. Bhati was born several generations after Pradyumn, their descendants are all Bhatis. After losing Mathura, Bhatis lived near Lakhi Jangal and settled Bhatner. Their branches are ....Jarechas, Sarvahiyas, Chudasamas and Jadavs'. Nainsi, *Khyat*, II, 16.

centre of this world. Bhatner, Pugal, Bikampur, Barsalpur, Deravar, Maroth, Kehror, Aasnikot, Tanot, Ludrovo, Mamanvahan were fortified settlements in this region controlled by Bhatis or their subclans, and contested by other Rajput clans. This historical geography extends further south and west through allied Yadu clans like Jarecha, Chudasama and Sammas, with whom Bhatis not only share a history of origin, but also the mythical *jadavsthali*.[114] Nineteenth century accounts refer to old Charanic genealogies that trace the movements of Bhatis further west into Afghanistan, and in some accounts as far as Egypt and Syria.[115] Charanic genealogies of Jarechas and Sammas trace their origins from four brothers Aspat, Gajpat, Narpat and Bhupat.[116] As they considered themselves descendants of a common ancestor, these clans did not intermarry, though they moved in a common geographical space. Besides, the descriptions of long mythical journeys became necessary in order to explain the hybrid character of these clans. Not only did Muslim groups like Samma, Abohariya, Chakita, Bhutto, Langa, Channa, Mohiya, Lodaro claim to share ancestry with the Bhatis, but Jats like Saran, Moodna, Seora, as well as the trading community of Bhatiya, also link their origins to the Bhatis. This can be witnessed in the manner in which Nainsi moves from Jaisalmer towards Kutch with his descriptions of Rajput clans, beginning with a genealogy of the Sarvahiyas of Girnar, and an account of loss of Girnar. He follows it up with an account of the Raidhans of Bhuj, Kutch, describing the multiple conflicts between clans in Kutch, through narratives of Khengar, Mandlik, Hamir and Lakho. Thus the Bhati genealogical space covered the extent from Punjab to Kutch, which is the extent of the Thar Desert. Moreover, Bhati alliances extended both into Multan and Punjab as well. Located on the route from Ghazna towards Gujarat, forts and settlements in this region became the sites of conflict between Bhatis and Muslim armies from northwest. This spawned a politics of both conflict and alliance, as Bhatis forged martial and marital alliances with both Rajputs as well as other groups in Multan and Punjab. In Nainsi's geography, Ludrova at the northern extreme of the Thar desert appears to be the door to the southern space. So when Bhati Vijayrao Lanjo

---

[114] Nainsi, *Khyat*, II, 3.

[115] Tod, *AAR*, II, 172.

[116] Samira Sheikh, 'Alliance, Genealogy and Political Power: The Chudasamas of Junagarh and Sultans of Gujarat', *MHJ*, 11, 1, 2008, 29–61. These histories claim that descendants of Aspat accepted Islam and came to rule Sindh as Sammas, Gajpat conquered the city of Gajna, Bhupat established Bhatner, and his descendants were the Bhatis, and the descendants of Narpat were Chudasamas.

married into the house of Paramars of Aboo, he was referred to as the *uttara disi bhad kivaad*, the 'sentinel of the north direction'.[117]

Similarly, the Rathor historical geography is defined through their movement from Kannauj to Khed, on their way to Dwarka for a pilgrimage. On his way to Dwarka, Rathor Seha Setramot was drawn into a dispute between Solanki Mulraj of Anhalwar Patan and Jarecha Lakha of Bhuj in the thirteenth century. Rathor Seho managed to kill the Jarecha and win the battle for Mulraj, as well as negotiate a marital alliance with the Solankis. His descendents wrested Pali from Mers, Khed from the Gohils and eventually Mandor from the Eendas.[118] In this manner they became a part of the Rajput world of the Thar, expanding through relationships negotiated with clans like Bhatis, Chauhans, Jarechas and Sodhas. The expansion of Rathors into Jangal, in the fifteenth century as Bikawat Rathors established Bikaner further expanded the Rathor space into the northwest, bringing them into conflict with Bhatis. A marriage contracted between Bika Rathor and the daughter of Sekho Bhati was instrumental in recognition of Rathor authority over large stretches of semi arid land between the Ghaggar and Sutlej rivers. The establishment of Bikaner involved subjugation of Godara and Saran Jats as well as the Johiyas, who had exercised semiautonomous control in this region. By seventeenth century, the Rathor political space appears to be located contiguous to Bhati space, in fact dominating it to the extent that Marwar, which literally meant the larger desert, came to be identified as a Rathor realm. However, there remained a contradiction between Maru, the desert and Marwar, the Rathor political space. In fact, when Nainsi compiled his *Marwar ra Paraganan ri Vigat*, it contained only seven paraganas, Jodhpur, Sojhat, Phalodi, Merta, Siwana and Jaitaran, that had been granted to Jaswant Singh. Nainsi's description of Maru through its nine forts is not in ignorance of the fact that all the forts were never a part of the 'political' Marwar ruled by the Rathors. For Nainsi, Maru was both, a political as well as a social and geographical space at the same time.

From Marwar, Nainsi's *Khyat* moves southwards towards Gujarat, with a description of early political history of Gujarat in *Atha Gujarat Desh Varnanam*. The inter-relationships of Rajput clans through war or marriage, as mentioned in case of Siha Rathor, linked these regions with each other. Pilgrimages to sites like Dwarka and Girnar also created opportunities for Rajputs from

---

[117] Nainsi *Khyat,* II, 33. The phrase *uttara disi bhad kivad* continued to be used by the Jaisalmer Bhatis in the Jaisalmer Gazette in the late nineteenth century.

[118] Nainsi, *Vigat* I, 8–12.

western Rajasthan to forge alliances with the ones in Gujarat. Rajput groups in Gujarat and Rajasthan also appear to be connected by the location of Abu as a geographical as well as genealogical and historical connector. Abu was the acclaimed site of the mythical *yagna*, which gave rise to the Agnikula Rajputs, thereby legitimising the claims of mobile warrior groups that had risen to power in this region. Being the centre of Paramara power, it became a reference point for Rajput groups like Solankis, Sodhas, Sankhlas and Chauhans that controlled southwestern Rajasthan, Gujarat and parts of Sindh. According to Nainsi, the descendents of the Paramara king Dharanivarah spread from Vahadmeru (Barmer in modern south west Rajasthan) and established nine forts marking the boundaries of Maru, though other clans subsequently took these.[119] For example, he refers to Umarkot, the Sodha capital in Dhat, as one of the forts of the nine forts of Marwar. Even though the powerful Ranas of Umarkot had sheltered Humayun as he escaped India, and thus provided a safe birthplace to the future emperor Akbar, by seventeenth century the Sodha power had been substantially reduced by both Rathors and Bhatis. However, both Hindu Rajput clans of Rajasthan and Gujarat as well as Muslim Soomras of Sindh, married into the Sodha clans of Umarkot in order to forge alliances.

Therefore, for Nainsi, Derawar (Bahawalpur), Bhatner (Hanumangarh), Pugal, Jaisalmer, Jodhpur, Bikaner, Jalor, Nagaur, Sanchor, Bhuj, Kelakot, Umarkot etc. were all part of a continuous political space, the Maru, which could be viewed as a region through the mobility of these clans that circulated through the Thar Desert. In Nainsi's historical imagination, it is not the geographical unity of the Maru that constituted it as a region, but the fact of it being connected through networks of movements and alliances, whether martial or marital. Alliances between Rajput clans facilitated circulation of men and resources in this highly mobile region. Located in Jodhpur, Nainsi attempted to describe this mobile region through *bats* or *vats*, encapsulating the connected histories of these men and thus clans in this region, which circulated throughout the region. Nainsi's imagination of Maru emerged from the movements of Rajput clans and far exceeded both the geographical boundaries of the Thar Desert as well as the political boundaries of the Rajput states.

Boundaries of James Tod's Rajast'han

James Tod (1782–1835) arrived in Delhi in 1800 and in a few years he embarked on travels that took him through Agra and Jaipur to Mewar and eventually

---

[119] Nainsi, *Vigat*, I, 1.

Gwalior. Tod was appointed the Political Agent of the Western Rajputana States in 1818, and occupied the position till 1822, against the resentments of his superior Ochterlony as well as the Maharaja of Marwar, Man Singh. In this respect his career has some parallels with earlier chronicler of the Thar, Munhata Nainsi. James Tod was an antiquarian with a deep interest in the history of the regions he was expected to administer. He was also keen traveller with avid geographical interest and imagination. One of the earliest recognitions of his geographical knowledge was the extensive use of his map of Central India in the battles against Marathas. Throughout his administrative career in India, James Tod maintained meticulous records of his travels as well as put together extensive annals of Rajput states that he compiled after his retirement as the *Annals and Antiquities of Rajast'han*. His other travel narrative, *Travels in Western India*, was published posthumously in 1839.

Tod was appointed to the Sindhia court in 1806, and followed the course taken by the Maratha army between Mewar and Central India. He realized that most maps provided unreliable information as positions of places like Udaipur and Chittaur had been reversed in earlier maps. Having never travelled any further west than Mandor, Tod's methodology had also been to send parties across the desert. The first party sent in 1810–11, travelled through Gujarat, Kutch into Sindh and returned through Jaisalmer and Marwar. The second party explored the desert south of Sutlej. These parties collected geographical, statistical as well as ethnographic information on the places they visited. Tod interviewed 'natives of every *t'hul* from Bhutnair to Omurkote, and from Aboo to Arore' who were brought to him while he was at the Maratha court in Gwalior between 1812 and 1817.[120] He also relied on the knowledge of *kasids* and other public conveyors of letters, who detailed both the peculiarities of the route as well as the accuracy of their distance. In this manner he was able to arrive at a map of central and western regions that he defined as Rajputana.

Much attention has been paid to the impact of Tod's *Annals* on historical imagination about Rajputs and Rajputana.[121] In more ways than one, Tod was responsible for creating parallels between Scottish warrior traditions and the Rajputs, by evoking a Scythian past for both. But more importantly, it was Tod's imagination of Rajputs as a singular race that also provided impetus to

---

[120] Tod, *AAR*, II, 233.

[121] Norbert Peabody, 'Tod's Rajasthan and the Boundaries of Imperial Rule in Nineteenth Century India', *Modern Asian Studies*, 30, no 1, 1996. 185–220. Henceforth *MAS*. Frietag, *Serving Empire Serving Nation*, examines Tod's life in India and the impact of his work on imagination about Rajputs in great detail.

the idea of 'Rajputana' as the political space of the Rajputs. However, while for Nainsi, the sense of a singular space emerges from networks of political and social exchange, in Tod's 'Rajast'han' the Rajputs, while belonging to a wider Rajputana are firmly entrenched in their locales. So, Sisodiyas in Udaipur, Kachwahas in Amber, Rathors in Jodhpur and Bikaner and Bhatis in Jaisalmer appear circumscribed by the boundaries of their political space, with which they increasingly identified. The treaties signed by the Rajput states between 1812 and 1818 resulted in curbs on both political and social expansion as political as well as marital alliances could not be negotiated without British intervention.[122] This meant that 'Rajputana' was viewed as conglomerate or confederation of distinct Rajput states rather than as a shared Rajput space.

However, this conglomeration was located in a space that was barely familiar to either Tod or other Englishmen in this period. Thus, for Tod, his travels and time spent in Rajputana were to lead towards developing knowledge of a region, which while referring to a familiar political imagination, held no resonance in his own geographical imagination. For Tod and other Europeans like Adams, Boileau, Powlett and Fagan, who lived and travelled through Rajputana, this dry barren sandy tract was quite opposite to the land that they had come from and thus evoked both awe and revulsion at the same time. It is, thus, not surprising that for Tod, the *Annals* were first and foremost an exercise in geography, "the historical and the statistical portion being consequent and subordinate thereto".[123] Tod draws a very clear line between historical and antiquarian research and geographical research, finding the former to be far more fruitful than the latter.[124] Nevertheless, Tod sought to 'fix the geography' and fill up the 'blank' that resulted from the absence of maps, which rendered most of this territory a *terra incognita*.[125] Thus, his objective was to "make

---

[122] Tod views these treaties as having brought universal pacification in Rajputana. He writes, "Till this period, not a chief present had throughout his life ever laid his head upon his pillow without being prepared to be roused from his sleep by the cry of 'the enemy is at the gate:' some ancient foeman who had come to 'balance the feud', or marauding mountaineer or forest Bhil who had emptied his cow pen. All these sources of anxiety were now at an end." *Travels in Western India Embracing a Visit to the Sacred Mounts of the Jains and the Most Celebrated Shrines of Hindu Faith between Rajpootana and the Indus,* (First Published, Allen and Co, London, 1839), Munshiram Manoharlal, New Delhi, 1997, 3.

[123] Tod, *AAR,* I, 2.

[124] Ibid.

[125] Ibid.

the map, so perfect as the superabundant material at the command of author might have enabled him to do".[126]His historical imagination also compelled him to, "institute a comparison between the map and such remains of ancient geography as can be extracted from Poorans and other Hindu authorities."[127] For Tod names of most places held a reference to a geographical feature of some sort. Parkar, for instance, derived its name because it was beyond the Luni, the Salt River, 'beyond the *khar*', hence, Parkar.[128] Likewise, the name Tirruroe is derived from *tirr* "which signifies moisture, humidity from springs or the spring themselves, that rise from this *rooe*".[129] The Rann had a similar association with *aranya*.[130]What is interesting is that while drawing references from ancient texts, Tod was actually attempting to define and map 'Rajputana' in contemporary scientific terms. Not only does Tod provide classificatory contexts like landscape, climate etc., he also fixes 'Rajputana' into the grid of latitudes and longitudes, in accordance with modern scientific cartographic practice. Thus, in his 'improved map' many positions were affixed by a 'zealous and scientific' geographer.[131] What perhaps Tod was attempting to do was to feed into larger geographical and topographical mapping exercises being carried out by Rennell and Reynolds. The mapping of various regions was part of a larger political context of correlating geographical and topographical knowledge with ability to control indigenous political regimes. In case of 'Rajputana' it served to open up the possibilities of being able to navigate the way to regions contiguous to Sindh, a land not only far unknown as compared to Rajputana, but also one that was far more 'unsettled'.

Tod did not merely survey and map the Thar, but it actually acquired a geographical shape in his imagination, by his efforts to put the "geography of Rajast'han ...in a combined form".[132] Standing atop the Guru Shikhar in Abu, Tod imagined the "wide expanse, from the 'blue waters' of Indus to the 'withy covered' Betwa". In his chapter on Geography of Rajast'han, Tod makes the reader turn 360 degrees and locate the Vindhyas, Aravali as well as the table land of Malwa. Saving the reader, "a painful journey over the T'hull" Tod turns

---

[126] Ibid.

[127] Ibid.

[128] Tod, *AAR,* II, 238.

[129] Ibid, 242.

[130] Ibid., 238.

[131] Tod *AAR*, I, 6. Tod refers to Captain Dangerfield, who carried out extensive surveys in Malwa.

[132] Ibid., 2.

westwards and introduces the reader to "that extensive plain of ever-shifting sand, termed in Hindu geography *Maroost'hulli* corrupted to *Marwar*".[133] Unlike the eastern parts where flora catches his imagination, Tod describes the western stretches through the desolation he encounters. Observing the Rann of Kutch, he writes, "…nothing meets the eye but an extensive and glaring sheet of salt, spread over its insidious surface, full of dangerous quicksands…".[134] Tod compared the Thar to the hide of "tiger, of which long dark stripes would indicate the expansive belts of sand, elevated upon a plain only less sandy, and on whose surface numerous thinly populated towns and hamlets are scattered".[135] For him, this space was not merely defined through its solitude and desolation, but also through hostility engrained in this terrain. He noticed that, "throughout this tract, are little hamlets, consisting of scattered huts of the shepherds, occupying in pasturing their flocks or cultivating these little *oases* for food, long line of camels, anxiously toiling through often the doubtful path, and the Charun conductor, at each stage tying a knot on the end of his turban".[136] The caravans could come across, "lying in ambush, a band of Sehraes, the Bedouins of our desert, on the watch to despoil the caravan".[137]

In his sketch of the Indian Desert, Tod attempts to mark the boundaries of the Indian desert, and correlate it to its political history as well as its human geography. This part of his text is not his personal narrative, but one put together like a map, from the reports of various missions mentioned earlier. In this section, Tod attempts to describe the desert either through natural divisions or through the divisions employed by the local people. This results in a curious mix of western methods of approximation of the region along with native political geography. Tod compares the Indian desert with the African desert, drawing parallels between the Nile and the Indus, geographical as well as civilizational. Tod dwells on the antiquity of references, unraveling etymologies, linking Rann with both Greek *Erinos* and Sanskrit *Aranya*. Thus, the Thar acquires not only an antiquity, but also a continuity with older world geological processes, as Tod imagines, "the chasm, now forming this rich valley (of Indus), must have originated in a sudden melting of all the glaciers of Caucasus, whose congregated waters made this break in the continuity of Maroost'halli, which would otherwise be united with the deserts

---

[133] Tod, *AAR*, I, 13
[134] Ibid., 14.
[135] Tod *AAR*, II, 234.
[136] Ibid., 236.
[137] Ibid.

of Archosia".[138] When he attempts to draw linguistic parallels between Indian *bah*, and Arabic *wadey*, or socio-historical ones between Indian *sehraes* and Arabic *bedouins*, he manages to place the Indian desert in the larger context of desert civilizations. Interestingly, considering that Tod's chapter on feudalism compares the Rajputs with European Anglo-Saxon knights it becomes clear that Tod views Thar from two different civilisational paradigms. The first is that of Rajput kingships firmly located in forts and capital cities, defined by valorous, chivalric ideals that Tod attributes to Rajputs. The other is of a dangerous shifting world, best represented by the *seah-kotes*, or the mirages of winter castles that Tod encounters several times in his travels. Tod's Rajputs do not appear to belong to this unstable world, which is represented by treacherous shifting routes and unsettled desert wanderers. In a sense these worlds are incommensurable in Tod's narrative. Tod, on the one hand is caught in the romance of unbound desert, is on the other entangled with the task of identifying domains of separate Rajput principalities, drawing boundaries on the map, as well as settling boundaries for the Rajput states.

It is in this context that Tod's understanding of the Thar completely deviates from Munhata Nainsi's.[139] For Tod, the kingship was located in specific courts, which paradoxically were becoming increasingly ineffectual in the nineteenth century. For Nainsi, kingship was located in the inter-relationships that had to be negotiated in these open spaces, away from the courts and capitals, where Rajput as well as non-Rajput groups wielded considerable authority. In the peripatetic, ambulatory world of Thar, the authority did not necessarily and totally rest with ones who controlled the forts located in towns, the nodes of power, as described earlier in the chapter, but with ones who controlled the routes. The power relationships in the mobile world of Thar, were located, on its routes, as well as in fairs, markets and towns, where its occupational travellers, came together and exchanged mobile wealth. In this mobile world the histories and identities of these occupational travellers were flexible, often indeterminate and interchangeable, as we would see in subsequent chapters. These two ways of imagining the Thar also lead us to understand the polity of the region in different ways, as we will see in the next chapter.

---

[138] Ibid., 234.

[139] In his memoirs of *Travels in Western India*, Tod however explores continuities between Rajputana, Kutch and Gujarat through Rajput networks, quite like Nainsi.

CHAPTER

2

Mobility, Polity, Territory

Rakhaich asked, "Who is the other palace for"? Then one *Apsara* said, " There is yet no one deserving enough for that palace, but one who dies while avenging his father's death, and dies in front of his *dhani's* eyes, while fighting for his cause will get to live in that palace".[1]

The political history of the Thar has largely been understood through the history of its political segments Jaisalmer, Jodhpur and Bikaner. As these states were assimilated into the Mughal Empire in the late sixteenth century, the study of polity in this region largely became the study of Mughal-Rajput dynamics. However, as we saw in the previous chapter, Thar was a fluid mobile region, geographically, economically, culturally, as well as politically. Munhata Nainsi's imagination of a 'Rajput' political history in the seventeenth century, revealed a fluid polity that relied on kin and clan relationships operating across the Thar through martial and marital alliances. However, by the nineteenth century, the political geography of this region was increasingly understood through the idea of boundaries, rather than frontiers. The subsidiary alliances between Rajput states and the British in early nineteenth century accentuated the binary between the idea of boundary and frontier. While the boundary became the site of legitimate political endeavors, the frontier became the 'outlawed' space. Frontiers were often contested non-agrarian terrains like hills, forests or deserts that increasingly came to be seen as hostile spaces by the nineteenth century. But frontiers were also spaces where complex political processes led to redefinition

---

[1] Nainsi, *Khyat*, I, 256–257.

of social and political identities. As the previous chapter explored the idea of Thar as a geographical frontier, this chapter examines the idea of Thar as a political frontier. It seeks to examine the idea of the Thar as a frontier defined through the mobility of its travelers and the relationship of this mobility with political control. It puts forth the argument that the Thar, for several centuries, had been a site for complex political negotiations, where authority was defined through control over access to mobile resources, whether pastoral wealth or trade. By the nineteenth century, discomfort with multiple levels of control led to certain kinds of controls being rendered legitimate and others illegitimate.

The dominant understanding of the nature of political control in the Thar region has been one where fort-centric Rajput lineages are understood to have exercised control over land as well as networks of circulation. However, my understanding of the nature of control in this region is that it was defined through negotiations and confrontations between various groups circulating mobile resources. The emerging fort-centric sedentary Rajput polity in medieval Thar had to co-exist with a mobile polity where control in outlying regions rested not just with a wide range of Rajput groups, but also others that controlled mobile resources like herders, carriers, traders and bards. Further, the ability of these groups to control mobile resources emanated from their ability to control the circulation of these resources. However, the emergence of sedentary Rajput polities also meant a shift towards a territorial polity where the basis of control began to shift from mobile resources to land. The assimilation of Rajput lineages of the Thar into the Mughal Empire in the sixteenth century and the imposition of British indirect rule in the early nineteenth century represent two important stages of this shift. This chapter attempts to trace these shifts and the manner in which these were reflected in the political culture of this region.

Historicizing Itinerancy: The Itinerant Warriors of the Thar

It is difficult to visualize a singular Rajput polity in the Thar region before the fifteenth century. The Thar was ruled by several Rajput lineages that had barely established themselves and were still in the process of negotiations with pre-existing political formations. As discussed in the introduction to this work, the process of emergence of Rajputs as a political group was closely tied with their emergence as a social group. However, while with the acquisition of political power, the ranks of latter (caste) closed by the fifteenth century, the ranks of former (status) remained open till much later, as I will go on to show.

I begin with a few descriptions culled from Munhata Nainsi's *Khyat*, where he discusses the establishment of Rajput clans in the Thar region. Nainsi, while chronicling histories of Rajput clans in the Thar, also traces their movements across the region, as seen in the previous chapter. In tracing early Rajput groups what Nainsi appears to be focusing on is the manner in which itinerant warrior groups move across ecological and political frontiers in order to establish themselves. These journeys appear to be undertaken in search of land, water, militaristic adventures, even as pilgrimages. The descriptions of these journeys, not only indicate the kinds of spaces that these groups travelled across in their journey to 'Rajputhood', but also the processes by which it was achieved.

Describing the journey of Sisodiyas to Mewar, Nainsi writes,

> An ancestor of the Sisodiyas, who had a *thakurai* in Nasik Tryambak, had no heir. He was advised to pray to Aambai Devi of Idar. On his way to the temple Grasias and Kamthalias attacked him.[2] The king was killed and the pregnant queen sought shelter with a Brahmin in Nagda. After her son Vijaydatt was born, the queen decided to commit sati and handed over the son to the Brahmin, who demurred, saying, 'He is the son of a Rajput, when he grows up he will hunt, acquire horses and enmity with the world around. I will lose my *dharma*'. The queen assured him, that her son and his descendants would be kings and that the descendants of the Brahmin will prosper in their rule.[3]

Nainsi describes the rise of Haras in Bundi in the following manner,

> In Bundi Minas were very strong. Haras, who faced *vikhau*[4] in Bhainsrod, had moved to Bundi. Minas wanted to marry the daughter of a Brahmin in Bundi. The Brahmin approached the Hara, who advised him to go ahead with the arrangements of the marriage, but killed the Minas while they were intoxicated. In this manner Haras took over Bundi.[5]

---

[2] *Grasiyas* have been understood to mean Rajputs who took up small patches of land in lieu of service (the basic term being *giras*, or a bite of food) particularly in Southern Rajasthan and Gujarat. However, in the present day context it represents a Bhil group that claims to have Rajput origins. Zeigler has translated Kamthaliyan as derived from *kamtha*, the frontier. Zeigler, *Action, Power and Service*, 34.

[3] Nainsi, *Khyat*, I, 1–2.

[4] *Vikhau* refers to periods of displacement faced by Rajputs.

[5] Nainsi, *Khyat*, I, 87.

The takeover of Sirohi by Deoras is described as,

> Deoras initially lived in Nadol, and later moved close to Abu, where five Deora brothers wandered from place to place in search of fortune. At that time Kanhadde ruled in Jalor. Then the Deora brothers decided to take Abu. They had five daughters each. So when a Charan of the Panvars came to them, they said, 'we have no land, but twenty five daughters, where will we find grooms?' The Charan set up an alliance with the Panvars of Abu. But the Panvars doubting their intention, asked for a surety in the form of a man. At the time of the wedding, however, the Deoras sent men dressed up as brides, who killed the Panvars and captured Abu.[6]

In a similar manner he describes the movement of Solankis to Gujarat,

> Two Solanki brothers had to leave Toda, when their stepbrother became the ruler, and they were disinherited. The elder brother was blind, and the younger a minor. They decided to go to Dwarka for a pilgrimage. On their way they reached Patan, where Chavdas ruled. The blind Solanki brother was able to identify the colour, breed and price of a mare, and predict that the mare would give birth to a one eyed foal. The Chavda impressed by the intelligence of the older brother, addressed him as *akalwant*[7]. The Chavda married his daughter to the younger brother and gave him many villages and cattle (employed him). Of this daughter Mulraj Solanki was born. Then the Solanki brothers decided to continue on their pilgrimage to Dwarka, leaving Mulraj behind in Patan. Now, Lakha Jarecha had already heard about the intelligence of the brothers. So he too married his sister to the younger brother."[8]

In another instance, Nainsi describes the rise of Chunda Rathor,

> Chunda worshipped the goddess with fervor, fasted and meditated for two days. The Devi manifested herself and talked to Chunda. She told him that four days from now, along the road at such and such place ten carts full of gold would

---

[6] Nainsi, *Khyat*, I, 123, 172–73.

[7] *Akalwant* literally means one with wisdom. Zeigler suggests that it could connote a Rajput sage who had gained the powers of insight, vision and ability to foresee the future through long meditation, devotion and self sacrifice. Such men were revered and respected as ascetics and acted as interpreters of omen, the *saganis*. Zeigler, *Action, Power and Service*, 33. The presence of such men correlates very well with the itinerant warrior ascetic idiom among the Rajputs.

[8] Nainsi, *Khyat*, I, 251–52.

come. Four armed men would guard the carts. The gold would be covered by
tattered cloth. You kill the men and bring the gold. Chunda killed the men
and buried the gold. Everyday more Rajputs with horses joined Chunda. Then
the goddess directed Chunda, 'I will make you a king (*gadhpati*)'.[9]

Table 2.1. Some Rajput Migrations according to Nainsi's Khyat

Name of the Clan	Claimed Place of Origin	Migrated From	Migrated To
Gohil	Plains of Central India	Pali	Northern Gujarat on the border of Saurashtra later called Gohilwara
Jarecha	Sindh	Sindh	Mallani, Kutch, Saurashtra
Tomara	Tanwarawati (Delhi)	Delhi/ Tanwarawati	Haryana, Punjab, Bikaner, Jodhpur, Sindh
Rathor	Kannauj	Central India	Marwar (Pali)
Paramara	Kamboja in the Hindukush	Alor in Sindh, Bhakar	Central India, Western Rajasthan, Mallani
Dabhi		Marwar	Saurashtra
Dahiya	Tryambakeshwar (Nasik)	Parbatsar	Thal, Nergarh, Pali, Mallani, Sanchor, Jaswantpur
Solanki	Badami	Gujarat	Marwar
Chauhans	Dhundhar		Nadol, Jalor, Dhundhar, Ajmer, Delhi, Haryana, Godwar, Harauti, Sindh
Chudasama	Sindh	Sindh	Saurashtra, Kutch
Bhati	Dwarka	Bhatner	Marwar, Jaisalmer, Multan Punjab, Sindh
Sisodiya	Nasik Tryambak	Gujarat	Mewar
Kachchwaha	Rohtas	Central India	Narwar, Amer
Chavda		Bhinmal	Saurashtra
Jhala	Kutch	Kutch	Mewar, Harauti

In all the above instances, the common element appears to be the itinerant
wandering adventurer in search of a foothold, but one who helps in upholding
of the social order. These itinerant adventurers pushed out of their original
inhabitations due to a variety of circumstances that could include famines and
floods, disinheritance due to expansion of clans, or search for employment,

---

[9] Nainsi, *Vigat*, I, 22–23.

found ample possibilities in frontier regions like the Thar. These 'immigrations' sometimes took place from far off regions like the Indo-Gangetic plains, Sindh, or even further from Baluchistan, and others from one part of the Thar to another. Within the Thar these groups of adventurers appear to move between regions with older established political centres, often towards Gujarat, as the frequently undertaken pilgrimages to Dwarka by a wide range of groups suggests, but also towards older power centres in south-eastern Rajasthan. The establishment of these groups was preceded by fierce struggle with similar itinerant groups, including groups like Mers, Minas, Bhils, but also others like Jats, Gujars, Rabaris, Charans, and Ahirs that traversed the Thar along with their cattle. Conquest, colonization, cult appropriation, formation of martial and marital alliances between these wide range of groups laid the path for the rise of some of these as sedentary aristocratic lineages, which was marked more often than not by the construction of a fort. Let us examine how these factors became operative in the case of two leading Rajput clans of the Thar that is the Bhatis and the Rathors.

Despite the long-standing feuds between Bhatis and Rathors, the patrons of Nainsi, in his *Khyat* Bhatis appear to occupy a place of pride. Not only does he devote a large part of his *Khyat* to the Bhatis, he also calls Jaisalmer a *chatra* over all the forts of *Hindusthan*.[10] Among all clans of the Thar, it is Bhatis who claim to have covered the largest space in the region in their migrations. They claim to be descendants of Krishna, *yaduvanshis* from Mathura with a long history of migration. They travelled into north and west directions settling in parts of Sindh, Punjab, Multan and Balochistan, in some accounts going as far as Egypt and Syria. Their itinerary carried them through, "Indraprastha, Surajpura, Mathura, Praga, Dwarica, Judoo-ca-dang, Behara, Gujni in Zabulistan; and again refluent into India, at Salbahana or Salpoora in the Punjab, Tunnote, Derawul, Ludrovo in the desert, and finally Jessulmer".[11] In local parlance the Bhatis recall having travelled from,

Mathura, Kashi, Pragvad, Gajni aru Bhatner,
Digam, Derawal, Lodrovo, nammo Jaisalmer.[12]

The older history of the Bhatis points to a tradition where they claim not only to have built the fort and the city of Gazni, but also settled Sialkot (Salbahnpur) named after Salbahan who conquered the whole of Punjab and

---

[10] Nainsi, *Khyat*, II, 15.

[11] Tod, *AAR*, II, 172.

[12] Hari Singh Bhati, *Pugal ka Itihas*, Bikaner, 1999, 9.

"the surrounding Bhomiyas acknowledged his supremacy. He had fifteen sons...all of whom, by strength of their own arms, established themselves in independence".[13] They also claim to be the ancestors of the Chagtai Mughals through an ancestor named Chakito who "became the king of Balich and Bokhara...of all from the gate of Baluchistan to the face of Hindust'han: and from him is descended the tribe of Chakito Mughals".[14]

These claims sound implausible, but considering the fact that this area was historically one that was criss-crossed by several groups, a part of such claims may well be justified. Bhatis interestingly also display wide social stratification. Apart from the various Hindu Rajput Bhati sub-clans, Jats like Saran, Moodna, Seora, as well as Muslim groups like Bhatti, Bhutto, Langa, Channa, Mohiya, Lodaro and the trading community of Bhatiya, all link their origins to the Bhatis. Besides, Bhati clan alliances extend to other Yadu clans particularly in Suarashtra, Kutch and Sindh like Chudasama, Jarecha, Sammas, Sarvahiyas and Raidhans. Tod notices this wide stratification in his *Annals* and writes,

"Bhati Abhey Rao brought the whole Lakhi Jungle under his control, and his issue, which multiplied, became famous as Aboria Bhatis. Sarun quarreled with and separated from his brother, and his issue, descended to the rank of cultivators, and are well known as Sarun Jats".[15] In accordance with the explanations presented for stratification among other Rajput groups, Bhatis too were 'polluted' by their contact with other groups in times of distress and were forced to eat "with husbandmen (Juts), and were married to their daughters. Thus the offspring of Kullur Rai became the Kullorea Jats: those of Moondraj and Seoraj, the Moodna and Seora Jats; while the younger boys Phool and Kewala, who passed off as a barber and potter, fell into that class."[16] This process continued and Makur's descendants "became carpenters and are to this day known as the 'Makur sootar'......descendants of Deosi became Rebarris (who rear camels) and the issue of Rakecho became merchants (baniahs) and are now classed amongst the Oswal tribe".[17]

In my understanding this is actually a reverse reading of Bhati social processes. In the desert, Bhati came to represent an assimilative category for

---

[13] Tod, *AAR*, II, 177.

[14] Ibid., 178. Also Nainsi, *Khyat*, III, 36–37.

[15] Tod, AAR, II, 179. Though according to Nainsi, Abohariya Bhatis are the descendants of Desal, who was the brother of Jaisal, who founded Jaisalmer. Abohariya Bhatis settled Abohar in Punjab. Nainsi, *Khyat*, II, 10.

[16] Tod, AAR, II, 180.

[17] Ibid.

varied groups controlling various parts and particularly routes through this desert frontier. Also, while the Bhati annals claim Bhatis having occupied wide territories, their actual spread occurred in contested territories, which meant they were being constantly pushed into the desert. In fact, the nine forts of Bhatis, Jaisalmer, Pugal, Bikampur, Barsalpur, Mamanvahan, Maroth, Derawar, Aasnikot and Kehror are all located in the heart of the desert. (See map) Increasing pressure from the Rathors of Marwar and later Bikaner limited them to their own territories from where they continued to make inroads, often unsuccessfully, into Marwar and Bikaner. The early dominance of Bhati Rajputs can be understood through some aspects of their clan history that Nainsi discusses. For instance, the incorporation of local goddess, particularly Charani goddess-worship traditions helped Bhatis not only legitimize their positions but also seek support from communities like Charans.

> Rao Tannu's son Vijayrao Chudalo was an *akharsiddh*[18]. Once a large army from Sindh attacked him. Vijayrao prayed to the goddess and offered his head if the army from Sindh retreated. Since Vijayrao was successful in the battle he went to the temple at midnight and proceeded to behead himself. The goddess stopped him, considering the gesture to be enough. Vijayrao then asked the goddess for a proof of her presence, which she gave in the form of her gold bangle. That is how Vijayrao was called *Chudalo* (one with the bangle).[19]

These incorporations appear to also have been necessitated because of continuous onslaught by Varahas, a Panwar clan from Derawar. Varahas offered the hand of their daughter for Vijayrao's son Devraj, and used the stratagem of marriage alliance to eliminate the Bhatis. However, Devraj managed to flee and was raised incognito by a Brahmin. At this stage another significant event took place in the Bhati world, which is acquiring a Charan. The Bhati genealogies and *bats* used by Nainsi have all been compiled by the Ratnu Charans, who became a part of the Bhati world in the following manner.

> The Varahas approached the Brahmin and asked him about the child he had with him. The Brahmin said, he was his own son, and to prove it made one of his sons, Ratnu, eat with Devraj. The Varahas satisfied, left, but Ratnu excommunicated by his brothers left for Sorath and became a *Jogi*. After

---

[18] *Akharsidh* has been translated by Zeigler as "one successful in battle". Zeigler, *Action, Power and Service*, 67. However, it can also be seen to mean someone who can control the battlefield, perhaps through miraculous powers.

[19] Nainsi, *Khyat*, II, 17–18.

Devraj grew up and gained control of Jaisalmer, he recalled Ratnu from
Sorath and made him his Barhat. The descendants of Ratnu are the Charans
to the Bhatis.[20]

Devraj, when he grew up, began *chakri* with his maternal uncle Bhutto
near Derawar. When he asked for land, his uncle granted him as much land
as could be covered by a buffalo hide.

Devraj soaked the hide in water and cut it into stripes and took all land
around him. He then acquired a number of horses. Then he decided to build
an impregnable fort. He approached the goddess again, who blessed him with
an impregnable fort. The Derawar fort, on road to Sindh and Multan, is truly
an invincible fort.[21]

Later, not only did he thwart an attack on Derawar by his maternal uncle
but also stormed the fort of his Varaha father-in-law. In a massacre that
followed, the Varahas were wiped out and their women insulted, including
the mother-in-law who had saved Devraj. This granted him an uncontested
control of an invincible fort in the middle of the desert, which became the focal
point of further Bhati expansion. Munhata Nainsi recalls this battle through
an epithet in his *Khyat*, which means that Varahas can be good enemies but
Bhatis can never be good friends.[22] From Derawar, the Bhatis expanded their
fief into the Panwar territory and took Ludrovo by deceit again.[23] By the
12th Century Pugal, Chohtan, Rohri and Sukkur had been incorporated in
the Bhati dominion. By this time Bhatis were not only in conflict with the
Muslim cheifships of Sindh and Multan, but also with the Panwars, Solanki
and Sodhas.

Around this time the fort and Jaisal founded the city of Jaisalmer in 1151,
atop the Trikuta Hill near Ludrovo, as the latter was becoming unsafe and
was abandoned. Jaisal's father Dusajh had carlier gone to Gujarat and had

---

[20] Nainsi, *Khyat*, II, 19–23.

[21] Ibid.

[22] *Viras bhalo Varihan, mitt na bhallo Bhati*
   Je gun kiya Raway, te sab kallar jhalliya.

   Ibid., *25.* This opinion of Nainsi is no doubt a reflection of the Rathor-Bhati relations
   in the 17th c., but it also points out that unlike the idea of Rajput as the chivalrous
   warrior, deceit was a common strategy for Rajputs during conflicts.

[23] Ibid., 26–28.

become a *grasiya,* and thus, his younger brother Vijayrao Lanjo became the king, who was succeeded by his son Bhojde.[24] However, Jaisal conspired with Ghaznavide chiefs and Bhojde was killed in the subsequent battle. Following the death of Bhojde, Jaisal became the Bhati chief. Jaisal was followed by his son Salbahan II, who completed the construction of the Jaisalmer fort. He was killed in a battle with Baloches while attempting to regain Derawar that had been taken by them.[25] It is only by the beginning of the 13th century that a more recognisable chronology of the Bhati kings becomes available, as it is corroborated through their skirmishes with the Delhi Sultans and later the Mughals. This is also the time when they came into contact and conflict with Rathors, who established themselves first in Pali followed by Khed, Mandore and finally Jodhpur. This phase also marks the decline of the mighty Bhatis who were now reduced to areas surrounding Jaisalmer as they also became subordinate allies of Marwar. The emergence of two powerful Rathor kingdoms in the neighbourhood, destabilized the polity of Jaisalmer.

The first Rathor chieftain in the Thar is believed to be Siho Setramot, who Nainsi calls Raja Singhsen of Kannuaj, a descendent of Jaichandra, thus creating a link between Rathors and the Gahadwala dynasty of Kannauj. However, the Bithu inscription of VS 1330 refers to Siha as the son of Set Kanwar, without a reference to the Gahadawala descent.[26]Nainsi recapitulates the account of how Siho came to be in the Thar region,

Raja Singhsen decided to go to Dwarka from Kannauj as a *kaapri* ascetic, as he wanted to repent his sins of fratricide. At this time Chavdas, who ruled in Gujarat and Lakha, who were the Jam of Sindh were enemies. As Sihoji reached Patan, Khetrapal the deity of Chavdas appeared in the dream of Mulraj Solanki and told him that Lakha is fated to die by the hands of Siho. So Chavdas invited Siho, but as he sat down to eat, he found that widows were serving him. When Siho asked he was told about the enmity with Lakha. Siho agreed to carry out the revenge and killed Lakha. He was given a daughter by the Chavdas and returned to Kannauj. One night in Kannauj his Chavadi wife dreamt that three cubs came out tearing through her abdomen. Then Raoji told her she would have three leonine sons who would win over large spans of land. Some time later these sons were born.[27]

---

[24] Nainsi, *Khyat*, II, 34.

[25] Tod, AAR, II, 197.

[26] Dasaratha Sharma, *Rajasthan through the Ages,* Vol I, 1966, 687.

[27] Nainsi, *Khyat*, II, 266–275.

Asthan, born of Chavda wife of Siho, was raised in Patan after the death of Siho. But as he and his brothers grew up, they became aware of their status in their maternal uncle's house. They decided leave Patan to seek their fortunes in Pali, where, "*lakhesuri* (millionaire) and *kodidhwaj* (billionaire) people reside(d)".[28] Pali came under the *thakurai* of Kanha Mer, who terrorized the inhabitants, the Paliwal Brahmins, as "he kept all newly wedded girls in his house for three days".[29]

> Asthanji was staying in the house of a Paliwal Brahmin, who had a daughter old enough to have been married. When asked, the Brahmin told Asthanji that the girl could not be married because of the Mers. Asthanji assured the Brahmin and asked him to go ahead with the marriage. As soon as the nuptials were conducted, the Mers tried to take away the girl, who sought shelter with Asthanji. His men chased away the Mers, who returned and attacked Pali, and took away all its cattle. Then Asthanji challenged the Mers, killed Kanha, and took all his cattle.[30]

The Paliwal Brahmins invited Asthan to accept the *rozgar* (employment) of guarding the town of Pali.[31] Eventually, similar invitations came from all over the villages around Pali. The terms of employment changed to tribute and the overlordship of Rathors was established in Pali.[32] They colonised land in about 50 villages and sought *bhog* (tribute) from another 60 villages. Sonig, one of the brothers of Asthan went to Idar and evicted Bhils.[33] Around this time Gohils were the overlords of Khed, (near Balotra) and Asthan sought matrimonial alliance with Pratapsi of Khed. Asthan and his brother Sonig connived with the Dabhis who were the *pradhans* in the area, and attacked Gohils taking them unawares. Gohils fled to Jaisalmer, then to Sorath and eventually Palitana, Sihore.[34] Dabhis too were evicted by the Rathors who then took 140 villages of Kodhena from the Bhatis.[35] As Nainsi describes,

---

[28] Nainsi, *Vigat*, I, 9.

[29] Nainsi, *Khyat*, II, 277.

[30] Ibid.

[31] Nainsi uses the term *rozgar* as "we are going to Gujarat to seek *rozgar*"...so Paliwals asked him to accept *rozgar* in Pali. Nainsi, *Vigat*, I, 10. However, there is no reference to this in the *Khyat*.

[32] Nainsi, *Vigat*, I, 10–11.

[33] Nainsi, *Vigat*, I, 12–13.

[34] Nainsi, *Vigat*, I, 14.

[35] Ibid.

once Asthan was able to take over Khed, he gradually marginalized the Dabhis with whom he initially shared the *hasil*.[36]

Rathors seized Mahewa (Mallani) and Barmer from Parmars and Bhinmal from Sonagaras. Rathor Raipal also attempted to control Mandor but the presence of Alauddin Khalji in Siwana and Jalor proved to be a deterrent to the Rathors.[37] Subsequently Rathors expanded their territories under Rao Kanhadde, Jalhansi and Chhada. As is evident in this period, territories frequently changed hands between various Rajput clans who were yet in the process of defining their own territories. Chhada was followed by Teedo, Salkha, Malo, Chunda and Rinmal all of whom engaged in clashes with neighbouring Rajput clans and also expanded the Rathor dominions either through conquest or through matrimonial alliances. This is also a period of flux or *vikhau*, as between Chhada and Chunda, there appears to be confusion regarding line of inheritance. It appears that Malo Salkhavat, who later became a Nath *Jogi* with the name Mallinath, wrested the throne from the incumbent Tribhuvansi.[38] Malo's brother Viram Salkhavat moved to Maheva, where he was killed by Johyas in a battle regarding theft of cows. Viram's son Chunda was being raised incognito by his Mangaliyani mother, when a Charan, Alha Rohadiya recognized his royalty.

> Chunda herded the cattle of the Charan. One day when he got tired while herding, he fell off to sleep. Alha happened to come that way, and saw the boy sleeping with a serpent shading his head.[39]

The Charan equipped him with a horse and weapons and presented him to his uncle Malo, who granted him a distant *thana* of Salodi. Chunda was not satisfied with it and undertook raids on the road to Phalodi. Mandor was given as dowry to Chunda by the Eenda Rajputs in 1395 and eventually became the capital of the Rathors.[40]Chunda Viramot (r.1385–1424) inherited the Rathor mantle and was instrumental in the move of the Rathors from the marginal Mahewa belt to Mandor. When he sat on the throne, "he called all Rajputs and assured them. Since he had a lot of gold, he did not ask the Rajputs for

---

[36] Ibid. The Gohil daughter-in-law implored that since they had killed her clansmen for the land, why did they need to share it with the Dabhis?

[37] Ibid.

[38] Nainsi, *Khyat*, II, 283–284.

[39] Nainsi, *Vigat*, I, 21.

[40] Ibid., 25–26.

anything for two years. He settled deserted villages and took some in *khalsa*. He then divided the *chaurasis* (group of eighty four villages, an administrative unit) into two and took villages with good *hasil* (revenue) into *khalsa*."[41]

The years following the reign of Chunda again witnessed a struggle for inheritance as instead of the elder Rinmal, Kanha became heir and the disinherited Rinmal sought refuge in the court of Sisodiyas.[42] Subsequently, Rinmal returned to Janglu, with the help of Charani goddess Karni, and carried out incursions into Bhati territory and occupied Bikampur. He also defeated Hasan Khan of Nagaur and added Nadol, Jaitaran and Sojat to the Rathor lands.[43] Rinmal's intervention in Mewar affairs led to his assassination in Chittor and for a short period of time Mandor was taken over by the Sisodiyas and his heir Rao Jodha had to seek shelter in the village Kasano in Janglu.[44] Around this time the Rathor strongholds also came under attack from the Delhi Sultans especially as Nagaur, Jalor and Siwana became garrisons of the Sultanate. Years later, Jodha succeeded in forging alliances with Deoras, Eendas, Sankhlas and Bhatis from Janglu, Pugal and Jaisalmer. Subsequently, Merta, Phalodi, Pokhran, Bhadrajun, Sojhat, Jaitaran, Siwana, Nagaur and Godwar were added permanently to the Rathor territory, thus making Marwar, the most powerful kingdom in Rajputana.[45] The table below lists the alliances forged through marriages contracted by Jodha and the expansion undertaken by his sons.[46]

Table 2.2. Marital Alliances of Rao Jodha

Jodha	Wives	Natal clan	Sons	Territories occupied or colonized
1	Jasmade Hadi	Hada	Satal	Jodhpur (Also founded Satalmer)
			Suja	Jodhpur (After the death of Satal)
			Neemba	(Died Young)
2	Champa Sonagri	Sonagra Kheeva Satawat	Varsangh	Merta
			Dudo	Merta

Contd.

---

[41] *Nainsi, Vigat*, I, 25.

[42] Dasaratha Sharma, *Rajasthan through the Ages*, I, 641.

[43] Nainsi, *Vigat*, I, 29.

[44] Nainsi, *Vigat*, I, 33.

[45] Nainsi, *Vigat*, I, 38.

[46] Nainsi, *Vigat*, I, 39. Another son Joga was declared unsound of mind.

Contd.

Jodha	Wives	Natal clan	Sons	Territories occupied or colonized
3	Sankhli Narangde	Mandalsinghot Sankhla	Bika	Bikaner, Janglu
			Beeda	Ladnu
4	Jamna Hulni	Banveer Bhojawat	Bharmal	Kodena
5	Bagheli	Baghela Urjan Bheemrajot	Sivraj	Siwana
6	Pura Bhatiyani	Bhati Bairisal Chachwat	Karamsi	Khinvsar

At this juncture, as can be seen from the table, all sons of Jodha sought to expand their fortunes in different directions, but one of them Bika, succeeded in founding a new state altogether. The Rathors of Bikaner were a subsidiary line of the Jodhpur Rathors. Rao Bika (r. 1465–1504), a son of Rao Jodha, ventured out in the latter's lifetime to conquer more land and create a separate niche for himself. Bika, "under the guidance of his uncle Kandul, led three hundred sons of Seoji to enlarge the boundaries of Rahtore dominion amidst the sands of Maroo. Bika was stimulated to attempt by the success of his brother Beeda who had recently subjugated the territory inhabited by the Mohils".[47] The kingdom of Bikaner was established in 1465, in the area that was traditionally known as Janglu.[48] The fort-city of Bikaner was founded in 1488, in a place that was strategically located on a route that connected Punjab, Multan, Marwar, Delhi and Sindh. The control of this area seemed to have passed from one group to another and at the time of establishment of Bikaner. Johyas, Chauhans, Sankhlas, Parmars, Bhatis, Bhattis, Khichis, Chayals, Kyamkhanis and Jats were dominant groups and controlled large portions of this arid region.[49] Several of these groups had accepted Islam but still seemed to maintain old associations, even continuing to enter into matrimonial alliances with their Hindu counterparts. In the 15th century, towards north and west of Janglu, Bhatis were still very strong and controlled all major forts including Pugal and Bhatner.[50] Jats in this region saw Bika as a possible buffer between them and the Bhatis. Bika is believed to have intervened in ongoing disputes

---

[47] Tod, *AAR*, II, 137.

[48] Nainsi, *Vigat*, III, 19.

[49] Ibid.

[50] Nainsi, *Khyat*, III, 36–37.

between Saran, Godara and Punia Jats, and eventually subjugated all of them. Nainsi describes Bika's role in a dispute between Godaras and Sarans as follows,

> The Saharans said, we will not be able to win, as Bikaji backs the Godaras. The Jats of Bhadang went to Narsinghdas Jatu, who brought his forces. When one hundred and forty Godaras were killed, they approached Bikaji, and told him that your Jats are being killed by Narsinghdas. At midnight half of the Jats of Bhadang approached Bikaji and told him that we will help you kill Narsinghdas. They took him to where Narsinghdas was asleep and Bikaji killed him. The Jat forces, fled, and Bikaji took all their cattle. Bikaji also killed Sonhar Jat at Harani Kheda.[51]

Another important group to aid Bika in the establishment of his kingdom was the Charans, particularly the Charani goddess Karni, who became the tutelary deity of the Bikawat Rathors. Karni had become a part of the Rathor world through her marriage to Depaji Rohadiya whose clansmen were hereditary Charans of the Rathors. In past, she had supported claims of Rinmal Rathor over Kanha, as well as helped Jodha in establishing Jodhpur. She guided Bika through his negotiations with various groups, and helped him establish himself first in Kodamdesar and then in Bikaner. In the process of his ascent, Bika realised the necessity of marrying into a clan of local significance, specifically that of Sekha Bhati of Pugal. Sekha, who was also a devotee of Karni, was reluctant to make this alliance, as Pugal was the frontier between Bhati and Rathor spheres of influence. At this juncture Karni stepped in, to negotiate this marriage, and forged an alliance between her devotees. She also settled a border dispute between them by deciding that Dhineru tank be considered a no man's land between the Bhatis and Rathors, selecting it as the site of her death. The role of Godara Jats and Charans in Bika Rathor's ascent was acknowledged by their presence at his coronation in Bikaner. While Karni was represented by her stepson Punyaraj (and in future by his descendants), the honour of anointing the king of Bikaner was reserved for Godara Jats.[52]

From the reign of Rao Bika itself, Bikawat Rathors had posited themselves as the true heirs of the Rathor legacy. After the death of Jodha, Bika also managed to get the ancient Royal emblems and heirlooms of the Rathors that

---

[51] Nainsi, *Khyat*, III, 15.

[52] Tod, *AAR*, II, 141. Also Harald Tambs-Lyche, 'Marriage and Affinity among Virgin Goddesses' in *The Feminine Sacred in South Asia*, Manohar, New Delhi, 2004, 63–87, 71.

they claimed to have carried from Kannauj, as he had been promised by Rao Jodha.[53] Rao Bika expanded the Bikaner state in the northwesterly direction and got involved in a battle with the Mohils in which his uncle Kandhal was killed.[54] His successor Rao Lunkaran (r.1505–1526) got in conflict with Chauhans of Dadreva, Kyamkhanis of Fatehpur and Chayals of Chayalwara. He also added Didwana, Baggar and Singhana to his kingdom and laid a siege on the fort of Jaisalmer.[55] He was killed when the Bhatis formed an alliance with the Jam of Sindh, Johiyas and Mohils.[56] Rao Lunkaran's son Jaitsi (r.1526–1542) suppressed the Muslim chiefs of Dronpur and the Johiyas and in 1527, he occupied the strong fort of Bhatner, from where all routes passing through the desert could be controlled.[57] He also got the Bidawat Rathors to accept the supremacy of the state of Bikaner and fixed an annual tribute.[58] Thus by mid sixteenth century, Rathor houses of Jodhpur and Bikaner had become firmly entrenched in the desert and had successfully challenged the supremacy of Bhatis in this region.

However, through the first half of the fifteenth century, Bhatis continued to defend themselves, as well as make inroads into Rathor territory. At the turn of the century Rawal Vairsi (r.1396–1448) recovered the fort city of Bikampur, which had been an old Bhati holding. Vairsi was followed by Chachiga Deva II (r. 1448–1497) who shifted the capital to Maroth, with the view of recovering the territories lost to Multan. But, Langas, Johiyas, Khichis etc.,

---

53 Powlett, *Gazetteer*, 9. Bika had been promised these heirlooms in exchange for Ladnu and for the promise that the Bikawats would never covet Jodhpur. These heirlooms included the throne, *chattra*, a dagger gifted by Harbhuji, an image of Nagnechiyanji, installed in Bikaner now, a *nagara*, a mare named Dal Singar and ancient cauldrons. Bika laid a siege of Jodhpur till he was handed these.

54 Nainsi, *Khyat*, III, 21–22.Kandhal was Bika's uncle who settled Bikaner with him. He initially participated in battle fought over a dispute regarding a Saran Jat woman Maliki between the Saran and Godara Jats, when the former sought the help of Narsingh Jatu promising him territory. Maliki, a spirited Jat woman is believed to have settled the village of Malkisar in Bikaner. Kandhal was killed by deceit by Sarang Khan, when engaged in a battle with Mohils. Nainsi, *Khyat*, III, 13–15, 21. Powlett *Gazetteer*, 6–8.

55 Powlett, *Gazetteer*, 11.

56 Ibid.

57 Nainsi, *Khyat*, III, 16–18. Bhatner was first taken by Khetsi Aradkamlot. Bhatis lost it after 10 years to Kamran, who lost it to Rao Ahmad Chahil. It, along with Nohar was reclaimed by Jaitsi and remained with Rathors of Bikaner till Shah Jahan turned it into *khalsa*.

58 Tod, *AAR*, II, 143.

who were old enemies of the Bhatis allied to fight against Chachiga Deva.[59] He formed an alliance with "...Shoomar Khan, chief of the Seta tribe.... whose granddaughter, Sonal Devi he married. The father of the bride, Hybat Khan, gave her with *daeja* of fifty horses, thirty five slaves, four palkis and two hundred she camels".[60]

The above discussion on the early history of Bhatis and Rathors in the Thar indicates the flux in which these groups existed and also points to the gradual change in identity. As is evident varied groups sought newer opportunities, lands and pastures in the frontiers of Thar and willingly participated in ongoing skirmishes in exchange of employment. One of the factors that bound these groups together was their mobility. These itinerant warriors entered into confrontations as well as negotiations with groups in control, which could have been older *kshatriya* clans, Mers, Bhils, Minas, Langas, Jats, or a number of groups with mixed identities like Charans. After conquest and colonization, cult appropriation became a significant vehicle for assimilation with existing groups, as did the institution of hypergamous marriages. Local *shaktic*, particularly *charanic* goddess-worship traditions, as well as *shaivite* traditions were woven into dynastic myths, which were emerging around this time and helped in the consolidation of community identities. Kin networks proved effective in extending control over vaster spaces and resources, with younger kin members moving further to conquer and colonize. The flux thus generated, led to the emergence of newer identities, in which control over resources determined not just the political but also social position of the groups. The new emerging powerful groups created strongholds by way of forts, either older or newly constructed, which became means of extending of political control. In their journey towards aristocracy, Rathors had begun a new spate of fort construction from the mid fifteenth century onwards, as is visible from the construction of impressive capitals with massive forts in Jodhpur (1459) and Bikaner (1486). Meanwhile, the control of older frontier forts like Bhatner, Derawar, Pokhran, Pugal, and Maroth etc. constantly shifted between the Bhatis and the two lines of Rathors. These forts, as political and administrative centres, served the purpose of administering the land around them as well as controlling pastures and routes through the desert.

Thar became the arena for such political upheaval because of its existence as a frontier that offered the possibilities for spatial and vertical mobility. Secondly,

---

[59] Ibid., 205–206.
[60] Ibid.

though the idea of colonization appears to suggest prima-facie the colonization of land, I would suggest that in the dry and arid Thar, it was also about control of space through which mobile resources like cattle and merchandise circulated. That the war-booty that the itinerant warrior fought for was land as well as cattle is evident from the numerous instances of cattle raids or *dhads* as part of warfare in the region. While the traditions patronised by the sedentarised Rajput courts portrayed them as territorial rulers, a close reading of Nainsi's *Khyat* and *Vigat* indicates that not only did the Rajput lineages of the Thar have a nomadic past, but mobile resources like cattle, addressed as *dhan, vitt* and *mal* (all connoting wealth), continued to be exchanged between Rajput clans, even till the late seventeenth century.[61] For instance, when a panegyric was composed about the 13th century Bhati King Salbahan of Jaisalmer, cows, buffaloes and camels were accounted for as his wealth.[62] Cattle and pasture lands continued to be a primary reason for battles between Rajput clans who undertook cattle raids or *dhads*. The lore of Pabuji (14th century) centres on the theft of camels from Dhat, across the Thar desert. The conclusive battle in the epic took place between the Rajputs Pabuji Rathor and Jindrao Khichi as Jindrao spirited away the cattle of Charans.[63] The cause of battle between Viramde Rathor and Johiyas in 1383 was the theft of cows by Viramde Rathor, who as Nainsi reports repeatedly engaged in cattle raids or *dhads* across the Thar desert.[64] In the reign of Rawal Bhim of Jaisalmer (r. 1577–1613), sons of Uhad Gopaldas a Rajput, looted the cattle of Pokhran and when people of Pokhran challenged them, the Bhatis helped them in recovering the cattle.[65] In fact, as late as 1650, when a treaty regarding the fort of Jaisalmer was negotiated, Bhati Ramchandra agreed to give up the fort of Jaisalmer and move to Derawar, only if he was allowed to take sufficient numbers of horses, camels and cattle.[66] Thus while on the one hand Rajput polity in the Thar was becoming increasingly sedentarised, control over mobile resources still

---

[61] In fact exploring the structure of military labour market in northern India, Kolff argues that "until as late as the nineteenth century, there was no lack of men opting for a life spent as errant soldier, migrant labourer, or pack animal trader". Kolff, 'The Rajput' in *Naukar, Rajput and Sepoy*, 74.

[62] Nainsi, *Khyat*, II, 37–38, 'Kavitt Bhati Salvahan ra'.

[63] Nainsi, *Khyat*, III, 58–80.

[64] Nainsi, *Vigat*, I, 16–17.

[65] Nainsi, *Khyat*, II, 99.

[66] Nainsi, *Khyat*, II, 108.

remained important for consolidating political fortunes. This can be seen in the manner in which rise of Chundo Rathor is described,

> Day by day circumstances became more favourable. Rav Chundo performed exceedingly great devotions to Chavnda Deviji (the tutelary deity of Mavar Rathors). Ten times more (traffic) began to pass along the roads. (Rav) Chundo continued to acquire wealth. A gathering of warriors and horses continued around him. Chundo adopted the behavior (of a ruler). He had (large amounts of) cooked food prepared. And he had offerings given (to the Charans and Brahmins).[67]

By the fifteenth century, while Rajput clans including Bhatis and Rathors increasingly began to view themselves as landed lineages, there also remained a persistent need to control mobile resources. As numerous pastoral and trade routes crossed these regions, the control of frontier forts like Bhatner, Pugal and Pokhran was significant for any of these clans. While the agricultural productivity of Ganga-Jamuna basins in northern India fostered agro-centric political structures, in the arid Thar, land was largely significant as passage and pasture. The dilemma of the Rajput ruler as 'annadata' or giver of grain and sustenance in this arid terrain is solved when the terms of references used for cattle are seen. Cattle is repeatedly addressed with terms like *dhan, vitt* and *mal*, all connoting wealth. By seventeenth century grazing taxes like *pancharai, ghasmari, singhoti, jhumpi* etc. formed an important part of the revenue collected by these states. In addition commercial taxes levied on transit of commodities like *dan, mapa, rahadari, pesar, nikal* too contributed significantly to revenue.[68] Thus while the identity of the Rajput ruler evolved around the idiom of ruler as the master of land, in actual terms the state relied on mobile resources.

Thus, even as clans like Bhatis and Rathors emerged as sedentary landed aristocratic clans, they still were forced into conflicts and negotiations with powerful mobile groups in order to control access to mobile resources. However, as sedentary rulers their genealogical traditions not only embed the 'Rajput' in the land, they also invert the narratives of their social contact and coexistence with groups that controlled these resources, as well as of their

---

[67] Nainsi, *Vigat*, I, 22. Translation Norman Zeigler, 'Evolution of the Rathor state of Marvar: Horses, Structural Change and Warfare' in *Idea of Rajasthan*, II, 197.

[68] B L Bhadani, *Peasants, Artisans and Entrepreneurs: Economy of Marwar in the Seventeenth Century*, Rawat Publications, Jaipur and New Delhi, 1999.

own itinerant pasts. The possibility of Rajputs having been cattle rearers, traders as well as rustlers is obfuscated by positing Rajputs as protectors of cattle rearing communities. The Rajput *bhomiaji* or *junhjar*, who laid down his life while protecting cattle became a dominant motif, with cattle protection being viewed as *dharma*. Mallinath, Gogade, Pabuji, Harbhuji, Ramdeoji are examples of Rajput folk deities revered by cattle rearing communities, as protectors of their flocks. The distinction between the protector and the protected indicates the emergence of a hierarchy between Rajputs and cattle rearing communities. This points towards a very complex picture of social and political structures in the Thar in the fifteenth century. On the one hand Bhatis and Rathors continued to deal with mobile groups, on the other they also attempted to align with the idea of Rajput that was consolidating, with land and territoriality as its basis.

The 'Long' Sixteenth Century and the Evolution of a 'Rajput' Polity

The sixteenth century appears to have been a period of significance in the formulation of a Rajput polity with rise of considerable internal differentiation among the clans, with shift from clan based polity to monarchical polity. By this period while on the one hand a clearer political map with strong polities like Mewar, Amber, Marwar, Bikaner and Jaisalmer emerged, on the other hand, relationships of ruling houses with their cadet lines also became complex. In the initial phases of their expansion, the cadet lines had maintained a sense of kinship towards the ruler of Jodhpur with *bhai-bant* (division among clansmen) being the norm for division of resources among clansmen. The principle off *bhai-bant* operated on the mutual recognition of authority of both the clansmen and the chief, with the former being recognized as *bhai-bandh* (kinsmen).[69] When a new area was brought under control, it was assigned to clan members who were regarded as the sons and brothers of the same patriarch. In principle, the *bhomiya* when from Rathor clan, was the *bhai-bandh* of the *tikayat* (anointed chief). The non-clan members were subjugates who at points did exert *de-facto* control and but were gradually brought under Rathor control. However, in the reign of Rao Malde (r.1531–1562), one of the most powerful rulers of the Jodhawat Rathor line, the shift from redistributive overlordship to centralized state brought about a transition in the relations of the Jodhpur

---

[69] G D Sharma, *Rajput Polity*, 5–13.

seat and clansmen, from *bhai-bandh* to *bhai-bandh-chakar*. In the reign of
Malde, even the Rathor sub-clan members could be reduced to the position of
a *chakar* (serviceman), with the *tikayat* in Jodhpur being the supreme. Malde's
efforts to establish himself as the supreme Rathor are reflected in his dealings
with Viramde of Merta who was his clansman, but whom Malde considered
to be a *bhai-bandh-chakar*. When Viramde captured Ajmer in 1556, Malde
demanded that he should be handed over the control of Ajmer as he was the
tikayat of Jodhpur and as Viramde was the *bhai-bandh-chakar* of the *tikayat*,
he could not hold the fort of Ajmer without his consent.[70] On Viram's refusal
Malde forced him out of Merta and further bestowed Rian in the Merta
territory to a grandson of Varsangh, another clansman.[71] This implied that
even though Viramde was a Rathor clansman, his authority over Merta had
to be validated by Malde. This also meant that the nature of control shifted
from patrimonial to prebendal in which *patta* had to be granted by the ruler.
The *patta*, which signified a written deed meant a land grant in return for
obligations of military service, which was transferable in nature.[72] The grant
of *patta*, to non-clan members became a way of garnering support, to balance
the claims of clansmen. However, while this polity saw the rise of the Rajput as
monarch, it also witnessed the rise of the recalcitrant *thikanadar* who continued
to control his ancestral villages or *bapoti* as well as mobility through them.

Rao Malde's reign was also one, which was successful in terms of territorial
expansion of the Rathors. Even before he inherited the throne of Jodhpur,
Malde was highly regarded as an intrepid warrior in Marwar and outside. He
can be regarded amongst the most important rulers of Rajputana and during his
rule Marwar rose to an indisputable position in Rajputana. He was considered

---

[70] Nainsi, *Vigat*, II, 53.

[71] Nainsi, *Vigat*, II, 42. This was considered justified as Merta had been settled by Rao
Jodha's sons Varsangh and Dudo by settling their kinsmen and Jat peasants in the
navikhati or unsettled lands. Merta before colonization by Varsangh and Dudo had
been unbroken stretch of land that had emerged as a strong Rathor settlement as well
as an important market town. While the descendants of Dudo considered Merta as
their inheritance (*bapoti*), in Malde's time the claims of the central Rathor authority
on the throne of Jodhpur over rode all other kinds of claims of clansmen. Even in case
an inheritance was accepted it was done only on the behest of the Jodhpur *tikayat*. This
led to the Mertia Rathors extending overtures to Akbar and their eventual absorption
as Mughal mansabdars.

[72] Zeigler, 'Evolution of the Rathor State of Marvar', 199.

to be ruthless, and according to Dayaldas, was guilty of parricide.[73] Malde attempted to fill the power vacuum created by the death of Rana Sanga. The "death of Rana Sanga and the misfortunes of the house of Mewar left Malde to the uncontrolled exercise of his power, which, like a true Rajput, he employed against friend and foe".[74] He intervened in Mewar affairs in favour of Udai Singh, whose throne had been usurped by a clansman Banbeer. He also decided to expand westwards and plundered Jaisalmer. Rawal Lunkaran of Jaisalmer was forced to sue for peace by giving him a daughter Umade Bhattiyani who earned the epithet of *roothi rani* due to her strained relationship with Rao Malde.[75] His reign saw Marwar emerge "...as a true state with a defined and institutionalized locus of authority in person of the ruler, who controlled the issuing of sanctions and the enforcements of regulations".[76] Zeigler points out that the reason for dominance of Marwar in this period was the increased supply of horses, which allowed the rulers of Marwar to equip outposts of the kingdom with mounted soldiers. As he points from an illustration from Nainsi's *Khyat*, the first recorded instance of establishment of stables, with a new category of salaried soldiers called *chindhars* at the outpost of Sojhat, is from Malde's father Rao Gango's reign (r. 1515–1532).[77] These deployments significantly increased in Rao Malde's reign, when due to a transitional phase in north Indian polity, the supply of horses to western India increased significantly.[78]

It was in Malde's reign, that Rathor polity came in contact with and was later subsumed by the upcoming Mughal state. Following his defeat in the battle of Bilgram, Humayun crossed the desert of Thar. Even though Malde

---

[73] Nainsi while dealing with Malde's accession does not mention parricide, Nainsi, *Khyat*, III, 94-95. Tod mentions that the "Yati's roll says that Ganga was poisoned; but this is not confirmed by any other authority", Tod, *AAR*, II, 119. Powlett quotes Dayaldas Sindhayach to state that Gango was murdered by Malde, Powlett, *Gazetteer*, 16. Ram Narayan Dugar's translation of the *Khyat*, edited by G H Ojha states that Gango was pushed from a terrace by Malde, G H Ojha (Ed.), Ram Narayan Dugar (Tr.), *Muhnot Nainsi ri Khyat*, Vol II, Rajasthani Granthagar, Jodhpur, (1934), 2010, 117.

[74] Tod, *AAR*, II, 19.

[75] Nainsi, *Vigat*, I, 47. Umade Bhattiyani never forgave Rao Malde for his amorous adventure with a *davri* (maid) on the wedding night itself. She lived all her life in separate palaces, adopted Rao Ram a son of Malde by another wife, and committed sati after his death.

[76] Zeigler, 'Evolution of the Rathor State of Marvar', 200.

[77] Ibid., 196. Also, Nainsi, *Khyat*, III, 84.

[78] Zeigler, Ibid.

had made him an offer of support in 1541, eventually he did not aid Humayun and the latter went to Umarkot.[79] At this juncture, the Rathor power was at its height in the desert, with Malde expanding into neighbouring territories and undertaking large scale fortifications.[80] At the end of the reign of Rao Jaitsi (r. 1526–1542) of Bikaner, Malde besieged Bikaner and Jaitsi's successor, Rao Kalyanmal (r. 1542–1574) had to leave the city of Bikaner and go to Sirsa where he spent the next two years of his life under the patronage of Shershah.[81] He was able to regain the *gaddi* of Bikaner with Shershah's help following the battle of Samel in 1544.[82] Realising that Malde posed a threat to his imperial intentions, Shershah engaged and defeated Malde. Having lost Jodhpur, Malde was forced to seek shelter in Siwana. Following the death of Shershah in 1545, Malde managed to retrieve most of the territory lost to Shershah, including Merta by 1557.[83] However, by this time continuous warfare had debilitated Malde and his adventures, within and outside Marwar, left him untrusted and isolated.

Besides, with the ascension of Akbar on the Mughal throne, the power dynamic shifted very quickly in Marwar. By 1557, Ajmer and Nagaur were lost and soon after Jaitaran and Parbatsar were annexed by Akbar in 1562. According to some versions, Malde sent his son Chandrsen to meet Akbar, but refused to seek an audience himself. As Tod puts it, "Malde succumbed to necessity: and in conformity with the times, sent his second son, Chundersen with gifts to Akbar then in Ajmer".[84] However, this did not improve Rathor-Mughal relationship, which was also marred by the memory of Malde's refusal to aid Humayun during his flight. With the death of Malde in 1562, the Jodhawat Rathors were soon reduced to the position of subordinate allies in the Thar. At the time of his death Rao Malde's territories included Jodhpur, Sojhat, Jaitaran, Phalodi, Siwana, Pokhran, Jalor, Sanchor, Merta , Barmer, Kotra, and some parts of Jaisalmer, but these were soon taken over by Akbar as Malde's sons were caught in a battle for succession.[85]

---

[79] Nainsi, *Vigat*, I, 60.

[80] Nainsi, *Vigat*, I, 45. Malde fortified Pokhran and Siwana and repaired forts of Jodhpur, Nagaur, Ajmer, Merta, Jalor and Sojhat.

[81] Powlett *Gazetteer*, 16–17, 19.

[82] Ibid.,20–21.

[83] Nainsi, *Vigat*, I, 63.

[84] Tod, *AAR*, II, 22.

[85] Nainsi, *Vigat*, I, 66.

Table 2.3. Malde's Wives and Sons[86]

Malde	Wives	Natal clan	Sons
1	Kachchwahi Lalachchde	Ratansi Shekhawat	Rao Ram
2	Umade Bhatiyani	Rawal Lunkaran Bhati	
3	Norangde Jhali		
4	Sarupde Jhali	Jhala Jaita	Mota Raja Udai Singh
			Chandrasen
5	Heera Jhali	Jhala Rai Singh	Raimal Malevot
6	Rambhawati Hadi	Hada Suraj Mal	Vikramaditya
7	Taankni Jaman	Kisah Kelhanot	
8	Lachch Ahadi	Ahad Preethiraj	Bhojraj
		Gangawat	Ratansi
9	Jamwali	Jagmal Surjawat	
10	Sonagri	Sonagra Akheraj	
11	Dhar Bai	Bhati Preethiraj	

At the time of Malde's death the heir designate Chandrasen who was the third son of Malde, was in Siwana.[87] Rao Malde's death led to a war of succession between Chandrasen and his brothers Ram Singh, Raimal and Udai Singh, who united against Chandrasen and sought Akbar's help in getting Jodhpur.[88] The city and fort of Jodhpur were occupied in 1565 and Nagaur in 1570.[89] This was a point of departure from pre-existing polity as now the claimants to the Marwar throne became *mansabdars* at the Mughal court. Instead of the Rathors of Jodhpur, Raja Rai Singh of Bikaner was granted Jodhpur from 1574 to 1577.[90] This further marked the rift between the Rathors of Jodhpur and Bikaner. Meanwhile Chandrasen resisted accepting Mughal suzerainty and shifted the center of his activities to Bhadrajun, which he was forced to evacuate in 1571.[91] He was also forced to handover the fort of Pokhran

---

[86] Nainsi *Vigat*, I, 55. Apart from these he also married a Bhan, a Chauhan, a Jadam, a Sodhi and a Jasad wife, whose names or father's names are not mentioned. The fact that they are identified only by their clan names is suggestive of the fact that these marriages were solemnized to strengthen interclan relations.

[87] Nainsi, *Vigat*, I, 66.

[88] G.D. Sharma, *Rajput Polity*, 11. Nainsi, *Vigat*, I, 23–24.

[89] Nainsi, *Vigat*, I, 68.

[90] Ibid.

[91] Ibid.

to Bhati Har Raj of Jaisalmer in exchange for a sum of one lakh *phadiyas*.[92] Chandrasen spent the last five years of a life as a wanderer, attempting to wrest his capital back. He resorted to pillage in imperial territory and two of his sons had died in battles with each other while the third accepted service in the Mughal army and was killed in the battle of Datani.[93] His brother Rao Ram was granted Sojhat in 1563, but Jodhpur remained in *khalsa*. Rao Ram died 9 years later.[94] However, it was only in 1583, that Udai Singh (r.1583–1595) was recognised as the ruler of Marwar, and a part of paragana of Jodhpur granted to him.[95] With this a long process of assimilation of Marwar into the Mughal empire began.

The assimilation of Bikaner into the Mughal empire followed a less contested route, but one that deepened the strife between the two Rathor states. The Bikaner ruler, Rao Kalyanmal (r. 1542–1574) attended the court of Akbar in 1570, unlike his Jodhpur brethren, Malde, and created firm ties with the Mughal court. At this occasion, two of his nieces were married to Akbar. After his death, his son Rai Singh was bestowed the *tika* of Bikaner. He was also granted the title of Maharaja and cemented his ties with the Mughal throne by marrying his daughter to Jahangir.[96] He maintained excellent relations with the Mughal court and spent most of his adult life fighting in Mughal campaigns in Kabul (1581), Bengal (1584), against Baluches (1585) and Jani Beg of Thatta (1592). Since the battle of succession had not been resolved in Jodhpur, Akbar assigned the *jagir* of Jodhpur to Rai Singh, which in principal should have been assigned to Chandrasen. This also went against the promise that Rao Bika had made to his father Rao Jodha, that Bikawat Rathors would never lay a claim in the Jodhpur territory. But, since the Mughal advent in the region had foreclosed any previously existing principles and norms, Rai Singh became the assignee of Jodhpur.[97] The wave of Mughal conquest extended to Jaisalmer and by 1577, Rawal Bhim of Jaisalmer is also known to have held a *mansab* of 1500 in the Mughal court.

The first phase of assimilation of the Rajput kingdoms into the Mughal Empire, was a phase of politico-administrative transition. The Rajput kingdoms were 'assembled' into a single *suba* of Ajmer, within which the

---

[92] Ibid., 70. Also, Nainsi, *Vigat*, II, 297.

[93] Nainsi, *Vigat*, I, 87.

[94] Ibid., 71.

[95] Ibid. Motaraja Udai Singh of Marwar also married his daughter to Akbar in 1586.

[96] Powlett, *Gazetteer*, 30.

[97] Ibid., 27.

kingdoms were further categorized as *sarkars* and *paraganas*. Motaraja Udai Singh of Jodhpur (r. 1583–1595) entrusted civil and military administration to a *pradhan* who came either from the cadet line of Pokhran or Ahuwa. Apart from this a non-Rajput Diwan, usually from the Oswal Jain community, was appointed to offset the influence of the Rajputs. Similarly *bakshis, kiledars,* and *kotwals* were appointed to undertake various duties at the court.[98] The power and influence of *thikanadars* began to erode in the new structure and as *pattadars,* they were expected to reclaim their *pattas* in succession and pay a *najar* (tribute) to the Jodhpur king as the Rajput ruler himself did to the Mughal emperor. In effect, the administrative and redistributive structure of the Jodhpur state became similar to that of the Mughal Empire. Around this time, the Jodhpur court began emulating the courtly practices of the Mughal court and instead of kin based placements, newer classifications came into existence, like the *rajvis,* who did not pay *rekh,* the *sardars* and the *ganayats.* Nainsi lists various rituals that were practiced in the darbar to characterize the mutual relationships of the ruler and the *sardars.*[99] This ritualization indicates that the administrative as well as political structures in Jodhpur were now affirming to the Mughal prototype.

However, this also means that in the Mughal political imagination, rather than being viewed as distinct political entities, the Rajput states were seen in the context of *jagirs* that could be allocated to the Mughal *mansabdars* that the Rajputs had become. While this led to gradual administrative changes, it also meant that the Mughal emperor acquired an arbitrative position in the intra-Rajput relationships. This is clear from the case of Udai Singh, who was granted the *tika* and *jagir* of Jodhpur in 1583 after it had remained in *khalsa* for almost 19 years, without a Jodhawat Rathor at the helm. For three years, in fact, Jodhpur had been granted to Rai Singh of the Bikaner branch. Udai Singh, who came to be known as *Mota Raja* (the senior king), unlike his predecessors who were called *Raos,* accepted the sovereignty of the Mughal Emperor and remained in the service of the Mughal court as a *mansabdar* and Jodhpur became the *watan jagir* of the Jodhpur Rathors. Several other territories were granted to him as *tankhwah jagir,* which underlined the subordinate status of the Jodhpur kings vis-à-vis the Mughal Emperor. This was further highlighted by the fact

---

[98] Nainsi, *Vigat,* II, Appendix 6, 482–483.

[99] Nainsi lists 42 ritual honors called *kurabs,* like '*uthan ro kurab*', which meant that the king stood up to receive a sardar only on coming or both while coming and going, '*paag khidkeeyaan tatha lapeto davi bandh ro inayat ro kurab*' or wrapping of the turban on the left side', etc. Nainsi, *Vigat,* II, 484–485.

that further successions in the line that is of Sur Singh (r. 1595–1619), Gaj Singh (r.1619–1638) and Jaswant Singh (r. 1638–1678) had to be ratified by the Mughal Emperors.[100] Karan Singh of Bikaner (r.1631–1669) was stripped of the throne and his *mansabs* in favour of his son Anoop Singh (r. 1669–1698) by Aurangzeb.[101] Besides, Sur Singh (r.1612–1631) of Bikaner, Jaswant Singh of Jodhpur and Anoop Singh of Bikaner became Rajput rulers who died outside their kingdoms while discharging administrative duties rather than in battle grounds in their own territories.[102] These examples strengthen the idea of the Rajput chief as loyal Mughal *mansabdar* in service of the empire.

The introduction of newer political and administrative norms created contradictions at several levels for the Rajputs themselves. In his seminal work on Rajput kinship, Zeigler lists the ways in which Rajput loyalties were defined in the pre-Mughal period. The Rajputs tied in relationships like *bhai-bandh* (brotherhood) and *saga* (relations formed through matrimony), operated through varied levels of loyalties. Ties of blood or marriage ensured the much required support in events of war, expansion or expulsion, and also commanded the primary loyalty. As Zeigler points out, that for semi-independent brotherhoods of far western desert stretches of Marwar, kinship remained a dominant institution even when Jodhpur was incorporated into the Mughal empire.[103] Given the ecological continuity, I extend this argument to include clans in Bikaner and Jaisalmer as well. However, towards central Marwar, wealth and proximity to power was a greater determinant in power sharing. Outside of brotherhood, service relationships or *chakri* determined the flow of loyalty, which could include clan members or Rajputs from other clans. Mughal policy of indirect rule in Rajput states impacted the internal relationships of Rajput clans by weakening kin relationships and strengthening service relationships. The service relationships or *chakri* became mirror images of Mughal *jagirdari*, whereby the *chakar* swore his allegiance, military support as well as attendance to his *dhani*, in return for a service tenure, or *patta*. While in the early period of Rajput expansion, territories were redistributed on the basis of kin relationships, later *pattas* could be given even in the absence of any such relationship. Towards the end of the seventeenth century, these *pattadars* were frequently transferred, just like the Mughal jagirdars. Within the older

---

[100] G D Sharma, *Rajput Polity*, 15.

[101] Powlett, *Gazetteer*, 37–38.

[102] Ibid., 32.

[103] Zeigler, *Action, Power and Service*, 143–145.

clan holdings or *thikanas*, the *thikanadars* themselves became granters, granting informal *pattas* and retaining *chakars*.[104]

However, implementation of the *pattadari* tenures often faced problem as the authority of the granter could be challenged, since older clan members held patrimonial rights to their ancestral villages or *janam-bhom*. Zeigler points that increasingly the Mughal emperor was seen as the source of authority required to enforce these grants. However, towards far western stretches of Pokhran and Mallani, the grant of *jagir* or *patta* remained theoretical until the grantee was able to enforce his right. Raja Jaswant Singh on being granted the paragana of Pokhran, had to appeal to Shah Jahan through his aunt Bai Manbhavati to allow the conquest of Pokhran,

> The *paragana* of Pokaraṇ is part of my Imperial jāgīr. But I have no authority over the area. For many years, Rāvaḷ Manohardās, who was my sagā, held it, and for this reason I made no complaints. Now, however, Bhāṭī Rāmchaṃd Siṃghot has succeeded to the throne of Jaisalmer. He is someone whom I have no reason to leave in control of Jaisalmer. If you allow me, I will attack Pokaraṇ and assert my authority.[105]

The assimilation of Rajput kingdoms into the Mughal empire paradoxically strengthened the regionality of the Rajput states, where the position of the ruling line became as unassailable as that of the Mughal emperor. This becomes quite visible in the case of relationship between Jodhpur Rathors and Mertia Rathors who descended from Dudo and Varsangh, two sons of Jodha. Under Viramde Dudavat, Mertias had become a very powerful branch whom Malde could not get under his *chakri*. In an argument between Malde and Jaimal Viramot in 1553 CE, the *pradhans* of Jaimal had challenged Maldeo by retorting, "who has the authority to give or take Merta? He who has given you Jodhpur has also given Merta to us".[106] Mertias were responsible for the entry of Akbar into Marwar and continued to be in Mughal service thereafter. However, when Akbar granted Merta to Raja Sur Singh in 1604 the Mertias protested, till they realized the emperor was unwilling to heed to their request. Besides,

---

[104] Zeigler, 'Some notes on Rajput loyalties during the Mughal period', in *The Mughal State*, Muzaffar Alam and Sanjay Subrahmanyam, (Eds.) OUP, 2006, 168-210.

[105] Nainsi, *Vigat*, II, 298. Zeigler, 'Some notes on Rajput loyalties', 193. Rao Chandrasen during his exile in 1575–76 CE had given away Pokhran to Bhatis in return for 1 lakh *phadiyas*.

[106] Nainsi, *Khyat*, III, 117–18.

Sur Singh himself confirmed the Mertia villages to them. This eventually led to cordiality in the relationships between Jodhawats and Mertias in the reigns of subsequent kings Gaj Singh and Jaswant Singh. In this manner, it can be seen that assimilation with the Mughal empire transformed political and social structures in the Thar.

Here, it also becomes important to delve into the differences that existed between the polity of Jodhpur, Bikaner and Jaisalmer. Considering the history of the Bhatis in the 15th and the 16th centuries, it emerges that the polity of the Bhatis was oriented towards the north and the west. Unlike Rathors of Jodhpur, who were involved in the affairs of Dhundhar and Mewar, the preoccupation of the Bhatis of Jaisalmer and Rathors of Bikaner remained with Sindh, Multan, Punjab and Umarkot. Bhatis came in conflict with Marwar and Bikaner over control of areas like Bikampur, Pokhran and Maroth, which were either important settlements on routes passing through the desert, or were good pastures, or both. Jaisalmer held an important position as an entrance to northern India from Sindh and Multan, in the early years of Sultanate, which is reflected in the epithet used for Vijay Rao Lanjo, that is the 'sentinel of the north direction'. But as strong polities like Punjab, Gujarat and Malwa as well as Jodhpur and Bikaner emerged, Jaisalmer was politically marginalized. This marginalization was further accentuated with the north-centricity of the emerging Mughal empire. With Delhi emerging as the centre, the desert including Jaisalmer became peripheral to the emerging Mughal centric polity of the Thar. While rulers of Jodhpur and Bikaner became active participants in the Mughal empire, the Jaisalmer rulers, though became *mansabdars,* were never too engaged with the empire. The association of Jaisalmer house with Hindu and Muslim Rajput groups of the frontier region like the Sammas, Soomras, Sodhas, Jarechas and Chudasamas etc. located the focal point of their polity quite away from the Mughal power centre Delhi, and Jaisalmer continued to remain a frontier region. Recalcitrant *thikanadars* became the norm in Jaisalmer as they robbed caravans and plundered territories. Despite being incorporated into the Rajput states and thus the Mughal empire, the outlying areas of the Thar desert remained 'uncontrolled' regions, often labeled *zortalab.* The Mughal accounts for Punjab and Sindh also are replete with instances of the ways in which frontier tribes were settled through coercion or co-option. In Punjab Jats, Khokhars, Bhattis, Khattars etc. were pastoral groups but were also seen to be involved in plunder.[107] Similarly in Sindh, the

---

[107] Chetan Singh, "Conformity and Conflict: Tribes and the 'Agrarian System' of Mughal India', in *The Mughal State*, 421–448.

Mughal state appears to have been involved in constant struggle with groups like Sodha, Sameja, Nahmardi, Kalmati, Bareja etc.[108] Annals of the Rajput states of the Thar also reveal instances of groups like Bidawats, Sodhas as well as Rathor and Bhati *thikanadars* engaging in cattle theft and highway robbery. The arid stretches of Jaisalmer, Bikaner, Marwar, Sindh, Punjab came to represent the hostile frontier that sheltered the despoiler. The Rajput and Rajput-like groups operating on these frontiers and making their living by controlling access in the frontier came to be regarded as freebooters and wayfarers. It is this frontier that marked the boundary between the noble Rajput warrior and the freebooter, which is also the distinction between the 'Mughal' Rajput and the others. This distinction was further reified by the manner in which historical traditions in the courts of Rajput states of the Thar recapitulated Rajput history through genealogical accounts.

The Making of the 'Rajput': Genealogy as History/Genealogy as Polity

The making of the Rajput in the Thar was a long drawn and gradual process that took several centuries. While conquest, colonisation, cult appropriation, martial and marital alliances had important roles to play, the development of historical traditions played an equally important if not even more significant role. Thar had a very long tradition of literary-historical composition by Jains, Charans and Bhats, in Prakrit, Apabhramsa, Dingal and Braj. Patronised by Rajput courts these traditions played a very important role in defining the idea of Rajput as they stressed on values like bravery, chivalry and generosity as Rajput attributes. One important aspect of these literary-historical traditions was the Rajput genealogies that were commissioned by the courts and preserved with great diligence. The *pidhivalis* or *vanshavalis* compiled by Bhats and Charans were seen by Zeigler as the "corpus of sounds that trace descent and heredity, and which legitimize a proper order among castes and men".[109] They were elaborate accounts of association by male blood to a particular ancestor, tracing the origin of the group itself.

In the first part of this chapter, I discussed the question of origin of Rajputs and the obscurity that surrounded the initiators of these clans. However, from

---

[108] Sunita Zaidi, 'The Mughal State and tribes in seventeenth century Sind', *IESHR*, 26, 1989, 343–362.

[109] Zeigler, 'Marvari Historical Chronicles: Sources for Social and Cultural History of Rajasthan', *IESHR*, January–March, 1976, Vol. XIII, No. 1, 218–250, 228.

the seventeenth century onwards very detailed genealogical information became a part of these traditions. Genealogies are believed to have been memorized and recited by specialized Bhats before they were committed to writing. According to Tessitori, some kind of written genealogies must have become available by early fifteenth century, though these genealogies had unexplained gaps.[110] He points out that rather than beginning with Siho, the founder of the Marwar branch of Rathors, an initial genealogy, complied by end of sixteenth century, begins with Salkho (mid fourteenth century), who was eighth in descent from Siho. He also mentions that a poetic composition *Rao Jaitsi ro Chhand*, composed about in about 1535 by a Charan, Vithu Sujo Nagarajot earlier does not mention any ancestors before Rao Salkha. David Henige has also pointed out that in case of Jodhpur genealogies a gap of eighty years between the Gahadwala ancestor Jaichandra and Rathor Siho, followed by seven ruling generations in seventy four years are difficult to explain, unless one considers the possibility of king lists having been converted into ascendant genealogies.[111] Therefore, it is clear that even in mid sixteenth century, when Marwar branch of Rathors had not only become fairly established, the Bikawat Rathor line had also separated, there was still no clear recapitulation of a long illustrious line of ancestors that connected them to the Gahadvalas of Kannauj. Jan Vansina explains the gaps in genealogies by pointing out that genealogies were never meticulously remembered, and since they are always constructed in a time much later than the one they refer to they are very likely to be misconstrued.[112] Genealogical reconstruction was done if it had a practical use.

If we go back to the Marwar expansion discussed in the last section, it becomes clear that a sedentary Rathor polity began to consolidate only in early fifteenth century with the rise of Rao Chunda. Before this the Rathor history appears to be a history of wanderers. In my understanding an important element at this stage appears to be references to Brahmins, Bhats and Charans as recipient of Rajput largesse and protection. In Chunda's instance, the discovery of his royalty by a Charan is particularly indicative of importance of communities that could keep records of the lineages. At this time oral as well

---

[110] L P Tessitori, A Progress Report on the work done during the year 1917 in connection with the Bardic and Historical Survey of Rajputana, *JPSAB (New Series,)* Vol. XV, No: 41 1919, 21.

[111] David Henige, 'The Jodhpur Chronicles', *The Chronology of Oral Tradition: Quest for a Chimera,* Clarendon Press, Oxford, 1974, 205.

[112] Jan Vansina, *Oral Tradition as History,* University of Wisconsin Press, Madison, 1985, 178–182.

as written accounts of the exploits of Rajput warriors began appearing in the form of Dingal *bats*.[113] These *bats* were often highly inspirational narratives of individual Rajputs, embellished with poetry. The individual character of early *bats* suggests that the accomplishment of individual warriors though celebrated, were not yet being woven together to tell dynastic accounts. The systemization of these records does not appear to have taken up till about the late sixteenth century. Two factors can perhaps be seen as important in systematization of genealogies, the first, literacy and the second, the necessity of establishing Rajput claims in the context of a group of outsiders that is the Mughals. While earlier genealogical accounts would have been oral, and thus multiple, the possibility of writing them allowed the "coalescing of variant accounts into a single standardized version".[114] The forms that brought the oral and written available materials together were *Khyat* and *Vigat*, which came into use by late sixteenth century, with the earliest available compendia being *Munhata Nainsi ri Khyat* and *Marwar ra Paraganan ri Vigat*, compiled in late seventeenth century.[115] The *Khyat* and *Vigat* included *pidhivalis* and *vanshavalis* of all Rajput clans in the Thar region. They attempt to assimilate oral *bats* regarding individual Rajputs and attempt to arrange them into a chronological order, filling gaps in the narrative.

It has been suggested that these genres of literary-historical writing in Rajputana developed in response to the Mughal historiographical traditions.[116] The assimilation of Rajput states into the Mughal empire also facilitated the movement of court historians and scribes between the Mughal court and Rajput court. Besides, Abul Fazl's compilation of *Ain-i-Akbari* drew on accounts summoned from the *subas*. The fact that from the late seventeenth century onwards several Rajput chiefs moved into imperial courts and further off into distant paraganas in imperial service, must have impressed upon them the importance of Mughal historical traditions as well as record keeping. The impact of Mughal documentation is clear in the multitude of administrative records that become available from late seventeenth century onwards in Rajput states.

However, while literacy was a significant factor, what was more important was the political need for a clear dynastic order, both for Mughals as well

---

[113] Agarchand Nahta, 'Rajasthani Baton evam Khyaton ki Parampara', in *Parampara*, (Ed.) N S Bhati, Rajasthani Shodh Sansthan, Chopasani, 11, 1961, 114–124.

[114] Henige, 'The Jodhpur Chronicles', 96.

[115] Sitaram Lalas, 'Madhyakalin Rajasthani Gadya Sahitya', in *Parampara* (Ed.) N S Bhati, Rajasthani Shodh Sansthan, Chopasani, 15–16, 1963, 237–253.

[116] V S Bhragava, *Rajasthan ke Itihas ka Sarvekshan*, College Book Depot, Jaipur, 1971.

as for Rajputs themselves. With the assimilation of Rajputs states into the Mughal empire, the confirmation of succession in Rajput states became the prerogative of the Mughal emperor. Thus genealogies became important in resolving succession disputes and hereditary claims. However, Zeigler, quite rightfully points out that these genealogies were as much for internal circulation as for external.[117] The Mughal historiographical approach to Rajput history has led the post Mughal conquest history of this region to be understood only in the context of Mughal-Rajput relations, thus a history of co-option or resistance to Mughals. This perspective, thus, fails to take into account the fierce internal struggle in the Rajput polity as well as one between Rajputs and other communities in this region. The emerging genealogies as well as historical literature shaped the identity of Rajput as against the non-Rajput. This literature posited the Rajputs as protectors of *dhrama*, ones who prevented the breakdown of social order. So, whether by Asthan Rathor acting against Mers, because by forcing sexual union with Brahmin women, Mers were causing a breakdown of moral order, or the assurance given by the Sisodiya queen that her son would safeguard the Brahmins, the Rajputs become representatives and guardians of a moral order. This is implicit in the accounts of Rajputs succeeding because of being blessed by goddess or Shiva. Later genealogies become even more blatant not only by tracing direct descent of a line of kings, but by including accounts of divine origin. The account of origin of Rathors in the *Rathaur Vans ki Vigat evam Rathauraon ki Vanshvalli*, attributes the origin of Rathors to the *amsa* of Indra which impregnated Javansat, the king of Chandrakala, and a child was born by the splitting of his spine, thus called Rathor.[118] A detailed genealogy of Rathors of Bikaner that perhaps was added to Nainsi's *Khyat* in the eighteenth century, refers to Rathors as *Suryavanshis*, begins with Aadi Narayana, and includes Ikshvaku, Harishchandra, Dasaratha and Rama.

Taken further, these genealogies also become a way of distinguishing Rajputs from other groups that fail to develop such detailed Brahminical genealogies. These groups like Bhils, Mers, Minas, Jats, Gujars, Raikas or Charans[119] were groups that had either been subjugated or had shared the itinerant space with these clans. While on the one hand Rajputs developed

---

[117] Zeigler, 'Marvari Historical Chronicles', 235.

[118] Ibid., 109–110.

[119] Most of these groups do have mythologies of origin which link them often to Shiva, and claims to lost kinships as I point out in the next chapter.

detailed genealogies, on the other, most of these groups evolved mythologies of a Rajput past, their own genealogies of sorts. In fact, by claiming to carry the Rajput *amsa*, while appearing to reiterate having been Rajput once, what these groups underline is the fact that at some point in time Rajputs were one of them. Thus, what these genealogical traditions do is to distinguish Rajput itinerancy from other kinds of itinerancies in the region. Rajput itinerancy is mythologised, and later seen in connection with Muslim rule, is projected to have taken place in order to restore a weakening moral order. For example, in Rajput histories, cow protection is seen as an act that is restorative of the moral order. An important aspect of this process is the way in which Rajput traditions often portray Rajput heroes and martyrs as protectors of cattle herding communities, rather than as cattle herders, traders or rustlers, as all these other groups were. The denial of a connection between itinerancy, cattle raids and the Rajput was the result of evolving 'genealogical orthodoxy' from the late sixteenth century onwards. This 'genealogical orthodoxy' was very interestingly manifested in Rajput-Mughal marital alliances, which were extremely important in defining the 'Rajput of the empire'.

A 'Mughal' Rajput or a 'Rajput' Mughal?

Rajput histories of the Thar from the late sixteenth century onwards have been subsumed within the histories of the Mughal empire. They have thus also been histories of sedentarised Rajput groups in collaboration with or in opposition to the empire. Rajput-Mughal polity became the frame to understand Mughal policy's relationship with all local polities because it generated fierce loyalties like that of Kachwaha rulers of Amber as well as equally fierce opposition from other Rajput houses like the Sisodiyas of Mewar. Some interpreters of Mughal-Rajput relationships have interpreted this binary in a rather communal context where opposition to Mughals and especially to Rajput-Mughal marriages was seen to represent a Hindu opposition to a Muslim empire. However, the study of Rajputs of the Thar in the seventeenth and the eighteenth centuries opens up several possibilities in the understanding of this binary. The first is to understand the ways in which Rajput polity remained complex and multilayered, even though its interactions with the Mughal empire significantly altered the nature of this polity. The second is to see how the constructions of the loyal 'Mughal' Rajput were used to posit the 'noble' Rajput against the 'unsettled' challengers, like Marathas, Jats, Satnamis, Sikhs, Meos among others in the Mughal period as well as later in nineteenth century. In the seventeenth and the eighteenth centuries, while the 'Rajput king' became the model adopted by some like Marathas, Jats and Sikhs, others like Meos, Bhils and Satnamis (all with a claim to Rajputhood) were increasingly 'outlawed'. This then creates a binary of a very different kind, in which it is difficult to decide the location of true Rajputhood, with the noble feudal Rajput warrior or with the proud rebellious outlaw. This binary also presumes a sharp break in understanding Rajputhood. If, the pre-Mughal Rajput was an itinerant warrior, who/what was the 'Mughal' Rajput? In case of the Thar, this distinction becomes pronounced with the identification of the rebellious warrior with the adversarial geography of the desert, *dharati dhoran ri* with the true Rajput belonging to the frontier.[120] Interestingly as Mughal *mansabdars*, many Rajputs kings were sent to the frontier outposts of the empire, like Kabul, Multan and Deccan, where they fought the challenges to the empire. Within the Thar, however, while the Rajput kings on the one hand adapted to the Mughal political culture, on the other, they had to continue to deal with a kin based political culture.

I would suggest that Rajputs managed to justify Mughal intervention in their political and social exchanges by turning Mughals into their kin by marriage, the *sagas* or *ganayat*, and thus making them a part of the kinship structure itself. Mughal-Rajput marriages played the most important role in this process. In the Rajput political culture, marriage played a very important role in cementing ties between clans. But marriages were also a way of garnering political support as well as ending feuds between clans. The marriage of a Chavda daughter with Siho Rathor, while seeking Siho's help in an ongoing rivalry with Lakho Phoolani, cited earlier in the chapter, can be seen as an example. The *Khyats* are replete with instances where marriages were contracted in order to end feuds or as terms of treaties, often with the losing party giving away a daughter.

The other side of such alliances is that marriages could enhance the wealth and status of either side, and as Ziegler explains could be seen as a way of entering the Rajput fold. For example, Rana of Mewar addresses Rao Rinmal as *moto sago* (powerful relative by marriage) and asks for a daughter, to make him a Rajput (*mhanu Rajput karo*).[121] Zeigler cites the example of betrothal

---

[120] Interestingly this association is very compellingly brought out in the 1997 Bollywood War film, Border, about the 1971 war with Pakistan. One of the protagonists, who is a Rajput from Jaisalmer, leaves his newly wedded wife to go to the war, in the true Rajput fashion. When the ugliness of the desert landscape is pointed out to him by another Rajput soldier from the hills, he points out that a son loves a mother even if she is ugly.

[121] Nainsi, *Khyat*, II, 333.

of daughter of Mamdan Sodha of Umarkot with Rathor Kumbho Jagmalot, who the Sodhas address as a *vado Rajput,* a great warrior.

> Kumbha said, 'Ranaji has made me a Rajput (*monu Rajput kiyo*). I have now become a Rajput. He has now given me prestige and made me important'.[122]

Such alliances became even more necessary in times of distress, when political gains and losses could be calculated.

> Rana Udai Singh of Umarkot came to Bhadrajan and accompanied by Rao Chandrasen went to Jaisalmer, and sent a message to Bhatis, 'give me a daughter in marriage (*monu parnavo*)'. Bhatis did not agree. Rana stayed there for fifteen days, but Bhatis did not give him a daughter. Then while coming back, Rao Chandrasen married his daughter Bai Karmeti to the Rana."[123]

This support could be sought from all allies, even Muslims. While attempting to settle a dispute between Rao Gango and his uncle Sekha, the latter's ally Hardas Uhar sought support from Sarkhel Khan of Nagaur and Daulat Khan of Ajmer, by promising daughters in marriage. On hearing this Sekha said to Hardas,

> 'Re Hardas, Whose daughters will we give in marriage? I do not have a daughter, neither do you'. Hardas said, 'Whose daughters? Heads will fall from the blows of swords. If we are victorious, then there are many Rinmalots (descendants of Rinmal) from whose daughters we will take two and marry; if we die fighting in the battle, who will perform the marriages? What is there to worry about?' [124]

Zeigler suggests that marriages with Muslims were accepted because there was no contradiction between being a Muslim and a Rajput. Rajput as a functional category could include anyone who fulfilled the moral requirements of being a Rajput in the Thar that is bearing arms, fighting and dying in battle, and possessing sovereignty. According to him, this category of Muslim Rajputs included Khanzadas of Nagaur, Pathans of Jalor, Sultans of Gujarat, Malwa, and the Mughals of Delhi.[125] In the early polity of the Thar, before divisions

---

[122] Nainsi, *Khyat,* II, 295.
[123] Nainsi, *Vigat,* I, 69.
[124] Nainsi, *Khyat,* III, 90.
[125] Zeigler, *Action, Power and Service,* 65.

of Hindu and Muslim become very clear, the circle of marital possibilities appear to include all warrior-like groups in the region. Rao Chunda's wife's sister was married to a Khokhar of Nagaur.[126] James Tod in his Annals of Jaisalmer elaborates upon the circumstances of the marriage of the Bhati Chief Chaichaga Deva, with Sonal Devi, daughter of Hybat Khan, who gave among other things two hundred female camels and fifty horses as dowry.[127] Rao Kelhana Bhati of Pugal married the daughter of Biloch chief Jam Ismail Khan in the mid fourteenth century. Daughters of Bukkan Bhati were married to a Muslim Rajput Depal-de Johiya and a Hindu Rajput Viramde Salkhavat[128]. Of several daughters of Rao Malde, one Kanakawati had been married to the Sultan of Gujarat and another to Haji Khan of Nagaur.[129] Viramde Jodhavat married his daughter to the Subedar of Ranthambhor and through him sought Shershah's help in regaining Merta.[130] All these alliances were possible because they were largely political alliances made for martial purposes, and thus, were seen as morally justified. The alliances with Mughals thus were no different from alliances that had been made in the past.

An interesting aspect of these marriages was the custom of *sala katari* or the custom of groom gifting clothes and a dagger to the brother of the bride, recognizing the gift of bride. In Rajput alliances this often took form of villages that were then granted to the brothers of the bride. The villages of Asop and Khinvsar were gifted as *sala katari* to Karamsi and Raypal Jodhawat by Salhe Khan of Nagaur on the occasion of his marriage to their sister.[131] Thus marriages were actually a form of very ritualized political exchange, which tied up various actors in the region in kin relationships.[132]

Seen in this context, Mughal-Rajput marriages acquire a dimension in which Mughals, as fellow 'Rajputs', could be allowed to become a part of Rajput social order. As *sagas*, they could be asked to intervene in ongoing disputes, as well as provided with support in their own battles. Besides, the grant of *jagirs* and *paraganas* could well be seen as *sala katari*, which was an accepted practice in

---

[126] Nainsi, *Khyat*, II, 310.

[127] Tod, *AAR*, II, 206.

[128] Nainsi, *Vigat*, I, 17.

[129] Nainsi, *Vigat*, I, 52.

[130] Nainsi, *Vigat*, II, 55.

[131] Nainsi, *Vigat*, I, 40.

[132] Interestingly, the ritual of Rajput marriage is itself shaped in symbolic language of battle. The groom dressed in battle paraphernalia, strikes the *toran* on the gate of the brides house with a sword, symbolizing the breaching of a fort.

the region. These alliances, from the sixteenth to the eighteenth centuries, were not seen as aberrations to a norm already in practice. Frances Taft points that the Mewar viewpoint, which saw these alliances as dishonourable, came to dominate the Rajput worldview only with the decline of the Mughal empire.[133]

However, these alliances did redefine both Mughals and Rajputs significantly. Taft lists 27 marriages between Akbar's and Farrukhsiyar's reigns. All these marriages were contracted with high ranking elite Rajput households.[134] Mughals did not contract similar alliances with other warrior groups like Jats and Marathas, despite that fact that Marathas were far more powerful in their own domain by the late seventeenth and eighteenth centuries. Nor did Rajputs marry their daughters to other Muslim groups or even other warrior groups after Rajput-Mughal marital alliances were contracted. This indicates that both Mughals and Rajputs went on to practice a kind of 'caste endogamy' that was inclusive of each other. I would further venture to say that Rajputs and Mughals began to mirror each other in their social mores. Harbans Mukhia points out that the Rajput custom of renaming wives or identifying them with their natal clans was carried into the Mughal *haram*. The Mughal *haram* itself became a secluded space like the Rajput *rawla* where female chastity had to be zealously guarded.[135] For Rajputs, marriages into 'Rajputised' Mughal household became acceptable as it was seen to carry the same kinds of values that were practiced in the increasingly elitised Rajput households. For Mughals, these alliances became acceptable as the Rajput was seen as the upper caste elite noble warrior that the Mughals (the 'Rajput' Mughal) strived to be. In fact, a number of Rajput princes were brought up in the Mughal household, in the tradition of *khanazads*, the title of *Mirza Raja*, being an apt example. Such was the sense of inclusion in the family that high ranking Rajput *mansabdars* like Jai Singh and Jaswant Singh were sent to quell rebellions by Mughal princes.[136]

---

[133] Frances Taft, 'Honor and Alliance: Reconsidering Rajput-Mughal Marriages', in *Idea of Rajasthan*, II, 215–241.

[134] Ibid., 218.

[135] Harbans Mukhia, *The Mughals of India*, Blackwell Publishing, 2004, 2008, 130–31.

[136] However, a very interesting act of rebellion was by Amar Singh Rathor, a stepbrother of Jaswant Singh, who was disinherited even though he was the elder son. This audacious Rajput appears to have won acclaim in Mughal politics and was granted Nagaur as *jagir*. According to Tod, he was rebuked and fined by Shah Jahan for having absented himself to hunt for a two weeks. Vernacular accounts however suggest that Amar Singh absented himself from court after learning of the defeat of Nagaur forces in a

In the Rajput world, caste endogamy and hypergamy created several kinds of hierarchies. *Sagas* came to be classified into clans with whom one way or two-way marriages could be contracted. The first were clans from whom wives could be taken, and thus hierarchically lower, and the second, to whom daughters could be given as well, thus higher or equal in hierarchy. In Rajput symbolism, Mughals saw Rajputs as one-way *sagas* from whom daughters could be taken but not given to. On the other hand elite Rajputs saw several of the Rajput clans as inferior clans, only taking wives from them, besides ruling out marriages with non-Rajput clans altogether. While in an earlier period, male Rajput ancestry could be considered enough for being accommodated in the Rajput fold, by the fifteenth century mixed ancestry was clearly frowned upon. The 13th century Jarecha chief Lakho Phoolani, who was the son of Jarecha Phool and an Ahir woman, was happily accepted in the Rajput fold and made the heir.[137] On the other hand, in mid fifteenth century Mewar, Chacho and Mera, uncles of Sisodiya chief Mokul, who were born of a *khati* (carpenter caste) concubine, are derogatively referred to as *khatanwala*. Mokul is observed as pointing out that it was not appropriate for those who eat barley (*jovala*) to mix with those who eat wheat (*gohuwala*).[138] In fact, categories like *gola*, *daroga* and *sipahi* Rajputs, who were the result of unions of high caste Rajputs with low caste women, became separate caste groups altogether who could not be included into the fold of Rajput *ganayats*. The 'genealogical orthodoxy' that Kolff points out, was not merely located in Mewar, with its refusal to become

---

boundary dispute with Karan Singh of Bikaner. When he sought permission to go to Nagaur he was refused permission as he was considered a *fisadi*. Losing his temper Amar Singh entered into an argument with the *bakshi* Salabat Khan and proceeded to murder him in the court. 'Rathor Amar Singh Gajsinghot ri Baat' in 'Rajasthani Sahitya aur Amar Singh Rathor', *Parampara*, Ed. Hukam Singh Bhati, 2002. In the following pandemonium he was treacherously murdered. In the battle that ensued between Amar Singh's forces several of his followers including Ballu Champawat and Devidas Rathor were killed. However, Shah Jahan continued with the grant of Nagaur to Amar Singh's son, Rai Singh. Tod, *AAR*, II, 34–35. Interestingly, this account does not find space in Nainsi's Khyat. However, Amar Singh's deeds find mention in several poetic and prose accounts like *Rao Amar Singh ri Bat, Shri Amar Singhji ro Uvako, Rao Amar Singh ra Jhulna, Amar Singhji ra Channd* etc. Amar Singh Rathor's rebellious character has also been immortalized by the puppeteers, who sing of his bravery in the *khayal* form.

[137] Nainsi, *Khyat*, II, 229. However, it must be mentioned that a number of later day Sindhi sources omit this aspect of Lakho's parentage and depict him as being an adoptive son of Phool.

[138] Nainsi, *Khyat*, II, 167.

part of Mughal-Rajput marital alliances, but in other parts of Rajputana as well, by creating categories within Rajputs themselves. Therefore, it is clear that purity of blood was clearly becoming a way of identifying the true Rajput, thus requiring elaborate genealogies. However, this notion of purity does not seem to have been applied to Rajput-Mughal marriages, at least in the sixteenth and the seventeenth centuries, as Mughals were seen as included in the upper caste categories.

It is through this process that the 'Mughal' Rajput came to be seen as the noble warrior, the prototype of Rajput that was to become the norm in future. It is in their capacity as 'Mughal' Rajputs that Rajput *mansabdars* in Mughal service were sent to confront the 'other' that is the Marathas in Deccan in the late seventeenth century. The chasm between the noble warrior and the brigand becomes clearer in the representation of Mughal forays into Maratha territory. In fact Shivaji's appearance in Aurangzeb's court highlights the relationship as well as the confusion between warriorhood and Rajput status. As someone who controlled most of Deccan, Shivaji's placement much below someone like Jaswant Singh was perplexing for Shivaji. In fact, Stewart Gordon points out that Shivaji may have been thought of as a Rajput, both by himself as well as Rajputs.[139] However, being snubbed by Aurangzeb as a *bhomiya*, Shivaji proceeded to 'recover' his genealogical origins in Rajputs of Rajputana, as well as go through a ritual Vedic coronation.[140] A verse composed at the time of his coronation in 1674 CE by Paramananda portrays Shivaji as a protector of gods, Brahmins and cows, an *avatara* of Visnu.[141] This kind of poetry in fact is no different from the heroic poetry being produced in Rajput courts around the same period. This, however does not impact the imperial view of Marathas, which continues to view Shivaji and his descendents as brigands, even though, in the eighteenth century the Peshwa court, as well as the emerging Maratha courts in Indore, Gwalior and Baroda embraced and replicated Mughal court practices. Despite their formidable military power, Marathas continued to be

---

[139] Stewart Gordon, *Marathas*, 88. Gordon quotes J N Sarkar to indicate that Rajput reports observed that Shivaji was a "genuine Rajput...marked with the characteristic qualities (or spirit) of a Rajput". (Jadunath Sarkar, *House of Shivaji*, 162, as quoted by Stewart Gordon).

[140] Gordon, Ibid. Interestingly, Zeigler points out that the genealogy that connects Shivaji to Rajasthan indicates Shahji Bhosle to be a descendent of Chacho Sisodiya, the half-Rajput uncle of Rana Mokal of Mewar, mentioned earlier in the chapter. Zeigler, *Action, Power, Service*, 54 (fn).

[141] James Laine, *Shivaji: Hindu King in Islamic India*, OUP, 2003, 23–25.

regarded as marauders, and along with Pindaris became a cause for alliance between the Rajputs and the British in the early nineteenth century. R P Rana points out that an underplayed feature of late seventeenth century and early eighteenth century Jat revolts in Delhi, Agra and Ajmer was the skillful use of fortresses or *garhis*, often seen as a Rajput symbol, by the Jat rebels.[142] A similar argument can be put forth for the Meos, who, as Shail Mayaram points out, continue to view themselves as Rajputs, and their resistance to Mughals as a sign of their Rajputhood.[143] Thus, it is clear that while caste endogamy imposed by genealogical orthodoxy closed the ranks of Rajput caste in Rajputana, the claim to Rajput 'status' remained open. This Rajput status continued to be defined by traits like resistance to authority and fierce independence, which were quite opposite to the idea of loyalty often discussed in Rajput-Mughal relationships. Resistance to the empire could be seen as a justification for seeking the Rajput 'status'.

However, for the Mughal period, the 'Mughal' Rajput warrior became the best representation of the empire, and became consequential in crushing revolts that these Rajputs also saw as a challenge to their own self identification. Was Rajputana then, both territorially and metaphorically, then increasingly being seen as the empire, and not the frontier? What impact did this understanding have on areas that were ecological and political frontiers? How did the dynamics of the 'Mughal' Rajput and the Rajput 'outlaw', the *barwuttea*, play out? While, at first glance, the political narrative of the sixteenth and the seventeenth centuries in the Thar may appear like the narrative of integration of these states into the Mughal empire, the existence of older clan hierarchies make it a much more complex process. The acceptance of Mughal sovereignty, while created another level of authority, it did not necessarily mean that the internal dynamics of Rajput polity were totally transformed. As Zeigler points out, the varying structures of kin relationships in Marwar produced differential loyalties.[144] Despite the unity imposed by the implementation of Mughal politico-administrative structures, kinship remained the dominant institution. In fact, he suggests that Rathor brotherhoods in far western Marwar remained outside the influence of powerful groups in eastern Marwar and had limited

---

[142] R P Rana, *Rebels to Rulers: The rise of Jat Power in Medieval India, c. 1665–1735*, Manohar, 2006, 183–184.

[143] Shail Mayaram, *Against History Against State: Counter perspectives from the Margins*, Permanent Black, 2004. Also Mayaram, 'Mughal State Formation: The Mewati Counter-perspective', *IESHR*, 1997, 34 (2), 169–197, 188.

[144] Zeigler, 'Some Notes on Rajput Loyalties during the Mughal period', 184.

contact with the Mughals. I would argue that within the Thar, semi-autonomy exercised by *thikanadars* in far western desert was also seen as a superior claim to Rajputhood. The co-option into the Mughal household, and thus subsequent pacification of Rajputs created categories of Rajputhood in Rajputana. While Rajput-Mughal *mansabdars* were seen as Rajputs through lineage, as well as their loyalty to the *dhani*, who was now the Mughal emperor, the semi-autonomous *thikanadars'* claim to Rajputhood came through their fierce independence. These *thikanadars*, never drawn into Mughal ambit, did not become a part of the Rajput-Mughal dynamics, as did the clans around Rajput power centres like Jodhpur and Bikaner. Even within the eastern Marwar, Rathor cadet lines like Champawats, Kumpawats, Mertias, Udawats, Jaitawats, and Karamsots had their own spheres of influence, as did other groups like Jesso Bhatis, Sankhlas and Sonagras. Apart from the Rajputs and the Mughals, there existed other political factors like the Nawabi of Nagaur, which was not a part of the Raj Marwar till the late eighteenth century. Established by the Khanzadas, who paid obeisance to the Sultans of Delhi and Malwa, it was directly incorporated into the Mughal Empire, except for short periods when Rao Chunda and Rao Malde conquered it. Nagaur was kept in *khalsa* or was granted as *jagir* to Rathor or non Rathor Rajputs or the Khanzadas at various times.[145] Similar was the case with Jalor, where a Nawabi was established after its conquest by Alauddin. This Nawabi was included into Raj Marwar by Rao Gango in 1516 CE. The Nawabi had close ties with the Sultans of Gujarat and even after the incorporation of Jalor with Raj Marwar the Muslim Diwans continued to court favours there. Apart from these established Nawabis there existed groups of Baloches, Sindhies, Pathans, Mers, Minas etc., who had contested claims over parts of the Thar. These claimants and new entrants like Marathas and Pindaris, created newer categories of 'rebels' in the eighteenth and the nineteenth centuries.

Post-Mughal Polity and the Rajput on the Frontier

The Rathor rebellion in late seventeenth century paved the way for weakening of Mughal control over the region with Rajput rulers actively participating in Mughal succession disputes.[146] In the eighteenth century, with the decline of

---

[145] Nainsi, *Vigat*, II, 413–416.

[146] After the death of Raja Jaswant Singh of Jodhpur in Kabul, Aurangzeb refused to recognize the posthumously born heir, Ajit Singh. Fearing for the infant's life a clansman Durgadas Rathor spirited him away, and he could become the ruler only

the Mughal empire, the polity of the Rajput states acquired a semblance of independence, which was fraught with contradictions. While the rulers saw themselves as inheritors of Mughal power, the *thikanadars* viewed themselves as sharers in this power. Given the absence of clear rule of succession, there were several claimants to power in all the Rajput states, which allowed nobility to gain an upper hand in the states.[147]

In Marwar, following the death of Aurangzeb, Ajit Singh managed to gain possession of important forts of Jodhpur, Sojhat, Pali and Merta. He became a major player in Mughal succession, marrying his daughter to Farrukhsiyar in 1815, and later conspiring with Sayyid brothers in deposing him. He was murdered by his son Bakhat Singh, at the instigation of powerful *sardars* and his other son Abhay Singh (r.1724–1749), who succeeded Ajit Singh.[148] In 1742–43, Marwar and Jaipur gained the joint custody (*shamlat*) of Salt lakes of Sambhar and Didwana that became the major attraction for Marathas and British in later periods. The years after the death of Abhay Singh saw dispute of succession between his son Ram Singh and nephew Bijay Singh, in which nobility openly sided with one claimant or the other. Bijay Singh (r.1752–1793) managed to dethrone Ram Singh after two years, but the enmity between them involved not just the powerful sardars of Ahwa and Pokhran, but also the Maratha Jayappa Sindhia. Between 1787 and 1790, the joint forces of Jaipur and Marwar faced Mahadji Sindhia and French mercenary De Boigne who forced to both states to part with Sambhar *shamlat* revenues. Bijay Singh's death led to another phase of succession dispute between sons, nephews and grandsons, most of whom were defeated, murdered or exiled. Even though Bhim Singh (r. 1793–1803) succeeded to the throne, he was challenged till his death by his cousin Man Singh.

In Bikaner, Sujan Singh's (1700–1735) reign was spent mostly trying to fight against the various attempts made by Ajit Singh, Abhay Singh and Bakhat Singh of Marwar who repeatedly attempted to take over Bikaner.[149] He also suppressed a rebellion by Bhatis and Johiyas and took over the Bhatner fort. He died in Raisinghpura in Bikaner in 1735.[150] He was succeeded by his son Zorawar Singh (1735–1745), in whose reign *sardars* of Bhadrajan, Mahajan

---

after Aurangzeb's death. Meanwhile, Jodhpur was taken away from the Jodhawats, leading to Rathor Sardars uniting for the cause of Ajit Singh.

[147] M S Jain, *Concise History of Modern Rajasthan,* 10.

[148] Tod, *AAR*, II, 70–71.

[149] Powlett, *Gazetteer*, 47.

[150] Ibid.

and Churu rebelled against him in connivance with Abhay Singh of Marwar leading to the plunder of Bikaner in 1739.[151] Zorawar Singh died in Anooppura in 1745 CE while on a campaign against Bhatis and Johiyas in Hissar. His cousin Gaj Singh (1745–1788) succeeded him and in his long reign he was able to repel the Bhatis, Johiyas and the Daodpotras of Bahawalpur. He was assigned the *paragana* of Hissar, which was proving to be unmanageable for the Mughal Empire. At the time of his death he had a *mansab* rank of 7000/5000 and the permanent *fauzdari* of Hissar-Firuza. He fortified the town of Bikaner as famine relief and was granted the right to issue coins in 1755.[152] Gaj Singh was succeeded by his son Raj Singh, who was poisoned by Surat Singh, who became the next ruler of Bikaner (1788–1828).[153] He was initially opposed by his nobles but he was able to repress the rebellion. He founded the city and fort of Suratgarh in 1799.[154] He led an army against the Bhattis of Bhatner but Bhatti Zabita Khan put a stiff resistance. But he later defeated Bhattis in Dabli and the fort of Fatehgarh was erected here.[155] In the same year, Bikaner was attacked by a mercenary army led by George Thomas, who managed to take Fatehgarh, but it was later recovered by Gaj Singh. In 1801 along with Khuda Baksh Daodpotra, Gaj Singh attacked Bahawalpur and won several forts like Phulro, Mirgarh and Maroth on the Delhi Multan road. In 1804, Gaj Singh attacked Zabita Khan and took over the fort of Bhatner and renamed it Hanumangarh.[156] On the other hand in Jaisalmer, Jaswant Singh (1701–1707), Budh Singh (1707–1721), Tej Singh (1721–1722) and Sawai Singh (1722) gradually lost their hold on areas surrounding Jaisalmer. Bhatis lost Pugal, Barmer, and Phalodi etc. to Rathors and Garah was lost to the Afghan chief of Shikarpur, Doad Khan, who named it Daodpotra.[157] Rawal Akhe Singh ascended to the throne in 1722, and during his reign, Derawal was lost to Doadpotra or Bahawalpur.[158] After his death in 1762, Moolraj III became the king, who was now addressed as Maharawal, even though Jaisalmer territories had largely shrunk by this time. Though Moolraj III ruled for 58 years, his reign is rather known for the intrigues of the Bhati nobles in Jaisalmer.

---

[151] Ibid., 49–50.

[152] Ibid., 65–71.

[153] Tod, *AAR*, II, 147.

[154] Dayaldas Sindhyach, *Bikaner ri Khyat*, Ed. Hukam Singh Bhati, Rajasthani Shodh Sansthan, Chopasani, Jodhpur, 2005, 118.

[155] Ibid.

[156] Powlett, *Gazetteer*, 75.

[157] Ibid.

[158] Ibid.

While the politics played out in the capitals has been well documented, what has not been paid much attention to is the impact which was visible in the outlying regions in the form of instances of robbery on the trade routes. The eighteenth century documents of the Jodhpur state are replete with instances of theft and robbery on highways through the desert by Rajput *thikanadars*. Examination of a number of petitions filed by traders in the late eighteenth century reveals that the trade routes were often open to the vagaries of local chieftains who made exactions that were considered unacceptable by the Rajput states. For example, the route from Marwar to Sindh was a disturbed one, as it was controlled not so much by the Jodhpur and Jaisalmer states as by the *thikanadars* of Pokhran and several groups that specialized in waylaying caravans. In 1765, a *katar* of 38 camels was waylaid by the *thikanadar* in Bhilada and the *patayat* of Pisangan was asked to initiate a settlement.[159] Another case of theft of camels was reported in 1766, from Dhadhmala in Kutch by Sadul Bhimsi of village Chohtan, Barmer.[160] In the same year, merchants travelling between Ajmer and Pali requested the creation of a new route as the *thikanadars* near Merta harassed the merchants.[161] A theft of camels from the stables of *darbar* of Bikaner was reported by the *kamdars* in 1771. A Raika was dispatched who reported that they had been sold in Pushkar.[162] In 1773, the merchants travelling to Sindh through Bikampur requested the creation of a new route as they were harassed by the *thikanadars* at Bikampur.[163] In the same year the *amil* of Nagaur was instructed to conduct enquiries in the matter of robbery of a caravan in village Jasrasar on its way to Jaipur from Bikaner by the local *thikanadar*.[164] A charge of levying a new *bulawo* was made against the *thikanadar* of Pokhran by Sahu Roop Chand of Umarkot in 1783, when the merchants were already paying *rahadari* and *dan* to the Jodhpur state.[165] Another complaint was made by Sipahi Dildar Khan at *sayer thana* of Phalodi, in 1799, regarding the robbing of his caravan of 130 camels by Bhati Maldeo of village Sahurekra. In case the merchandise was not recovered the value of

---

[159] *Miti Baisakh Vadi 7, Guruvar,* Jodhpur, *Sanad Parwana Bahi* No 2, VS 1822/1765 CE. (Henceforth, *S P Bahi*).

[160] *Miti Ashadh Sudi 15, Jama Khrach ro Navo,* Kotda, *SP Bahi* No 5, VS 1823/1766 CE.

[161] *Miti Chaitra Vadi 4, Guruvar,* Jodhpur, *SP Bahi* No 4, VS 1823/1766 CE.

[162] *Miti Bhadva Sudi 7,* Maroth *Kacheri, SP Bahi* No 11, VS 1828/1771 CE.

[163] *Miti Ashwin Sudi 14, Budhwar,* Phalodi *Sayer, SP Bahi* No 13, VS 1830/1773 CE.

[164] *Miti Kartik Vadi 9,* Nagaur *Sayer, SP Bahi* No 13, VS 1830/1773 CE.

[165] *Miti Ashwin Sudi 6, Somvar,* Jodhpur, *SP Bahi,* No 29, VS 1840/1783 CE.

the merchandise had to be paid from the income of the *sayer*, as was in the case of a cloth merchant in Bahawalpur in the same year.[166]

Here it becomes necessary to explore the relationships between martiality, Rajputhood and control over mobile wealth in this frontier region. Was martiality demonstrated through cattle raids or banditry on highways expression of warriorhood? Could these instances be understood in the context of a cultural system of exchange prevalent in frontier areas, where mobile wealth remained the accepted form of wealth, thus war booty? After all in dry arid deserts like the Thar it would have been moveable wealth that was more desirable than control over unproductive land. The looting of cattle was a form of warfare, locally addressed as *dhads*, involving rival Rajput groups and had been used to augment cattle wealth as well as to settle social and political disputes.[167] Theft of cattle was not something to which any social stigma was attached. David Gilmartin has demonstrated that in nineteenth century Punjab cattle stealing was undertaken by young men to show their prowess. It was a practice associated with protection of clan livelihood as well as clan honour. Village headmen as well as clan chiefs along with *rassagirs* extended protection to networks of cattle theft.[168] I would argue that in the Thar, the ability to carry out such thefts was closely associated with the ability to resolve

---

[166] *Miti Falgun Vadi 11, Budhvar*, Phalodi *sayer*, *SP Bahi* no 52, VS1856/1799 CE.

[167] This facet of rise of Rajput power has been explored by several researchers. See Nandini Sinha Kapur, 'The Bhils in the Historic Setting' in *Mobile and Marginalized Peoples: Perspectives from the Past*. (Eds.) Rudolf C Heredia & Shereen Ratnagar, Manohar New Delhi, 2003,160. Norbert Peabody has associated banditry, looting and lifting with internal political disorders between raja and chieftains, and within chieftains; Norbert Peabody, "Cents, Sense, Census: Human Inventories in Late Precolonial and Early Colonial India', *Comparative Studies in Society and History*, Vol. 43, No. 4 (Oct., 2001), 840. C A Bayly suggests a relationship between the local decline in agricultural production (occasionally exacerbated by famines) and 'banditry'. He writes: 'Mercenaries and tribal people such as Bhattis, Mewatis and the corporations of mercenary adventurers known as Pindaris had to turn to outside looting and plunder when rent extracted from their home territories was no longer sufficient to maintain their style of life'; C A Bayly, *Rulers, Townsmen and Bazaars: North Indian Society in the Age of British Expansion*, Cambridge, 1983, 91. Shail Mayaram has made a similar connection in the context of the Meos. Shail Mayaram, *Resisting Regimes: Myth, Memory and the shaping of a Muslim Identity*, OUP, Delhi, 1997.

[168] David Gilmartin, 'Cattle, Crime and Colonialism: Property as negotiation in north India', *IESHR*; 40; 2003, 36.

the disputes and recover the cattle. A similar association existed on the trade routes as well. The abilities to loot caravans and to extend protection to the caravans were closely related. These associations signified the true location of authority in frontier regions. The safety of cattle or caravans was not dependent upon the ability of the state to extend protection. It rather depended upon the ability of the state to negotiate with the people who controlled the region and thus routes through it. An example of control was the fact that *thikanadars* in the frontier zones like Pokhran continued to levy *bulawos* or protection taxes like *rukhwali ri bhachch*. This was noted by the British as well, who also considered the older system of *bulawo* necessary "for affording a ready system of local protection to travellers and merchandise and property in the cause of ensuring safe transit to person through tracts where both stand in need of armed escort".[169] For the Rajput *thikanadars* in the frontiers, there existed dissociation with core Rajput-Mughal polity. While the core polity became more and more agro-centric and territorial, the *thikanadars* continued to operate in older ways where political control was exercised through control over mobile resources. This fierce control that they continued to exercise, despite it pitching them in opposition to their clansmen and the empire, labeled them outlaws and bandits, as opposed to the loyal, chivalrous noble warrior. These oppositions carried into the nineteenth century when the desert came to be seen not just an ecological frontier, but also a political and civilisational one.

Over the long history of Thar, several kinds of groups had exercised such control, including Rajputs, as well as groups with a claim to Rajput past like Bhils, Mers, Jats etc., but also interestingly the bardic community of Charans, who were entrusted with the task of safeguarding caravans through the desert. Various stories about how Charan women found water in the desert, fed wandering armies, provided horses and oxen to armies are indicative of their control in the desert.[170] Besides, the location of Charani *sagati* temples on routes through the Thar, suggests that Charans exercised immense control on movements across the desert, whether of armies or of traders. The fact that Chundo Rathor had to be equipped with a horse and weapons by Charan Alha Rohadiya before being presented to Maloji is indicative of the fact that Charans had access to both these important martial requirements and mediated their supply. Charans could play out their multiple roles as bards, guarantors and guards only because they controlled the frontier space and the authority of emerging Rajput states in frontier areas was manifested through mobile groups

---

[169] Foreign Political, 17th December 1838, 37–49, NAI (Henceforth FP).

[170] Kamphorst, *In Praise of Death*, 253. Also, Tambs-Lyche, *Power, Profit Poetry*, 174–197.

like the Charans. This network thrived on circulation of mobile wealth like cattle or merchandise. As Rajput capitals like Jodhpur, Bikaner, Jaislamer, Umarkot etc. developed as centres of sedentarised Rajput aristocracy, the outlying fringes of these centres remained the domain of semi-sedentarised groups. Since, mobile groups had traditionally relied on circulation of mobile wealth; the tendency to control this circulation was also a reflection of political control in this region. The theft and robberies in this region thus should not be viewed as an indication of lack of control by sedentary states, but rather, as an expression of control exercised by the mobile groups. Several levels of negotiations can be witnessed in the processes through which the sedentary states sought to address these acts of defiance. Granting revenue free villages to Charans, as well as allowing *thikanadars* to levy protection taxes were both mechanisms to engage with mobile groups. The processes of state formation in the Thar region thus, charted an ambivalent course, aligning with larger sedentary state formation like the Mughal empire on the one hand, and, accommodating resistances to such structures on the other. The inability of the Mughal state as well as the Rajput states to extend control in the Thar desert, led it to be seen as a lawless zone, which was beyond the pall of the state.

'Lawlessness' has often been equated with absence of authority and thus, that of state. I would argue that in frontier zones like the Thar, we would need to understand multiple meanings of authority, rather than authority as something absolute. The geographical distance of the frontier in the Thar from the core areas of authority would have made imposition of Mughal administrative norms as well as laws difficult. But I would argue that rather than being a result of geographic isolation, it would have resulted from manner in which norms of power sharing developed in this region. As some mobile groups, including Rajputs sedentarised, others remained on the peripheries of this new emerging sedentary world. While Rajputs in core areas increasingly focussed on territorial polity, it did not mean that their older reliance on mobile wealth declined. The groups in the fringe areas of the Thar continued to be the means through which the sedentarised groups had access to mobile resources.

Rajputana Agency: Politics of Indirect Rule and the Making of the Rajput

The politics of British expansion in what came to be seen as Rajputana followed the various successes that the East India Company had met with Marathas, Bengal and Mysore. It was clear that the East India Company now wanted to be established as the paramount power in Rajputana, a term very clearly in usage by this time. The internal dissentions within the Rajput states had reduced the authority of the ruler to a mere ceremonial one, whereby the *sardars* had become recalcitrant and in some cases controlled the state.[171] In Marwar there arose a dispute between Bhim Singh (r. 1793–1803) and Man Singh (r.1803–1817 and 1818–1843) between 1793 and 1803 CE. In Bikaner, Maharaja Surat Singh (r. 1788–1828) also faced a near rebellion by sardars who brought in Jamir Khan Pindari in 1816 CE. The neighbouring state of Jaisalmer gradually weakened under a series of weak rulers. Moolraj III (r. 1762– 1820), who decided to call himself a Maharawal, appointed Swaroop Singh, a Jain *mutsaddi* to counter the intrigues of Bhati nobles. Instead, they conspired and killed Swaroop Singh and imprisoned Moolraj III. After long protractions, Moolraj III was released, the conspiring *sardars,* along with the Prince Rai Singh, were banished and Salim Singh, a minor son of Swaroop Singh, was made the *Diwan* of Jaisalmer. Maharawal Moolraj gave up the administration completely to Salim Singh. This was a period of terror for the kingdom as Salim Singh got most of the important *sardars* as well as the heir apparent Rai Singh and his sons murdered.[172] The rulers adopted the strategy of reducing their dependence on nobles by recruiting mercenary armies of Marathas and Pindaris, a move that often failed, because unlike the *thikanadars*, the mercenary armies had no attachment to the state beyond their upkeep, for which they were willing to resort to plunder.[173] Secondly, they

---

[171] Such instances were quite rampant throughout Rajputana in the nineteenth century and as Norbert Peabody suggests, through the example of Jhala Zalim Singh in Kota, who made explicit display of the control he exercised over Umed Singh, these instances decentred the traditional notion of warrior service or *chakri*. Peabody, 'From 'royal service' to 'maternal devotion' during the Jhala Regency: Local Politics at the end of old regime' in *Hindu Kingship and Polity in precolonial India*, Cambridge University Press, Cambridge, 2003, 112–146, 116.

[172] Tod, *AAR*, II, 214–216.

[173] According to Stewart Gordon, Pindaris were irregular Pathan or Afghan horsemen who accompanied Maratha armies in central India during the 18th century when the Mughal Empire was breaking up. (Stewart Gordon, *Marathas*, xiv). According to C A Bayly, 'Pindaris had to turn to outside looting and plunder when rent extracted from their home territories was no longer sufficient to maintain their style of life'. (C A Bayly, *Rulers, Townsmen and Bazaars*, 91). However, according to James Tod the term Pindari came to be used to distinguish an unlicensed freebooter and was 'synonymous with Khuzzak, Grassia etc'. Pindaris were loosely organized under self-chosen leaders, and each band was usually attached to one or other of the great Maratha leaders. The most redoubtable leaders among them were Amir Khan, Karim Khan, and Chitu, a Jat. Their special characteristic was that they received no pay, but rather purchased the privilege

appointed non-Rajput officials, like Bhandaris, Singhvis, Khatris, Brahmins as *diwans* and *bakshis* to counter strong *sardars*. Both, Marathas and Pindaris had advanced into the Thar through Gujarat and Malwa and had taken to conducting raids on the behest of one claimant of power or the other.[174] As mercenary 'freebooters' these bands sacked and plundered, and imposed heavy war indemnities on the Rajput states.

---

of plundering on their own account. The Pindari or the Third Anglo-Maratha war of 1817–18 was fought to disband and pacify Pindaris under John Malcolm. The Maratha chiefs of Gwalior and Indore were forced to sign treaties to disband Pindari troops and most Pindari leaders were put to death. However, one of the powerful Pindaris agreed to disband only on the condition of being granted the possession of Tonk. James Tod, *Origin, Progress and Present State of Pindarees*, Nagpur Govt. Press, Nagpur, 1920. The Krishna Kumari episode (1804) became the reason for entry of Amir Khan Pindari at the behest of the Jodhpur ruler Man Singh (r. 1803– 1817, 1818–1843) to intervene. The princess of Udaipur, who had been bethrothed to Bhim Singh of Jodhpur was offered in marriage to Jagat Singh of Amber, after former's premature death. Man Singh who considered this as an affront especially as Jagat Singh was supporting the cause of Dhonkal Singh a rival claimant to the throne, stopped the sending of *tika* to Amber from Udaipur. In the ensuing hostilities, Man Singh sought the help of Pindaris, which resulted in the death of Krishna Kumari, in order to avoid a battle between Jodhpur and Amber. Tod, *AAR*, I, 367–368. Man Singh also used Amir Khan's services to get the powerful *thikanadar* of Pokhran, Sawai Singh assassinated in 1809. The repeated overtures from the rulers of Jaipur and Jodhpur provided Marathas and Pindaris opportunities for plunder and loot. Amir Khan Pindari often visited Marwar and even captured Merta and Ghanerao and collected tribute from Man Singh.

[174] In 1755, during the reign of Bijay Singh of Jodhpur, Marathas under Holkar and Sakharam Bapu had captured Sambhar and had besieged Nagaur. Bijay Singh entered an agreement with Marathas and agreed to pay a war indemnity of 50 Lakhs and a tribute of 1,50,000 annually, even though the Marwar treasury was empty. To escape making the payment Bijay Singh extended overtures to Abdali who had defeated the Marathas in the battle of Panipat. Abdali offered protection from the Marathas, but he himself was gone by 1766 CE. In May 1769 CE, Mahadji Sindhia and Tukoji Holkar attacked Marwar and forced Rathors to pay tribute and arrears. Mahadji occupied Sambhar and demanded a fourth share of Godwar that had been captured by Bijay Singh. During this half century, the rulers of Jodhpur were engaged in various battles with the neighbouring areas. Battles were fought with the Jaipur state, with Sambhar and Didwana being contested territories. The Maratha incursions forced the rulers of Jodhpur and Jaipur to come together, and fight Mahadji Sindhia in Lalsot in 1790. Sindhia, with the aid of DeBoign's artillery easily defeated the joint forces of Jodhpur and Jaipur, who in accordance with the treaty of Sambhar were forced to pay 60,00,001 lakh rupees as war indemnity to Sindhia as well as *bharna* worth 3 lakhs and Sambhar remained under Maratha control. G D Parihar, *Marwar and the Marathas (1724–1843)*, Hindi Sahitya Mandir, Jodhpur, 1968, 82–107.

The beginning of nineteenth century witnessed the rise of recalcitrant nobility in all three Rajput states of the Thar. The recalcitrance of the thikanadars as well as the threats posed by the continuous exactions led to the states of Bikaner and Jodhpur seek British help, which resulted in the treaties of perpetual friendship, alliance and unity of interests with the British by 1818.[175] However, it needs to be pointed out that the subsidary treaties were as much a result of British need to limit the advance of Marathas and Pindaris, as of the Rajput rulers.

The process of treaties started in 1803 itself, when the British were involved in a series of Anglo-Maratha wars. Though Jaipur and Alwar signed early treaties in 1804, Man Singh of Jodhpur refused, only to seek a new treaty in 1805–06. According to these treaties the Rajput states agreed to act in accordance with British military advice, place their armies at British disposal and pay for British army. By 1816, the terms of proposed treaty included exclusion of all foreign influence and acceptance of British arbitration in all matters. By this time as a result of the third Anglo-Maratha war, defeated Maratha chiefs had signed treaties with British and Amir Khan Pindari had been disarmed and made Nawab of Tonk. Thus Marathas and Pindaris having been routed out, the Rajput states were not left with much choice and in 1818 the treaties of perpetual friendship, alliance and unity of interests were signed. The treaties required the Rajput states to cede all their external affairs to the British, supply troops and pay tribute. In 1832, in the durbar held in Ajmer, the creation of a separate agency for the Rajputana was announced, that was to be headed by the Agent to the Governor General (AGG), to be stationed in Ajmer. In 1897, the Agency was quartered and Western Rajputana States Agency was created to govern the affairs of Western Rajputana states.[176] Political Agents were appointed in different states in Rajputana, who reported the affairs of the states to the AGG and acted as a bridge between the native rulers and the British administration. A number of these men wrote their own accounts of their residence in Rajputana in shape of their memoirs or gazetteers.[177]

---

[175] Barbara Ramusack, *The Indian Princes and their States*, Vol 3, 78.

[176] The position of the Agent to the Governor General was occupied by Englishmen who had served as Residents and Political Agents in several states of Rajputana, and who exercised considerable influence in these states. This included Lt. Col. Sutherland, Col. Elliot, Lt. Col Keating, C K M Walters, Col. Trevor, Sir Robert Crossthewaite and Sir Robert Holland among others.

[177] Col James Tod, Major K D Erskine, Col C K M Walters, Col P W Powlett are important Political Agents whose accounts provide copious information about the Princely States.

The treaties imposed several conditions on the rulers, and created several contradictions as well. In exchange for personal insignia of sovereignty, the princely states accepted a position of subordinate cooperation and the British overlordship. Now the British governed not only the way the princely states interacted with each other, but also how they dealt with their own clansmen. Even matters like adoption and matrimonial alliances among the Rajput clans became subject to British approval. Thus, rulers were increasingly reduced to a rather ceremonial position vis-à-vis the British, though the creation of modern administrative structures as well as standing armies also reduced their dependence on Rajput *thikanadars*. This impacted the internal dynamics of the Rajput rulers, nobles and clansmen severely, and provided British administrators a greater opportunity to interfere in the affairs of the Rajput states.[178] The question of why Rajput states signed these treaties has been dealt with quite authoritatively by M S Jain who argues, that for the Rajput states treaties merely meant transfer of tribute from Marathas to British, though on even more stringent terms.[179] In absence of Mughal paramountcy for about half a century, it was the rise of 'turbulent and uncontrollable' nobility that made rulers accept subordinate treaties.[180] The treaties made it possible for British to choose to arbitrate in internal as well as external conflicts of the Rajput states. An interesting aspect of this arbitration was settlement of boundary disputes, particularly between Marwar, Mewar, Bikaner and Jaisalmer. Boundaries were measured, fixed and marked by poles under the supervision of British officials.

Some of these men themselves took interest in acquainting themselves of Rajput History and contributed to some of the early British understanding of Rajputs.

[178] Ian Copland, *The British Raj and the Indian Princes: Paramountcy in British India, 1857–1930*, Orient Longman, Bombay, 1982, Michael Fisher, *Indirect Rule in India: Residents and the Residency System*, OUP, Delhi, 1992 and Barbara Ramusack, *The Indian Princes and their States*, Cambridge University Press, 2004 extensively explore the nature of paramountcy in Princely states. K N Pannikar, *British Diplomacy in North India: A Study of the Delhi Residency*, 1803–1857, Delhi, 1968, unravels the establishment of the Residency system in northern India. Sukumar Bhattacharyya, *The Rajput States and the East India Company from the close of 18ᵗʰ Century to 1820*, Munshiram Manoharlal, New Delhi, 1972, discusses the impact of policies of early administrators of the Esat India Company in Rajputana. A C Banerjee's *The Rajput States and British Paramountcy*, Delhi, 1980 explores the working of the Residency system in Rajputana. P R Shah, *Raj Marwar during Paramountcy: A study in Problems and Policies up to 1923*, Sharda Publishing House, Jodhpur, 1982, sheds light administrative policies in Marwar under indirect rule.

[179] M S Jain, *Concise History*, 23.

[180] Ibid.

However, these were not easy alliances, as the rulers continuously asserted themselves. For instance, Man Singh of Marwar sheltered Appa Sahib Bhonsle in 1829 and refused to attend Ajmer durbar in 1832. His association with the Naths, against whom there were several accusations of criminal activities created discord between him and the British. The British administrators encountered the "disorganized state of Marwar",[181] and the "plundering and predatory incursions carried out by the natives of Jodhpur ...on the frontiers of Marwar, a malaise slowly penetrating the interiors as well".[182] In response, the British army under Capt. Sutherland marched against Man Singh and captured Jodhpur fort in 1839, which Man Singh ceded easily. Even though Man Singh of Marwar has sometimes been projected as last Rajput resistance to British, the fact is that he always stopped just short of entirely antagonizing the British. His relationships with his *thikanadars* were strained to such an extent that he thought it prudent to maintain ties with the British. His successor Takhat Singh (r.1843–1872), who was adopted from the Idar line, and was the chief of Ahmednagar, also had fraught relationships with Marwar nobles as he tried to bring in the Ahmednagar nobles with him. A few years after his acsension, Takhat Singh began to levy succession fee, impose heavy fines as well as confiscate jagirs. The anger of the *thikanadars* of Gular, Alneawas, Asop and Ahuwa was expressed in the form of revolt of against Takhat Singh, which has been seen as participation in the 1857 revolt though there is no single view on the nature of participation of the Marwar nobles.[183] However, when Takhat Singh revoked *jagirs* of several *thikanadars*, the latter asked the British to arbitrate. Maharaja was threatened with deprival of all authority and forced to cede administrative powers to a council of ministers. In the nineteenth century thus, the Rajput states of the Thar were reduced to a subordinate position in a way that these had not experienced even in the period of Mughal dominance.

Interestingly, while the Rajput rulers were continuously made to agree to humiliating conditions, on the other hand their proud ancestry and legacy of bravery was also espoused. The history of Rajputana was constructed through frames of warriorhood, whereby the Rajputs emerged as proud warriors more than anything else. Drawing upon older texts James Tod (also Alexander

---

[181] FP, 25th April 1838 (97–99), NAI.

[182] FP, 25th December 1838, (10–24), NAI.

[183] While a number of local narratives in Rajasthan hold Thakur Kushal of Ahwa as a rebel in 1857, Jain holds that the rebellion by *Thakurs* in 1857 was directed at Takhat Singh rather than the British.

Forbes) constructed the Rajput through two interesting frames. The first was of a Hindu history of Rajputs, whereby valor was represented through idioms of fort protection against Muslims. The second, was through their relationship with Anglo-Saxon traditions, as against upstarts like Marathas. In the nineteenth century, the East India Company had to engage with several groups like Sikhs, Marathas, Afghans, Gorkhas, Hyder Ali and Tipu Sultan of Mysore, and the *talukedars* of Awadh, among others. The Rajput states of Rajputana are the interesting absence in this list. The treaties with the Rajput states were not military treaties forced after wars, but treaties that were carefully negotiated after long deliberations. Yet, in the larger discourse of martiality, where groups are identified on the basis of their martial characteristics, Rajputs were the foremost entry, as the most martial of the martial races. The discourse of martiality was parallel to the discourse on criminality, as both were seen as inbred characteristics.[184] In case of Rajputs, warriorhood along with traits like chivalry and loyalty were seen as their innate characteristics, which as Tod tried to demonstrate could be a result of their old Scythian origins.[185] It is in Tod's work that a new genealogical orthodoxy can be seen as emerging, as Tod obsessively treads the racial terrain. He researched and put together a corrected list of genealogies of 'thirty six royal races'.[186] Racial purity along with resistance to Muslim domination became the standard for adjudging the hierarchy of Rajput clans, as also 'Rajputhood'. This is apparent in his *Annals*, which he begins with Mewar, declaring, "with exception of Jessulmér, Mèwar is the only dynasty that has outlived eight centuries of foreign domination".[187] His castigation of Kachchwahas of Amber for making a marital alliance is scathing, "the name of Bhágwandas is execrated as the first who sullied Rajpoot purity by matrimonial alliance with the Islamite".[188] This assumption is carried forward in other accounts as that of William Crooke, who writes of Rajputs as, the "proud tribes of Rajputana...who by careful marriage regulations have preserved the purity of their blood".[189]

---

[184] Metcalf, *Ideologies of the Raj*, 125–127.

[185] Tod, *AAR*, I, 471.

[186] Ibid., 68–69.

[187] Ibid., 173.

[188] Tod, *AAR*, II, 286.

[189] William Crooke, 'Rajputs and Mahrattas', *The Journal of Royal Anthropological Institute of Great Britain and Ireland*, Vol. 40, Jan-Jun 1910, 39–48, 40–41.

James Tod's Annals of the Rajputs, turned the Rajputs from "fierce warriors into inspirational national heroes".[190] Freitag's work demonstrates how Tod's *Annals* became instrumental in reconstruction of Rajput identity through their translation into Indian languages like Bengali, Marathi, Hindi, Urdu and Gujarati in the late nineteenth and early nineteenth centuries.[191] In the same period Ramya Sreenivasan points out the proliferation of Bengali narratives inspired by the Padmini narrative, including Tod's version in his *Annals*.[192] The relationship between racial purity and warrior ethos was highlighted through the narratives of sack of Chittaur in the *Annals* as well as popular works like that of Gabrielle Festing. In fact, the "Tale of the Virgin Princess", the tragic story of Princess Krishna Kumari of Mewar, was cited as another example of the warrior ethos that resided even in the women of the Rajputs.[193]

However, the Rajput subservience to both Mughals and Marathas in the past centuries came in conflict with this projected fierce independent warrior ethos. In this narrative Rajputs emerged as fallen princes, who at once displayed both valor and ruin. In order to restore their former glory, the anarchy in their courts, in the form of disaffection of nobility had to be dealt with. But this had to be done while maintaining the dignity of the princes, who should not be made to feel that their religion or power was being subverted.[194] Even though Tod's political work brought Rajputs under the British, Tod in fact felt that having being rescued from Maratha thralldom, Rajputs were being reduced to the same deleterious effect because of the conditions that treaties imposed on them.[195] The British had to be responsible for the Rajput system of governance, which was patrimonial and feudal, and thus better than both Mughal 'despotic' and Maratha 'predatory' forms. [196] However, given my understanding of Rajput 'familial' relationship with Mughals as well as the

---

[190] Freitag, *Serving Empire Serving Nation*, 31. Interestingly, the last translation that Freitag mentions is in 1996, which demonstrates the fascination with Tod even in late twentieth century.

[191] Ibid., 174–179.

[192] Sreenivasan, *The Many Lives of a Rajput Queen*, 214–227.

[193] Gabrielle Festing, *From the land of Princes*, Smith, Elder and Co., London, 1904. The Princess consoled her mother, 'I fear not to die! Am I not your daughter?' Tod, *AAR*, I, 368–369.

[194] Freitag, *Serving Empire Serving Nation*, 87.

[195] Norbert Peabody, 'Tod's Rajasthan and the Boundaries of Imperial Rule in Nineteenth Century India' *Modern Asian Studies*, Vol 30, No.1 (Feb 1996), 185–220, 207.

[196] Ibid., 210–211.

'predatory' past of Rajputs, these distinctions are amorphous. Marathas could claim admiration for resisting Mughals, but their rise from the humble Kunbi stock ruled out any possibility of comparison between them and Rajputs. For Tod, the Maratha court was 'nomadic', the Maratha governance 'lawless'. The difference between Rajput and Maratha was that of a noble warrior and an itinerant predator.

What the British attempted to do in nineteenth century Thar was to streamline the Rajput polity. For Tod, Rajput predation could be justified on account of their misfortunes. He points out, "the Sodha ......with his poverty has lost his courage, retaining only the merit of being a dexterous thief and joining the hordes of sehraes and kossa who prowl from Doadpotra to Guzzerat."[197] The 'lawlessness' of Bhati subclans like Maldote, Kelhan, Varsang, Pohar and Tejmalot, who Tod compares to Bedoiun, Kazak or Pindari, becomes excusable due to their circumstances. In Tod's estimation, the purpose of British presence in the Thar was to enable the British to turn the 'lawless' Rajput into the 'noble' Rajput, by creating a unilineal vision of Rajputhood. In this vision both a genealogical orthodoxy as well as fort protection against Muslims became ways of identifying the Rajput. In a common Rajput history of 'Rajputana' the discordant narratives could be woven together to create a seamless narrative of a race with an illustrious past.

As Thar was an ecological frontier, where nomadic and agrarian resource bases interacted, it also became a political frontier where nomadic and sedentary polities interacted. Therefore, while sedentary state systems evolved with Rajput fortress cities at their centres, the Rajput polities were also forced to accommodate mobile polities in the fringe areas, as these frontiers were the domains of mobile groups controlling circulation of mobile resources. It is not surprising, that in the nineteenth century, most of these groups were accused of lawlessness and banditry, and some were later notified under the Criminal Tribes Act of 1871. Not only was the mobility of such groups severely restricted, the community of Charans was characterised as 'begging bards' singularly responsible for practices like female infanticide among Rajputs. The establishment of indirect rule in Rajput states in the nineteenth century also radically altered the internal structure of Rajput polity in the Thar. As the British found it easier to negotiate with the ruling clans, the position of the cadet lines was substantially reduced. For them, the multiplicity of control as it existed in the fringe areas of the Thar added to the incomprehensibility of

---

[197] Tod, AAR, Vol. II, 257.

the Thar, which was increasingly viewed as a hostile region. The ordering of Thar thus required the imposition of clear, identifiable boundaries as well as the suppression and control of mobility of itinerant groups.

In the seventeenth and eighteenth centuries, however, it had been actually only through such groups that the authority of the state, whether Rajput or Mughal could be manifested in the Thar. Thus, within the Thar and particularly in its outlying fringes there were several nodes of authority. The areas lying between these multiple nodes of authority rather than being studied through binaries of state and non-state should be seen as zones of negotiated authority, where a range of power structures developed. In the 19th century, not only was the mobility and control excercised by such groups suppressed, the ruling Rajput states of the Thar were also reduced to subordinate positions in ways that they had not experienced even in the period of Mughal dominance. On the other hand, orientalist writing constructed the 'Rajput' as a genealogically superior race. They romanticized the Rajput and essentialised chivalry, valor and pride as Rajput values, which were as much constructs of historiography as criminalizing of certain communities was.

3

Itinerants of the Thar:
Mobility and Circulation

If peripatetic myths do not transcend the ideals of a settled order, they do not unquestioningly accept the terms of that order either. Marked by a deep ambiguity, they embody the predicament of those who are both inside as well as outside settled society, those who are seen as a normal part of the social landscape, yet strangers----accepted as well as censured.[1]

In the contemporary Thar region, a number of mobile groups continue to follow very old routines of migration, despite drastic social and political transformation, particularly over the past century. Even as the Indian states of Rajasthan and Gujarat rapidly urbanize, it is still common to come across large herds of sheep or camels using the roads meant for modern transport, or to find the carts of Gadiya Luhars parked in the heart of the city. These communities form a part of the long history of this region and represent its lived past. They also represent the struggle and the dynamism for survival in a system that acknowledges their existence with increasing unease.

The previous chapter explored the rise of Rajputs as territorial rulers and the resulting political marginalization of itinerant groups as itinerancy posited a challenge to the emerging territorial political culture. However, while

---

[1] Neeladri Bhattacharya, 'Predicaments of Mobility: Peddlers and Itinerants in Nineteenth Century Northwestern India', in *Society and Circulation Mobile People and Itinerant Cultures in South Asia* 1750–1950, (Eds.) Claude Markovits, et al Permanent Black, Delhi, 2000, 163–214.

itinerancy was marginalized in the political structures, itinerants of several kinds continued to travel on, and exercise control over networks of travel and circulation on the frontiers of the Thar. They also continued to challenge social boundaries created by the emerging genealogical orthodoxy and thus in a sense represented the social frontiers of the Thar. A perusal of complex mythologies and histories of these groups questions the fixity of caste identities. In fact, the oppositions between 'mythologies' and 'histories' are pointers towards the fluidity and ambivalence of caste and occupational identities, and the processes through which they were essentialised.

In the Thar, the existence of a vast population of peripatetics throughout its human history and the political compulsions of settlement, signify a struggle that cannot merely be understood by placing settlement and mobility in bipolar opposition. Mobility formed the core livelihood practice of the region, as its fragile ecology demanded a constant circulation of people and resources. Pastoralists, traders, carriers, peddlers, travelling artisans, bards and genealogists, ascetics and also some scattered hunting and gathering groups were occupationally mobile groups in the Thar Desert. These groups moved across the desert carrying out their occupational roles that made them essential to the Thari social structure. Mobility was dependent upon the existence of unoccupied stretches of land that could be used in multiple ways by peripatetic groups, particularly the pastoralists. Cultivation in the Thar was largely incidental, dependent upon rain and diligent water harvesting, precluding the possibility of sustained dependence on agrarian production. Besides, the groups that had been settled and were farming on a fairly continuous basis also did not remain entirely sedentary. Peasant mobility and frequent attempts at desertion of land particularly in times of distress, which were fairly regular, remained a recurrent phenomenon in the Thar region. The poor quality of land and inadequate quantity of produce led to it being deserted by a peasantry that remained semi-sedentarised at best. Also, it would be difficult to find communities that did not indulge in animal husbandry and cattle rearing though they might not follow annual cycles of migration. Though it has been increasingly discouraged by princely as well as post-colonial states, mobility is clearly an inherent part of Thari life. Therefore, sedentary/mobile or peasant/nomad have remained rather fluid categories in the Thar.[2]

---

[2] Chetan Singh and Shail Mayaram have pointed out in their respective areas of work that these categories were often very flexible. See, Chetan Singh, 'Forests, Pastoralists and Agrarian society', in *Nature, Culture, Imperialism*, 21–48. And Shail Mayaram, 'Mughal State Formation: The Mewati Counterperspective', *IESHR*, 1997, 34 (2), 169–197, 174.

However, research into the 'pasts' of mobile groups was indeed a difficult enterprise. While the state of Rajasthan appeared to be immersed in its martial past, a land of wars, forts and of kings, of heroic poetry and epic adventures, there appeared to be no 'history' that most of its itinerants could allude to. There, of course are myths, of origin, of migrations, of 'fall from Rajputhood', but hardly any written accounts to speak of. As I began looking for 'historical' sources for studying mobile groups like pastoralists, peddlers, mobile artisanal groups, wandering bardic communities, I was struck by the lack of information. Munhata Nainsi's 17ᵗʰ century gazetteer, *Marwar ra Paraganan ri Vigat* referred to Rajputs, Brahmins, Banias, Charans, as well as an assortment of resident artisanal communities like the Kumhars, Bambhis, Raigars etc. Infrequently, it recorded the presence of Raikas, Gujars, Banjaras and Lowanas, but communities like Sansis, Ganwarias, Kalbelias, Gadiya Luhars, Nats, Sewags, Pancholis were completely missing from the *Vigat*. The detailed Jodhpur *Sanad Parwana Bahis* of the late eighteenth century, which recorded petitions from all over the Jodhpur state also rarely mention these communities. It was only in the colonial ethnographic accounts in the nineteenth century that these communities found a mention. They were also recorded in the early census operations, which aimed to identify these communities in order to regulate their movements. Another set of records that mentioned these communities were 'criminal' records or ones that recorded the processes of 'settlement of criminal tribes'. Ethnographic accounts, censuses and state records are in real sense records of prejudices against these communities whereby most of them were mentioned as habitual criminals, descriptions that form the basis of knowledge about these communities even today. As a historian with little 'historical' evidence to rely upon, I tried to look at colonial ethnography against the grain, and in the light of new ethnographic research undertaken in recent years. Also, it is while I began engaging with oral epic traditions, performed by low caste bardic communities in Rajasthan, that I realized that versions of popular epics that were circulated among mobile communities were, in a sense, their claims on histories of this region. It was not difficult to see that these renderings, as well as origin myths, simultaneously referred to a past as well as a present location of these communities, the movement between the two signifying the process of their marginalization as well as their survival.

This chapter has three objectives. Firstly, it seeks to describe the Thar Desert as a mobile space where a wide range of mobile communities circulated. It examines the world of itinerants and the processes of production and circulation they were engaged in. These descriptions rely on vernacular accounts like the *Khyat* and *Vigat*, records of the Rajput states, British records and administrative accounts, censuses, ethnographic accounts, both colonial and contemporary, travelogues as well as oral epic traditions. Secondly, it seeks to explore the ways in which the movement of these groups contributed to the formation of a circulatory regime, whereby people, cattle, commodities, information, lore etc. circulated in the fluid mobile region of the Thar.[3] While the circulation of Rajputs had led to the making of the Thar as a 'political' region, the circulation of these communities defined the 'regionality' of Thar in more than one ways. Thirdly, this chapter argues that mobilities of Rajputs, and that of other itinerant groups, like herders, traders, carriers, artisans, genealogists, bards, that together constitute the social frontiers of Thar, are historically connected.

The Travelers

There were several kinds of mobile communities in the Thar region that undertook long and short journeys within and outside the desert. These groups maintained their own sense of space or time that was specific to the group in question. For instance, for a pastoral nomad it referred to the time that it took him to move with his herds in a seasonal manner. On the other hand, for a Bhat it could mean the time between two visits to a particular village community to record births, deaths, marriages and other events. Similarly, for a long distance trader, a sense of space, referred to the long routes on which he carried his merchandise, while for the petty peddler it meant the villages around himself where he peddled his ware. As discussed in the first chapter, the settlements in the Thar region formed several overlapping networks of travel, which were utilized by the mobile groups, where they encountered each other and circulated their particular wares or services. In this manner they constantly moved and circulated on these networks, thereby carrying out their particular occupational roles. This was certainly not purposeless wandering as mobility and circulation of men, merchandise, ideas and traditions were essential features of the Thari society. Though it is not possible to study these communities in a linear manner, I have delineated them into four categories:

A. Pastoralists
B. Traders, Carriers and peddlers
C. Itinerant Menial and Artisanal Groups
D. Genealogical, Bardic and Minstrel Groups.

---

[3] Neeladri Bhattacharya refers to similar circulatory regimes in Punjab composed of traders, peddlers, artisans and ascetics, which fed into larger trading networks extending into Kabul and Tibet. Neeladri Bhattacharya, 'Predicaments of Mobility' in Nature, Culture and Imperialism', 49–85.

Even a categorized description will point out how these groups actually functioned in relation to each other and performed specific tasks in the society, though often merging into one another. In the following sections, I also explore the dynamic relationships that existed between these groups as well as the sedentary and the mobile communities, and how at times these roles got blurred.

Pastoralists

Pastoralism has been widely practiced in the arid and semiarid tracts of the Thar Desert from times immemorial. According to the 2012 Livestock census, pastoralists in Rajasthan rear 16.36% of India's sheep, 13.32% of India's goats and 64.01% of India's camel population. Rajasthan has the second largest sheep and goat population in India.[4] Eleven arid districts of Rajasthan contribute 78.86% of total camel population of India.[5] However, while numbers are available for cattle and livestock, no clear estimates are available for the number of people involved with their rearing. In absence of reliable community wise data after independence, it is difficult to approximate the proportion of communities engaged in animal husbandry or livestock rearing in India. On the basis of a rough estimate, it is believed that about 200 'tribes' are engaged in pastoralism in India.[6] Ilse Kohler defines pastoralists in the Indian context as, "members of ethnic caste or ethnic group with a strong traditional association with livestock-keeping, where a substantial proportion of the group derive over 50% of household income from livestock products and their sale, and where 90% of animal consumption is from natural pasture or browse, and where households are responsible for the full cycle of livestock breeding".[7] Pastoral communities can be either nomadic or semi-nomadic, depending on the nature of their herds. Also, peasant households also usually raise a small number of milch cattle, oxen, goats as well as camels for agricultural needs. These are usually reared both on common and individual resources. Among the communities engaged in mobile pastoralism and animal husbandry, there exist several kinds of distinctions based on the kind of animal husbandry and live stock rearing

---

[4] Livestock Census of India, 2012, 39.

[5] National Research Centre on Camel, Bikaner, Vision: 2030, published July, 2011, 2.

[6] Vijay Paul Sharma, Ilse Kohler-Rollefson and John Morton, *Pastoralism: A Scoping Study*, Centre for Management in Agriculture, IIM, Ahmedabad and League for Pastoral Peoples, Natural Resource Institute, University of Greenwich, UK, 1997, 2.

[7] Ibid.

practiced. Besides, often groups engaged with pastoralism combine animal rearing with other itinerant pursuits like trading or blacksmithing.[8]

There are several communities traditionally engaged with animal husbandry and livestock rearing in the Thar. While Raikas, Gujars and Sindhi Muslims are seen as nomadic pastoral groups, agricultural groups like Jats, Bishnois, Sirvis and Charans are agro-pastoralist groups combining agriculture with pastoralism. Of all these groups the one that has continued to practice nomadic pastoralism is the Raikas, also known as Rabaris in Gujarat. They are traditionally camel rearers and herders but some sections of Raikas have increasingly turned to sheep pastoralism. According to Vinay Srivastava, Raikas, "….not only manage the camels owned by others, but also have their own herds. There may be individual camel specialists in other castes, but the Raikas have the reputation as the foremost camel-breeders of Rajasthan, even if they have kept sheep for time immemorial".[9] The 1891 census of Marwar describes them as "old inhabitants of Marwar who were engaged in stock breeding since the time of Shah Jahan."[10] However, the Raikas had been living in this region for a much longer period. Raikas claim to belong originally to Baluchistan, where there still is a temple of the Charani Goddess Hinglaj who they worship. Enthoven traces their origin to Marwar and Sindh, as well as their regular pilgrimages to the temple of Hinglaj in Baluchistan.[11] The initial migrations in the folkloric memory of the Raikas refer to flights to Sindh and Kutch undertaken to escape Muslim persecution.[12] According to Sigrid Westphal-Helbusch, the significant migrations of Raikas took place between 1100–1400 CE, when they moved from Marwar to Sindh and Kutch. The migrations of Raikas in fact follow similar paths as that of Rajputs and Charans, two other migrant group in this region, indicating intertwined histories. Westphal-Helbusch ascribes the goddess worship traditions of Raikas to Charan influence.[13]

---

[8] Ibid.

[9] Vinay Kumar Srivastava, *Religious Renunciation of a Pastoral People*, OUP, Delhi, 1997, 6.

[10] *Report Mardumshumari Raj Marwar*, 19.

[11] R E Enthoven, *Tribes and Castes of Bombay Presidency*, 3 Vols, 1921–23, Bombay, (Reprint Cosmo, 1975), II, 252.

[12] Ibid.

[13] Sigrid Westphal-Helbusch, *Hinduistische Veihzuchter im Noedwestlichen Indien, I: Die Rabari*, Duncker and Humboldt, Berlin, 1974. Since I was not able to access this work in English translation, I have depended upon Harald Tambs-Lyche's use of Westphal-Helbusch's work in his book, *Power, Profit and Poetry*, 154.

Nainsi's seventeenth century *Vigat* mentions Raikas as living in villages, but this is rather infrequent as, Nainsi tends to overlook them since he records the major or dominant castes only; he thus records them in very few villages.[14] B L Bhadani nevertheless calculates the population of Raikas in the seventeenth century through the Jalor *Vigat* and he estimates that in 500 villages in the paraganas of Siwana, Sojhat, Jaitaran and Jalor, Raikas inhabited 112 villages.[15] He further makes estimation on the basis of the *Mukata Bahi* of Jalor and on the basis of the taxes paid by the Raikas; there appear to have been 554 Raika males in 132 villages of Paragana Jalor. Therefore he calculates the total population of Raikas to have been 9390 in *paragana* Jalor in 1663 CE. [16] The Marwar census of 1891 estimated their numbers to have been 98,406, while the Census of Marwar and Mallani recorded 1,12,096 Raikas.[17] Vinay Kumar Srivastava estimates the population of Raikas in 1989 to be about 4,00,000.[18]

In their own origin myths, they consider themselves to have been created by Shiva and Parvati to take care of their camels. The first Raika initially called Chaamad, caught in the forest while herding one night, cooked and slept by the fire of a pyre. On finding him Shiva called him a ghost, and thus, Raikas refer to themselves as Shiva's *bhuts*. This myth also connects the origins of Raikas and Kachela Charans, as having originated from two sisters.[19] They also claim to have originated from Pinda, created by Shiva, and Rai, a nymph. The twelve daughters of the couple were married to Rajputs. The son, Samar or Sambal is regarded as the ancestor of all Raikas.[20] In a variant of the origin myth, Sambal was asked to move out of the heaven (hence *raah-bari* or Rabari) and his sons and daughters married Rajputs, who thus became Rabaris.[21] In Enthoven's account, Rabaris were Rajputs who were excommunicated (*raah-bari*) because of the marriage with *apsara*.[22] Also, like many other groups they claim to have sheltered Rajputs during the latter's annihilation by Parsurama.[23]

---

[14] B.L. Bhadani, Pastoral sector in the Economy of 17th Century Marwar', Paper presented at Second International Seminar on Rajasthan, Udaipur, 1991, 34–35.

[15] Ibid.

[16] Ibid.

[17] *Report Mardumshumari*, 567.

[18] Srivastava, *Religious Renunciation*, 7.

[19] *Report Mardumshumari*, 568.

[20] Srivastava, *Religious Renunciation*, 15–17.

[21] Tambs-Lyche, *Power, Profit and Poetry*, 155.

[22] Enthoven, *Tribes and Castes*, II, 253.

[23] Tambs-Lyche, *Power, Profit and Poetry*, 155.

Raikas, like most pastoral communities of the Thar claim a link to the Rajputs, as "it is said that all Rajputs were *maldharis* (cattle keepers) or vice-versa, it is asserted that all nomadic peoples have Rajput *ansa* (essence) in their veins".[24] Their deities are addressed as Raikanath and Momai mata (the mute goddess). Raikas also build small temples dedicated to Pabuji, a Rathor Rajput *junjhar*, who is supposed to have entrusted them with the task of herding camels.[25] The lyrical-visual narrative of Pabuji, *Pabuji ri phad* is recited by Bhil *bhopas* in Raika gatherings. Raikas are the patrons of the *bhopas* and treat the *phad* as a mobile temple.

While Raika myths link them to Rajputs, their present status is that of a marginalized nomadic group. They have traditionally preferred to use the term Raika, for themselves in reference to their nymph ancestor Rai.[26] However, they are often pejoratively referred to as *bhut*. Like many other nomadic communities they are also outcastes as they "act against the normal *reet*, they are called *raah-bari* or Rabari...they live in the desert ....never bathe.....do not know the rituals...and their faces are fearsome....that is why they are called *bhut*".[27] They are called so by "other sedentary castes, chiefly because the occupation of animal grazing keeps them away in forests, pastoral land, and the relatively uninhabited places, which are the favorite abode of the ghosts".[28] In recent years, Raikas insist on the use of the term Dewasi which means close to the gods. They prefer such honorisms as being addressed in this manner becomes a way of ascertaining a position vis-à-vis a lost past.

Apart from herding their own and other communities' herds, Raikas have traditionally acted as messengers and carriers and have been called *kasids*. In the Bhili Bharath, when Balo Himmat decides to join the Bharath, it is Raikas who are sent to fetch his wife Antra, so that the marriage is consummated before he dies in the battle.[29] In the epic of Pabuji, his sister Sonabai calls a Raika to deliver a message to her brother. Harmal Raika was sent by Pabuji

---

[24] Janet Kamphorst, ' The Deification of South Asian Epic Heroes- Methodological implications' in *Epic Adventures: Heroic Narratives in the Oral Performance Traditions of Four Continents* (Eds.) Jan Janson and H. J. Maiereds, LIT Verlag, Muenster, 2004, 95.

[25] Vinay Kumar Srivastava, 'The Rathor Rajput Hero of Rajasthan: Some Reflections on John Smith's The Epic of Pabuji', *Modern Asian Studies*, Vol 28, No 3, July, 1998, 589–614.

[26] *Report Mardumshumari*, 568–571.

[27] Ibid.

[28] Srivastava, *Religious Renunciation*, 13.

[29] Bhagwandas Patel, *Bharath: The Epic of Dungri Bhils*, Centre of Indian languages, Mysore, 2012, 301–322.

to locate the she-camels that he needed for the dowry of his niece.[30] In the 19th century, the Raikas seem to have held a particularly important position in the Bikaner state as the imperial graziers. They are believed to have come to Bikaner with Rao Bika and a line continued to tend the imperial herds till 1920s when the camel herds were disposed off because of the complaints that had been received regarding the imperial graziers "herding their animals in fields with standing crops, often with vengeful motives".[31] Raika are believed to be able to "identify camels and sheep even in the night and can trace their animals in large herds. It is said that they can even identify the offspring of their animals".[32] But in practical terms the animals are identified by the branding marks called *kheng*. Since Raikas do not own all the camels or sheep that they herd, there are actually two brand marks on an animal, the first of the owner and the second of the herder.

The Raikas are traditionally classified into two categories, the Maru and the Chalakiyas. The Marus rear only camels and the Chalakiyas rear sheep predominantly, though their herds may contain a few camels as pack animals. The Marus consider themselves to be superior to the Chalakiyas, as the latter herd sheep, which is considered inferior to camel herding. The Chalakiya women wear brass (*pital*) ornaments therefore they are also called Pitaliya.[33] They also do not wear the traditional head cloth called *ati,* and the men wear shorter loin cloth or *tevato.*[34] At present the category of Chalakiya or Pitaliya is no longer used by the Raikas themselves, and they instead categorise themselves on a regional basis, as Marus who belong to Marwar, Godwara Raikas who are from the Godwar region and Kachela of Kutch in Gujarat, who call themselves Rabaris.[35] These categories maintain ties of brotherhood but do not intermarry.

In Marwar, Raikas appear to have majority of their hamlets in Jalor, Jaitaran and Pali belt, from where they set out on their annual migrations.[36]

---

[30] *Report Mardumshumari,* 573–577.

[31] Vinay Srivastava, *Religious Renunciation,* 15.

[32] *Report Mardumshumari,* 573.

[33] Ibid., 568

[34] Vinay Srivastava, *Religious Renunciation,* 19.

[35] Ibid. Vinay Srivastava also claims that the distinctions made in the dressing patterns of the Maru and Chalakiya do not hold correct anymore. Besides Maru Raikas also rear sheep now. Purnendu Kavoori claims that it is Godawara Raika who traditionally reared camels and Maru who reared sheep, though he too states that these distinctions are no longer applicable. But the superiority claimed on the basis of camel rearing indicates the internal dynamics of the Raika community. Kavoori, *Pastoralism in Expansion,* 5.

[36] Ibid.

The permanent homes of the Raikas are located in these districts, from where they practice seasonal transhumance, returning to the point of departure. They usually own small landholdings, which are insufficient for subsistence and largely serve the purpose of maintaining ties with village community. In context of the necessity of rooted existence, these holdings also provide a means of identification. These plots are nominally cultivated and the Raikas depend on herding for their livelihood, which points to a basic contradiction in their existence. While through social and economic relations with agrarian castes, they experience the need locate to themselves in territorial terms, the need to migrate is "irreducibly, if variably constitutive of their identities".[37]

Purnendu Kavoori's work on social distribution of mobility in western Rajasthan explains that while most communities in western Rajasthan like Raikas, Gujars, Jats, Rajputs, Sindhi Muslims, Kyamkhanis, Meghwals etc. own some cattle, sheep and goats, it is Raikas and Gujars who are largely nomadic and transhumant. Raikas have the largest flocks in the region, as well as highest rate of landlessness, where as Gujars are small and medium land owners. He points out that while in dry and arid districts like Jaisalmer and Barmer, the pastoral households could own large holdings, in semiarid, or fertile belts like Pali Godwar, their holdings are minimal. Also, desert conditions mean that in order to spread the risk, all communities own herds as well as land, thus leading to heterogeneity in herd as well as land size. However, with intensification of agriculture in semi-arid areas while land size of castes like Rajputs and Jats go up, their herd sizes decrease, as they no longer need to keep herds as a gainful economic activity. In past such herds would have been entrusted to Raikas, thus circulating resources. Besides herd ownership by landed communities meant an interest in maintaining common resources like fallows and pastures. With decreased dependence on herding, landowners tend to increase acreage and eat into fallow time, as well as fallow land. Since Raikas have large herds they also have greater needs for commons, for which it is necessary to have proprietary rights within the village community. In absence of proprietary rights Raikas are often left with no option but to increase their nomadic pastoral cycles in pursuit of pasture.[38] Besides, the absence of significant land ownership also impacts the Raikas' ability to influence decisions regarding future of commons like *jods* and *orans*.

---

[37] Arun Agrawal, *Greener Pastures: Politics, Markets, and Community among a Migrant Pastoral People*, Duke University Press, Durham, North Carolina, USA and London, 1999, 20.

[38] Kavoori, *Pastoralism in Expansion*, 87.

Since large herds cannot be pastured in one place, so the Raikas traditionally follow three kinds of migratory patterns. We can attempt to understand the mobility of pastoral nomadic communities in context of the "strategy that Raika shepherds deploy to accommodate the spatial and temporal structure, intensity and unpredictability of the environmental variation".[39] However, Kavoori demonstrates that even in the years of normal to good rainfall short and long-range migrations take place on a regular basis.[40] Enroute the grasslands the herds seek pasture in forests as well as in croplands, where the herds manure the fields in return as they graze on crop stubble. Kavoori demarcates three patterns of seasonal migrations of nomadic pastoralists as well as farmers, through the Thar that historically have led to either grasslands within the Thar, or into Punjab, Multan, Sindh, Gujarat and Malwa. The first is the migration pattern of pastoralists that stay permanently in the pastures (*dang me rehna*). In this case the herds are never brought back into hamlets but migrate from pasture to pasture. The shepherds of these flocks are rarely able to get back to their hamlets.[41] The second category is of herders who migrate locally within western Rajasthan, within a range of 100 kilometers or so grazing on the single crop tracts.[42] The third pattern relates to those herders who practice transhumance proper to eastern and southeastern parts of Rajasthan, as well as into Haryana, Punjab, Uttar Pradesh, Madhya Pradesh and Gujarat. These herders migrate seasonally, and seek pasture in the forests and fertile croplands in these regions. With the development of modern travel networks comprised of railways and roads, changes in land use patterns due to spread of canal irrigation as well as increased urbanization, it is possible that there may have occurred some deviations from old routes. However, the direction of movement of herds has not varied much, not only in the Indian Thar, but also in Deccan where similar transhumance takes place, as shown by Sonthiemer in the case of Dhangars.[43]

---

[39] Arun Agrawal, *Greener pastures*, 23.

[40] Kavoori observes that, "livestock migration is also reported in years that were marked neither by scarcity nor famine. Thus in the year, 1895–96 which was not a famine year, 11,360 persons are reported to have migrated from Marwar along with 96,477 heads of cattle". Kavoori, *Pastoralism in Expansion*, 66–67.

[41] Srivastava, *Religious Renunciation*, 39.

[42] Kavoori, *Pastoralism in Expansion*, 71.

[43] G D Sontheimer, 'The Dhangar: A Nomadic Community in a Developing Agricultural Environment' in *Nomadism in South Asia*, Eds. Aparna Rao and Michael Casimir, OUP, New Delhi, 2003, 2008, 364–398.

The other traditional pastoral community of the Thar is the Gujars, also called Ahirs in Gujarat.[44] Though, Gujars at present own agricultural tracts in Rajasthan and rear milch cattle, they are still identified with their original profession of goat and sheep herding. Gujars claim origin from the aristocratic Gurjara-Pratiharas as well as to having old settlements in Gujarat and Punjab. Even in the nineteenth century this paradox in their position was apparent, as in the census of 1891, they are reported to be "inferior as they graze sheep and goats and do not engage in cultivation....but in older days they were counted among rulers. They like herding better than farming...they live outside the villages where there is fodder and water for their animals".[45] The debate on their origins was also referred to by Russell, who writes, "owing to their destructive appearance and their exploits as cattle raiders, the origin of Gujars has been subject of much discussion. In north India, Gujars are a pastoral caste and... the saying about them is that the Ahir, Gadaria and Gujar want only waste land for grazing their flocks".[46] Gujar origin myths claim to include Manu and Satrupa, as well as Shiva. One of the myths traces the origin of Gujars to Gaupal Gujar born of Shiva's semen consumed by a cow, as she grazed on grass on which it had been dropped by Narada.[47] However, Gujars in Rajasthan draw their social identity around the worship of their deity, Devnarayan, to whom they dedicate temples in their villages. Devnarayan is seen as an incarnation of Vishnu, and his narrative is recited in the form of a lyrical-visual narrative the *phad*. The narrative of Devnarayan apart from channelizing Gujar devotion towards a vaishnavite tradition also focuses on Gujar martiality that poses a challenge to Sisodiya kings of Mewar. At present even though Gujars are socially, economically and politically dominant as compared to other mobile communities, their social image remains inherently that of an illiterate 'uncultured' community.

Nineteenth century census and settlement reports mention many more pastoral groups. However, most of these groups have become agro-patoralists and no longer practice nomadic pastoralism. One such pastoral community was the Raths in Bikaner region. The Settlement officer of Bikaner P J Fagan calls Raths ruthless, as ".. they cultivate little or no land...and they own in some cases immense herds and it is difficult to gain any satisfactory idea of what the

---

[44] Tambs-Lyche, *Power, Poetry and Profit*, 151.

[45] *Report Mardumshumari*, 44–45.

[46] R V Russell, *The Tribes Castes of Central Provinces of India*, II, Macmillan and Co., London, 1916, 166–170.

[47] Aditya Malik, *Nectar Gaze and Poison Breath*, OUP, New York, 2005, 107.

number of cattle in this part really is. It fluctuates greatly and is dependent in any given year on the state of grass crop and the amount of water available for cattle in the wide natural depressions known as *tebas* or *johurs* or in wells. When grass and water fail the Raths will without demur desert their villages and take their cattle off to Sinde canals or to the Satluj, or in some cases they will go as far as the cis-Satluj districts of the Punjab".[48] The Ahirs were also a pastoral caste, who claimed to have come from Braj. In the 1891 census they were reported to be, "a wild caste ....and their occupation is to herd cows".[49]At present Ahirs are largely a landed community and also practice diary farming. The Bishnois were another agro-pastoral group that were beginning to combine their their pastoral pursuits with agriculture in the late nineteenth century.[50] The Ghosis were a Muslim pastoral group and were largely seen to be engaged in sale of milk and ghee.[51] The Ghanchis were traditionally professional oil millers but also reared cattle and sold milk and ghee."[52]

Pastoralists and Sedentary Communities

With this, I come to the question of the kind of relationships that exist between mobile pastoralists and the sedentary peasants. This relationship can be understood in two ways. The first is through the binary opposition in which mobile and sedentary are placed, and second by looking at these as complimentary parts of a system defined through specific ecological conditions. There are several reasons to look at mobility and sedentarism in binary opposition to each other. It is evident from the discussion above, that the position of the pastoralists in the rural society has remained marginal. The relationship between sedentary peasant communities and pastoralists remains complex. In the hierarchical structure of the society, the sedentary communities have always been upper caste land holding groups like Rajputs, Banias, Brahmins, Jats, whereas the mobile communities are placed lower in the caste hierarchy.

Land-holding peasants and pastoralists were historically subject to differential tax regimes. Migrant pastoralists paid heavy grazing fees in order to use the grazing available in the village on their annual migratory routes.[53]

---

[48] Fagan, *Settlement Report of Bikanir*, 9.

[49] *Report Mardumshumari*, 492–93.

[50] Ibid.

[51] Ibid.

[52] Ibid.

[53] Bhadani, 'Pastoral Sector' 13.

For example in Bikaner, cultivators were "not liable to any demand on account of the unoccupied waste, they might graze their cattle, …..*chaudhries* could not as a rule realize any dues except from outside cattle which came to graze in the village".[54] Fagan lists the rates as being, "levied on cattle brought to graze from outside the state; per cow 4 annas, per buffalo 8 annas, per camel Re 1, per sheep or goat 1 anna".[55]

Besides, the *jods* and *orans*, while considered to be common property of the village were largely controlled by land holding peasants of upper castes. In some instances, the *jods* were maintained through the grants given by the state. In 1771 CE, a *sanad* of Jodhpur states that a worker kept specially for the purpose of growing and taking care of grass in the *jod* in Aakoli, Sanchor was paid Rupees thirteen and he employed twenty to twenty five labourers, which was not sufficient. So it was ordered that he be allowed to employ more men and be paid more for the purpose.[56] In 1774 CE, the *joddar* and *tafdar* of the *jod* of village Lambiya of Merta were ordered to not let cows enter the *jod* till there was grass in sufficient amount. It was also ordered in the same *sanad*, that once there was enough grass, the villagers were to be given grass as available.[57] The disbursal of grass from these *jods* was also done on the orders of state officials as is evident from the same *sanad* that Bharat Singh Pemsinghot of Khamp Udavat was to be given 1000 cartloads of grass every year.[58] The name of the beneficiary suggests that he was a Rajput of high status. It is not known how much share could lower caste *pahis* or migrant pastoralists could get out of the state or community owned grasslands but compared to upper caste peasant proprietors it would have been considerably low. Also, the extent of land ownership would have been an important factor in decision making regarding the village commons thus excluding or marginalizing small landowners from the process.

In the contemporary rural social structure, the Raikas and Gujars rank lower than the landholding Rajputs, Jats and Brahmins as well as the Banias, though higher than the itinerant artisanal groups like Nats and Kalbelias. Therefore, their position in the village is a disadvantageous one. Since their landholdings are nominal, and returns uncertain they are as likely to be trapped in indebtedness as a peasant. Besides, the greater access and control that the

---

[54] Fagan, *Settlement Report of Bikanir*, 45.

[55] Ibid.

[56] *Miti Paush Vadi 12, Somvar, SP Bahi* No 11, VS 1828/1771 CE.

[57] *Miti Bhadrapad Sudi 7, Somvar, SP Bahi* No 14,VS 1831/1774 CE.

[58] Ibid.

upper caste peasantry can exercise over individual and common resources, while excluding the pastoral castes from them, force the nomadic pastoralists more and more into cycles of migration. So, while the migrant pastoral communities fulfill a role that is an essential part of the rural economy, yet their share in decision making regarding the common resources of the village is a nominal one. And, as the resources that they use are common resources, their marginality excludes them from any kind of "functional" decision making regarding those resources.[59]

However, in most arid and semi-arid parts of the Thar, pastoralism and agriculture also complement each other in many ways. For one, peasant and nomad, though defined through categories of community and caste, are not occupationally watertight categories. In this region due to aridity agriculture has historically remained marginal and dependent on infrequent rains. Therefore it has been necessary for the peasants as well to engage in livestock rearing as well as raising small herds of goats and a camel or two. As Fagan points out,

> "it is impossible to form any satisfactory estimate of the average profits derived from camels....the agriculturalist looks (at it) more as a member of his family than a dumb animal. He is used as a domestic beast of burden to carry water, grain etc. and as a pack animal to carry goods for hire: he is harnessed to the plough and is far more useful than a bullock in the sandy soil; he is used as a riding animal and his owner makes fair profit from the sale of the camel's young and his wool".[60]

Besides, peasants could easily turn mobile in case of failure of rains.

Co-existence between pastoralism and agriculture are the best adaptive techniques for optimum resource use. Pastoralists usually arrive in these villages on their routes of migration after the crops are reaped and before the plowing and planting begin. This means that there is a lot of stubble available for the cattle and sheep to feed on for which the pastoralists pay the peasants, apart from manuring the fields with animal dung. The rights of pastoralists to use the resources in this manner have been traditionally accepted because it has been a part of the composite rhythm of the rural society. Apart from

---

[59] This irony is displayed by Arun Agrawal in his article 'I don't need it but you can't have it: Analysing Institutional Conflicts between Farmers and Pastoralists', Pastoral Development Network, 36a, London, Overseas Development Institute, London, 1994, 36–55.

[60] Fagan, *Settlement Report of Bikanir*, 19.

manure, pastoralists historically also supplied pastoral products like cattle, ghee, wool, felt and hides etc. In past, carrier pastoralist communities also doubled as grain and salt traders. Therefore, it appears that the relationships between the nomadic pastoralists and sedentary peasants can be considered to be mutually dependent ones.

Besides, pastoralism historically contributed significantly to the economy of the region. Both peasants and pastoralists paid grazing taxes of several kinds which formed a significant part of the economy of the Thar. In the 17th century, *ghasmari* (a broad grazing tax on cattle feeding on grass and stubble) *pancharai* (on animals feeding on leaves) and *singhoti* (grazing tax on sheep and goats) were realized. Apart from these taxes like *korad* on stalk of dry *moth* or *til*, *bhuraj* etc. were also levied.[61] The collection of these dues was made through village headmen or *chaudhries* or through state officials in the *khalsa* areas.[62] It also appears that another grazing tax which was called *bhunga* was sometimes collected in *jagir* areas in Bikaner though it was not realized from the villagers in *khalsa* villages. It was so due to the absence of any kind of enumeration of cattle in Bikaner. Instead, the state took, "a portion at least of its share of the income of the waste by a house tax".[63]

Table 3.1. Pastoral Taxes in Paragana Merta, 1634 CE[64]

S. no.	Name of the Animal	Rate of Grazing per Animal (in dugani)
1.	Cow	5
2.	Buffalo	10
3.	*Jhote* (Calf Buffalo)	8
4.	*Baratho* (Cow-calf)	4
5.	Goat and Sheep	1
6.	*Jhumpi* (hut)	15

[61] Bhadani, 'Pastoral Sector', 11.

[62] Powlett, *Gazetteer*, 39.

[63] Fagan, *Settlement Report of Bikanir*, 45.

[64] Ibid. Based on Nainsi's *Vigat* B L Bhadani has calculated the *ghasmari* for the paragana Merta in 1634 CE. This also includes the tax on the semi-permanent structures that the pastoralists constructed in the grazing fields called *jhumpis*. On the basis of figure given by Nainsi, Bhadani also estimates the grazing tax on sheep to have been 1 *dam* per sheep in the khalsa villages of Sojhat. In Jodhpur, Jaitaran, Siwana and Sanchor grazing tax per camel was 8 annas. Besides, taxes like *ghora kamal* and *dhumalo* were also realized. Between 1653 and 1662 CE, grazing taxes appear to have formed about 6% of the total income of the Jodhpur state. Out of this about half came from taxes on cows and calves. In Phalodi though, they appear to have been 34.27% by the year 1692 CE.

Apart from the fact that pastoralists paid heavy taxes for grazing, pastoral products circulated widely on the commercial networks and formed an important segment of the commercial economy. The pre-dominance of pastoral products on trade networks is not surprising given the arid ecology of the region. The production of these commodities, however, was dependent on availability of resources like pasture and forage from commons. As long as pastoralism and agrarianism are understood as two complimentary systems and not as water tight categories, it is possible to see agricultural production, fodder production, cattle trade, and trade in As long as pastoralism and agrarianism are understood as two complimentary systems and not as water tight categories, it is possible to see agricultural production, fodder production, cattle trade, and trade in pastoral byproducts like wool, ghee or hides as components of the same precariously balanced system. Throughout the period that this book covers, cattle, ghee, wool, hides and felts and livestock formed the bulk of exports from the Thar region, engaging a number of communities that processed and traded in them. These commodities formed the bulk of exports from the Thar region, engaging a number of communities that processed and traded in them.

There existed a thriving commerce in cattle and livestock, which followed patterns of circulation that extended across the Thar. Local breeds of cows from Bikaner, oxen from Nagaur, horses from Mallani, Kutch and Punjab, camels from Umarkot, Sanchor and Kutch were traded far and wide. Cattle trade itself represented a continuum between agrarian and pastoral communities. Cattle traders on their cycles visited villages supplying villagers with necessary livestock, while buying excess. While preparing the settlement report of Bikaner in 1893 CE, Fagan observed that, "the heifers (*bachri, jhota*) are always kept; steers are kept for three or four years and then sold ungelt for Rs 15 or 20 apiece to travelling traders, or at the Gugano fair at Bahadra. The male buffalo calves (*jhota*) are sold for transport to the Manjha tract of the Punjab, where they fetch about the same or a little less than a steer".[65] In this manner, sedentary villages were incorporated in the circulatory regimes of the itinerant cattle traders, often Banjaras.

Other sites of exchange were the periodic fairs, which in the Thar, more often than not were cattle fairs. For instance, two fairs of Ramdeoji were held in Magh (January-February) and Bhadva (August-September) in all parts of the Thar, but the main fair held in Pokhran in Bhadva was attended by cattle traders from all over the Thar.[66] Another important fair was the Tilwara or Balotra

---

[65] Fagan, *Settlement Report of Bikanir*, 18.

[66] Munshi Hardayal, *Majmui Halaat wa Intijaam Raj Marwar Babat San* 1883–84, *Mutabik Samvat*, 1940, 45.

fair also known as *Chaitri ka mela*, which was attended by livestock dealers of
Marwar and other parts of Rajputana, Gujarat and Sindh.[67] It was an annual
fair, commencing a few days after Holi, and lasted a month and a half and
was attended by about eight thousand people, as noted by Boileau in 1830s.[68]
At this fair, "all sellers of goods and animals were expected to make a small
offering of either money or food at the temple and the general outcry though
out the fair was Jai Mallinathji".[69] The Mundhwa fair at Nagaur, which had
been instituted by Maharaja Bakhat Singh of Jodhpur, was known chiefly for
bullocks, which were brought from all over. It commenced with month of Magh
(January-February) and lasted six weeks; the merchandise of various countries
was exposed and purchased by the merchants of adjoining states.[70] *Tejaji ka mela*
in Parbatsar held in Bhadwa (August-September) likewise attracted commerce
from Jodhpur, Kishengarh and Ajmer. In the fair oxen from all over come to
be sold and merchants from far off come and buy them.[71] In Jaisalmer also
two religious fairs were held, which were attended by cattle traders. The first
of these was celebrated on the last day of the month of Baisakh or April, at
about ten miles from the city of Jaisalmer. The other fair was held in honour
of the pastoral deity Gogaji at the same place, in August or September of each
year.[72] Another fair dedicated to Gogaji was held at Gogaheri, on the ninth of
the Bhadva month.[73] In this fair good quality horses were purchased for royal
contingents like that of Juhar Singh of Ajmer, who bought 12 horses.[74] The
Jodhpur and the Bikaner *darbars* made large purchases at this fair. In 1767
fifty horses were purchased for the *darbar* and the purchase was exempted of
all taxes.[75] A similar exemption was granted to purchases made at the fair of
Kaparda near Jodhpur famous for its Jain shrines, held between *Chaitra Sudi
1 to Chaitra Sudi 15*.[76] The *Eklingji ra Mela* in neighbouring Udaipur state
facilitated good thoroughfare through the Jodhpur state. *Mata Sakambhari Ka*

---

[67] Adams, *The Western Rajputana States*,138.

[68] Boileau, *Narrative*, 115.

[69] Adams, *The Western Rajputana States*, 138.

[70] Tod, *AAR*, II, 129. The Mundhwa fair at Nagaur still remains one of the largest cattle
fairs in western India, where indigenous varieties of cattle is bought and sold.

[71] *Miti Ashadh Vadi 14*, Jodhpur *Kacheri*, *SP Bahi* No 18, VS 1834/1877 CE.

[72] Adams, *The Western Rajputana States*, 140.

[73] Tod, *AAR*, II, 362.

[74] *Miti Margashish Vadi 7*, Pokhran, *SP Bahi* No 6, VS, 1824/1767 CE.

[75] *Miti Chaitra Vadi 7*, Jodhpur, *SP Bahi* No 6, VS 1824/1767 CE.

[76] *Miti Falgun Vadi 6*, Jodhpur, *SP Bahi* No.2, VS 1822/1765 CE.

Mela in Sambhar was also a meeting point for merchants from Jodhpur and Jaipur states. In Mukam, Surpura, a fair was held in the honour of Jambhoji of the Bishnoi community. The trade in cattle was a highly prized one with state collecting dues on sale and transit. For the pastoral nomadic communities these fairs were meeting grounds of sorts where they not only exchanged cattle, but also news and pastoral practices. Since a lot of these fairs were held in honour of deities and heroes worshipped by pastoralists like Tejaji, Pabuji, Mallinath ji, Goga pir, and Ramdeo ji, they also became foci of expression of reverence for these pastoral communities that were considered outside the dominant social discourse.

A byproduct of large scale livestock rearing was manufacture of ghee, an important commodity on the circulatory regime in the Thar. Ghee was manufactured mainly by the Jats, Gujars, Ahirs and Ghosis in the villages. Ghee production was dependent upon rains and availability of forage, which affected milk production. The pricing of ghee was however controlled by whole sale merchants in the large markets. Ghee was also taxed on sale and was listed as a commercial item in *sayer jihati*. In Marwar in the seventeenth century there appears to have been tax called *ghiyayi*, realized in kind, for ghee produced from cow, buffalo and goat milk. For *paragana* Phalodi, these rates were 1 *seer* per cow and 2 *seers* per buffalo and 1 *chatank* per goat.[77] B L Bhadani estimates that on an average 324.7 seers of ghee was realized as tax from Phalodi.[78] Phalodi, Nagaur, Didwana and Sambhar were two large wholesale *mandis* of Ghee from where ghee was sent to far off places. Sale and purchase of Ghee was also subject to commercial taxes like *rahadari, dan, mapa,* and *kayali*. The transit tax on ghee was called *bad dan* in Jalor. In the year 1662, there appears to have been a tax realization of 12 *dams* per *kundi* of ghee, with a *kundi* containing 20 *seers* of ghee. Bhadani has estimated that 650 *maunds* of ghee was manufactured in the *khalsa* villages of *paragana* Jalor in this year, which suggests that the total manufacture of ghee in the entire *paragana* would have been much higher.[79] In Nagaur the state received *dalali* on the sale of ghee.[80] Like as in the case of salt, in Sambhar and Didwana the returns from the transactions in ghee were divided equally between Jodhpur and Jaipur states. A study of *arhsattas* of Didwana reveals that in seven months

---

[77] Bhadani, 'Pastoral Sector', 15.

[78] Ibid., 31.

[79] Bhadani 'Pastoral Sector', 30.

[80] *Miti Magh Sudi, 6,* Nagaur *Sayer, S P Bahi no 6 VS* 1824/1767 CE.

from *Chaitra* to *Ashwin* in the year 1708 CE the Jodhpur state earned rupees 11, 674 from the sale of ghee. A reciprocal amount was received by the Jaipur state. [81]

Table 3.2. Income from Ghee Didwana VS 1765/1708 CE

S.no.	Month	Cart loads	Wt (Approx) in Maunds	Total income In Rupees	Share of the Jodhpur State
1.	Chaitra	45	1270	317	158.5
2.	Vaishakh	20	625	156	78
3.	Vaishakh	185	6247	1561	780.5
4.	Jyeshtha	150	4977	1244	622
5.	Jyeshtha	135	4622	1186	493
6.	Ashadh	17	620	206	103
7.	Ashadh	110	3750	937	468
8.	Sravan	-	1219	443	221
9.	Bhadrapad	-	37687	12558	6279
10.	Ashwin	-	653	201	100.5
11.	Ashwin	-	15883	4358	2269
					11,674

The above estimate is only for a few months in a single *paragana*. So it can be inferred that over whole of Thar, the income from various kinds of taxes levied on ghee would have been very high making it an important commodity in the market. In the nineteenth century ghee was listed as one of the major exports of the Jodhpur and Bikaner states.

Hides and felts were other major pastoral by-products in the Thar. With a large pastoral economy it was inevitable that a good industry in leather should develop in this region. The Bikaner salt agreement of 1877, states that in Bikaner alone about 40,000 skins were tanned in a year. [82] The process of tanning and trade in leather was "carried on in most parts of Bickanir, especially in the northern pergunnahs of Sooratgarh, Sirdargarh and Hanumangurh, and the eastern pergunnahs of Renee, Rajgurh and Soojangaurh and Muggra in the west. Large numbers of cattle graze in these pergunnahs". [83]

---

[81] *Miti Jyeshtha Vadi 4 Arhsatta*, Didwana, VS 1765/1708 CE, *Basta* no.1, Jaipur Records.

[82] Bikaner Salt Agreement, Folio 174–176, 1877 CE, Salt, Bikaner, RSAB..

[83] Ibid.

The community involved with the production of leather was the Meghwals who were also known as *bambhis* or *dheds*. They claim descent from Brahmins and explain that they were ostracized from the community as they consumed the flesh of a calf during a famine. Others claim that they were created by god to carry dead cattle and accept their ostracism.[84] Munshi Hardayal in his 1891 CE census reports that the Meghwals consume the "flesh of dead cattle and tan the hide for the owner ...of which they either give half to the owner or pay for the half."[85]They are followers of Ramdeoji and therefore "there goes a saying in Marwar all that Ramdeoji found for followers are *dheds*".[86] He notes that, "the caste of Bambhis is low but is so necessary that people get them to work without paying and later pay them in kind".[87]Another form of Bambhis are the Raigars, "the difference being that traditionally the Bambhis skin cattle while Raigars tan it".[88] Meghwals in the nineteenth century had taken up weaving but a large majority of them were engaged in leather industry. The tax realized from Meghwals in Jodhpur state was called *siladibab*.[89]

The leather products of the Thar were exported all over. These were shoes, leather water vessels called *chaguls*, leather *hukkas*, camel and horse saddles, and bridles. The skin of sheep and goats was called *naree* while the skin of cows and buffaloes was called *goria*. The cheapest quality of salt obtained from Chappar or Lunkaransar was used to tan these hides. Cow or Buffalo hide could cost Rs 3 to 8 while Sheep or goat hide could cost 1 Rupee to 1 Rupee 8 annas. It was advised in the Bikaner salt agreement 1877 CE that the taxing of salt would lead to the increase in the cost of leather production.[90]

Wool was a byproduct of sheep pastoralism. While sheep and goats were reared for meat, wool also appears to have been an important product in the Thar region. Until the nineteenth century, wool production and use in the Thar was largely a rural and nomadic enterprise. The use of woolen garments was not very popular as the aristocratic class wore quilted cotton garments. Coarse blankets that were manufactured locally either by shepherds themselves or by professional weavers were used by common people. Tirthankar Roy examining wool production and distribution in colonial India, points out that

---

[84] *Report Mardumshumari*, 527.

[85] Ibid.

[86] Ibid. *Ramdeoji ko mile dhed hi dhed.*

[87] Ibid

[88] *Report Mardumshumari*, 540.

[89] *Report Mardumshumari*, 434.

[90] Bikaner Salt Agreement, Folio 174–176, 1877 CE, Salt, Bikaner, RSAB.

in the Thar region only a very coarse variety which was called carpet wool was produced.[91] The finer apparel grade of wool was produced in Tibet and Kashmir. Environmental conditions, indiscriminate inbreeding, coarse feeding and nutrition and long migrations affected the quality of the wool produced. The heat and aridity led to hard dry and frizzy wool, which was almost devoid of natural greases. The coarseness of hair produced a coarser yarn and hence a coarser carpet.[92]

Even today the indigenous system of wool classification is based on the processing criteria, the basic distinction being between washed and unwashed wool.[93] The production and circulation of wool involves three stages. In smaller quantities it is processed and woven by the shepherds themselves who sell their coarse produce on their routes. When the quantities of wool are large, the professional shearers are employed who are called *katariyas* who are generally Muslim Khatiks. They move about the region in shearing seasons in bands and move from one flock to another. These communities of Khatiks, who as the name suggests are butchers professionally. The Marwar census reports them having taken up wool weaving and, "they colour the wool black and red and make rugs etc from wool. They also make the *dhavalas* worn by Jat women and beautiful *bandanwars* of colourful wool".[94] The Khatiks "are often itinerant unlike the Bania, they manage to attract clientele from more remote villages, although a village may be part of the catchment of more than one trader".[95] The sale of wool is carried out through collective mechanisms with the *nambardar* of the group negotiating the price.[96] In case the wool is sheared on the migratory route then the shepherd settles his flock in an empty field after negotiating with the peasant, who allows this in return for his fields being manured. After the wool is sheared it is taken into the *mandis* and then exported to Punjab.

---

[91] Tirthankar Roy, 'Changes in Wool Production and Use in Colonial India', Gokhle Institute, 2002, 4.

[92] Ibid., 7

[93] Purnendu Kavoori explains that the unwashed wool is called *bina dhoh*, while there may be varieties of washed wool depending upon the water source at which it is washed, like pitcher, ditch, gutter, pond or river. Besides Chokhla quality of sheep wool is never washed while the Jaisalmeri cannot be sheared without a good wash as it is tough coarse and gritty. Kavoori, *Pastoralism in Expansion*, 119.

[94] *Report Mardumshumari*, 541.

[95] Kavoori, *Pastoralism in Expansion*, 117–18.

[96] Arun Agrawal, 'Indigenous decision making and hierarchy in migrating pastoral collectives: The Raika of Western India' in *Nomadism in South Asia*, 417–447.

In the seventeenth century there appear to have been two taxes on wool production, *ghora kambal* and *serino*. Nainsi provides with the figures of income in these heads from four paraganas in six years.

Table 3.3. Tax Realization from the Manufacture of Wool 1657–63 CE (figures in rupees)[97]

Paragana	Jama Asal	1657–58	1658–59	1659–60	1660–61	1661–62	1662–63
Jodhpur	571	488	491	494	494	500	397
Merta	7794	5101	452	5674	6015	6464	
Sojhat	374	122	133	134	134	133	133
Siwana	264	165	268	268	268	268	

As we can see the greatest amount of realization was done from Merta, which was strategically located near Bikaner which was a big wool *mandi* and Nagaur, which was a centre for manufacture of blankets. In 1767, we find merchants coming into these *mandis* to sell wool, like one Kaliram, who sold 12 bales of wool in the *mandi* of Phalodi.[98] We do not however get the idea of the amount of wool per bale, or whether the seller was a shepherd, shearer or merchant. He was however granted exemption in the *rahadari* in Phalodi.[99] Various kinds of blankets were manufactured in Nagaur that were either of pure wool or mixed with cotton. In 1783, a merchant from Gujarat ordered blankets from Nagaur with specification that "100 blankets should be ones with no cotton thread....and the rest 400 should be with cotton threads".[100] The rate of blankets was rupees ten per pair and they were manufactured in the *karkhanas* owned by the Jodhpur state, as well as by the individual weavers. The *karkhanas* purchased wool from open market as in 1781, when the Jodhpur state ordered the purchase of wool from Nagaur *mandi*, for which the money was to be paid from Nagaur *sayer*.[101] Though the rate of taxes extracted from merchants and manufacturers on wool and wool products are not available for the eighteenth century, we do get evidences of the exemptions granted to merchants of Nagaur on sale and purchase of wool.[102] The blankets or *looes*,

---

[97] Bhadani, 'Pastoral sector', 34–35.
[98] *Miti Ashwin Vadi 14, Phalodi Sayer, SP Bahi* No 6, VS 1824/1767 CE.
[99] Ibid.
[100] *Miti Margshish Sudi 7, Ravivar,* Nagaur, *SP Bahi* No 29, VS 1840/1783 CE.
[101] *Miti Magh Vadi 7,* Nagaur Sayer, *SP Bahi* No 28, VS 1838/1781CE.
[102] *Miti Kartik Sudi 12,* Nagaur, *SP Bahi* No 2, VS 1822/1765 CE.

thus manufactured were exported from Marwar, to areas like Gujarat, Sindh, Punjab and Agra. In fact, Tod remarks that blankets were one of the very few items exported from Marwar.[103] Wool manufacture and trade was thus an important part of the economy of the Thar involving shepherds, shearers, agents, carriers, manufacturers and traders, often in multiple capacities.

Traders and Carriers: The Commerce of Circulation

A dominant image that is formed when we refer to the Thar is of the Marwari trader. But the mercantile groups that operated within the Thar were numerous and their movements on the trading networks were in accordance with their specific roles in the processes of the market. The category of trader in the region could include the village *bania* or *sahukar*, itinerant traders and carriers like Banjaras, Charans, Lowanas, Gadeets, Bohras, Bhils or at times even Brahmins, as well as the 'Marwari' traders who appear to have been based in the market towns.

The Marwari was the most ubiquitous figure on the early modern commercial networks. Marwari is by no means an unsegmented category, but composed of many distinct trading castes involving social and economic hierarchies. The Mahajans or Banias were prominent traders of Marwar who participated in long distance and local trade, often respected, envied and derided at the same time for their business acumen as well as their shrewdness.[104] Tod surmises that, "nine tenths of the bankers and commercial men of India are the natives of Marrodes and these chiefly of Jain faith. The laity of Khartra sect send forth thousands to all parts of India and the Oswals so termed from the town of Osi, near the Looni, estimate one hundred thousand families whose occupation is commerce".[105] Their status in Marwar was perceived by Munshi Hardayal to be very high as, "they live in Calcutta, Bombay, Madras and Hyderabad…where they have separate bazaars called Marwar Bazaars……they are influential in Marwar as they work for the *ahalkars* and *jagirdars* and also because they are into moneylending.….peasants are indebted to them.….especially in west as the important *mutsaddis* and *musahibs* of the Raj are *Mahajans* and they lend money".[106] One gets an idea of their importance from Adams' observation

---

[103] Tod, *AAR*, II, 127.

[104] The honorific name of the Marwari traders was *Mahajan, Seth* or *Sah* while the pejorative one was *Kirad* or *Leda. Report Mardumshumari*, 420.

[105] Tod, *AAR*, II, 127.

[106] *Report Mardumshumari*, 420–421.

"much of the prosperity of Marwar, is due to the young Banias (dealers) born in the country, who on attaining manhood, go all over India and beyond its borders to engage in trade".[107] Jain Mahajans outnumber the Vaishnavs with Maheshwaris and Agrawal being prominent in latter.[108] Among Jains Oswals, Porwals, and Saravagis are prominent, though Srimals, Srisrimals, Bagherwals and Khandelwals are also found.[109] While Oswals could be found in "all parts of Marwar...Porwals were mostly confined to the paraganas of Bali, Jaitaran or and Sanchor....and Saravagis were found mostly in Maroth, Nawa, Sambhar, Parbatsar and Didwana".[110] Though there were distinctions between various *nyats*, they were bound by mores of social exchange like marriage that also took place between the Jains and the Vaishnavs.[111]

Even though the markets appear to be dominated by the *Banias*, it was actually a complex conglomerate of varied groups acting in varying capacities. Acting as receivers for grain or commodities like ghee collected as agrarian or commercial revenue, it was the traders who were in a position to supply much required wealth to the rulers. Therefore, they were intricately linked with both, the revenue networks as well as the commercial networks. Madhavi Bajekal points out in the case of eighteenth century Eastern Rajasthan that the merchants were forced to buy the grain collected as revenue or pay a tax called *parna* in lieu of doing so.[112] The most prosperous of merchants were wholesalers known as *kothiwals*, who controlled commodity trade at all levels.[113] The *kothiwals* had their shops and agents called *dalals* or *arhattiyas* in practically every town and village and dealt with a number of commodities. Below these *kothiwals* in the hierarchy were the *pattiwals*, retailers who conducted small time retail trade either on a permanent basis or in the weekly markets. A group of Muslim traders called *Turkia Bohras* were another group who had shops in the market towns They claimed to have "come from Siddhapurpatan in Gujarat... and were engaged in retail trade of *kirana* and items of European manufacture... and were petty moneylenders also".[114] Their shops were largely located in Jodhpur, Pali, Bhinmal and Jaswantpura.

---

[107] Adams, *The Western Rajputana States*, 437.

[108] *Report Mardumshumari*, 421

[109] Ibid.

[110] Ibid.

[111] Ibid.

[112] Madhavi Bajekal, 'The State and Rural Grain Market in Eighteenth Century Eastern Rajasthan', *IESHR*, 25(4), 1988, 443–473.

[113] Gupta, *Trade and Commerce in Rajasthan*, 138.

[114] *Report Mardumshumari*, 441.

The *bisatis* spread their ware on the pavement, their name derived from *bisat* or spread.[115] According to the census of 1891, the *bisatis* claimed to have been Saiyyads who were earlier dealers in horses and were called *saudagars*. They were called *bichchayatis* in the same context and they had shifted to selling *mal maniyari* which usually included glass and metal ware, cutlery etc.[116] In the nineteenth century the *bisatis* also took to selling items of European manufacture like buttons, needles and tapes in the fairs.[117] Another category was of the *pheriwals* who, as the name suggests, went on their *pheris* or rounds across the land. These were peddlers and hawkers, the merchants who went from village to village, selling and buying.

Apart from merchants there existed groups involved in the carriage of goods. These could be termed as the mobile traders of Marwar. The *Baldiyas* also known as *Banjaras* with their *tandas* or caravans of oxen were engaged largely in the transport of salt and grain. Their "tandas could include 40,000 oxen at a time".[118] Nomadic in nature their "houses were on bullocks and they led a wandering life with their families…".[119] The term Banjara originated from *banaj* which referred to trade. In popular parlance it is usually taken to mean a wanderer. They were honorifically called *nayaks* and pejoratively called *bigadu* as they were accused of despoiling crops on their way. They were suspected of stealing oxen and children on their way and selling them in far off places.[120] At present they are a diversified group and at times all mobile communities are clubbed together under this umbrella term. Historically, Banjaras were a numerically large community traversing a greater geographic area.[121] Large communities of Banjaras exist in Rajasthan and other parts of India particularly Andhra Pradesh, but they are no longer grain, salt or cattle traders. Instead, they make and sell hand embroidered garments, where perhaps the memory of bygone times is depicted in the motifs. The Banjaras were one of the largest communities that lost their livelihood when railways transformed the networks of commercial exchange. Though they are a very marginal community today, they are nevertheless engaged in circulation networks by their role in the trade in embroidered fabrics.

---

[115] Ibid.

[116] Report on Marwar Customs and Tariffs, 1895, 7, Customs, Jodhpur Records, RSA.

[117] W. Crooke, *The Tribes and Castes of North Western Provinces and Oudh*, Vol II, Calcutta, 1896, 115.

[118] Tod, *AAR*, II, 133.

[119] *Report Mardumshumari*, 444.

[120] *Report Mardumshumari*, 444–445.

[121] Radhakrishna, *Dishonoured by History*, 24.

Lowanas and Gawarias were two other groups engaged in similar trades. Of Lowanas, who lived in Pachpadra, Mallani and Sanchor the Report Mardumshumari, 1891 reports......."come from Sindh...their occupation is to carry merchandise on camels. Earlier when their were no rails in Marwar, they used to carry salt to Malwa, Gujarat and Sindh and brought back grain, rice and coconut but now because of rails their position has declined".[122] In 1837 the town of Balotra had a population of 625 Lowanas.[123] The Gawarias claim to "have come from Amritsar and Patiala...and some of have taken to carrying goods along with the Banjaras and lead a wandering life".[124] Bhils in Mallani were largely engaged in the carriage of salt from Pachpadra salt works. Brahmins of Barmer were involved in the trade of rice, sugar and cloth from Sindh. They had several hundred camels and, "brought the produce of Sindh from the town of Jeysulmer and extended their traffic as far as Guzerat and Kattywar".[125] Joshis who were Pushkarna Brahmins originally from Sindh and whose occupation was trade, took, "ghee, and gum to Gujarat, Jodhpur, Nayanagar and to Bhiwani and brought back goor, khand, coconuts and betelnuts".[126]

Another group that was engaged in caravan trade in the late eighteenth Century was the Charans who traditionally being cattle herders and traders were also engaged with carriage and commerce in this period along with their other role of bards.[127] While at present Charans in Gujarat are pastoralists, the Charans in Rajasthan are largely cultivators. The name Charan "generally held to mean a wanderer" was associated with graziers, cattle settlers and grain traders.[128] Charans in various parts of the Thar were found to be traders in grain and livestock and were closely identified with Banjaras. Tod referring to the general nature of carriage "identified Charans with the Banjaras, using the name alternatively".[129] They "may have united the breeding of cattle to

---

[122] *Report Mardumshumari*, 446.

[123] Boileau, *Narrative*, 225.

[124] *Report Mardumshumari*, 581.

[125] Walters, *Gazetteer of Marwar and Mallani*, 275.

[126] Ibid.

[127] Charans were a community with multiple roles. They were recipients of *sasan* land grants and were negotiators in affairs of Rajputs like marriages and disputes and were engaged to guard caravans and palaces. They also were poets and composers as well as genealogists. Apart from being engaged in carriage they also undertook trading and moneylending. These other roles of Charans have been discussed further in the work.

[128] Russell, *The Tribes Castes of Central Provinces*, II, 164.

[129] Ibid., referring to Tod. Also Tod, *AAR*, II, 501.

their calling of bard".[130] In course of time, "the carriers restricted themselves to their new profession....splitting off from Charans and forming the caste of Banjaras".[131] In Mallani, Charans were found to be "large traders in live and dead stock...possessed great privileges as a sacred race being exempted from local dues throughout Marwar".[132]Their familiarity with the desert terrain made them the protectors of caravans travelling across the formidable terrain. Charans, according to James Tod, were "bunjarees by profession, though poets by birth. It was the sanctity of their office which converted our 'burdais' into 'bunjarees' for their persons being sacred, the immunity extended likewise to their goods, and saved them from all imposts; so that in process of time they became the free traders of Rajputana".[133] The traditional role of Charans on the trade routes in the Thar was primarily that of protectors of caravans. Even in times and areas where caravans were routinely looted, Charans were seen as, "great traders...who...paid no dues and in troubled times when plunder was rife...although trading with thousands of rupees worth of property were never molested".[134]Tod saw them as guardians of caravans through desolate regions from Jalor, Bhinmal, Sanchor towards Gujarat, who could overawe lawless Rajputs, Bhils and Kolis.[135] When the caravans were attacked the Charans threatened the robbers with self immolation and sometimes the immolation of a whole body of women and children, for whose blood the marauder was declared responsible.[136] Charans were able to utilize their advantageous position and since they had "exemption from perpetual and harassing imposts...they gradually became chief carriers and traders".[137]

Charan origin myths also associate Charans with Shiva, as they claim to have been created by him to graze the bull Nandi. Charans claims that they were created to herd the celestial bull Nandi, after the Bhats could not graze him as they were scared of the lion of the Goddess Parvati. A myth recorded by Russell in *Tribes and Castes of Central India* traces Charan migrations along with Rajput migrations from north and central India.[138] This version matches the one available in Enthoven's *Tribes and Castes of Bombay Presidency*, according to

---

[130] Ibid., 258.
[131] Ibid., 259.
[132] FP October 1868, (69–80).
[133] Tod, *AAR*, II, 500.
[134] Walters, *Gazetteer of Marwar and Mallani*, 277.
[135] Tod, *AAR*, I, 554.
[136] *Report Mardumshumari*, 343.
[137] Russell, *Tribes Castes of Central Provinces*, II, 258.
[138] Ibid., 162.

which Charans are said to have "accompanied the Kshatriyas in their southward flight and took charge of carrying of supplies".[139] However, in the description provided in the Marwar census, 1891, they claim a divine origin along with the Brahmins.[140] Charan migratory history traces their movements between Baluchistan, Jaisalmer, Marwar, Gujarat and Kutch. These movements can also be traced through the location of Charani Devi temples on the migratory routes. Charani goddess myths, like that of Hinglaj, Avad, Khodiyar, Karni are spread all over the Thar, which in particular underline the association of these goddesses with protection to travellers on desert routes. Charans are believed to have three and a half subdivisions, Maru, Kachela, Sorathia and Tumbel, which maintain ties of brotherhood with each other.[141]

Among the trader-carrier groups another prominent group were the Gadeet. They belonged to Nagaur and were Muslims. They "were soldiers but later turned to carrying merchandise on oxcarts. Their carts have become useless since the railways came. Earlier, when the merchandise was offloaded from ships in Bhavnagar and Ghogha Bandar they brought the merchandise to Pali, from where it was sent to different places".[142]

The groups involved with carriage brought together the pastoral and commercial circuits that coalesced in periodic fairs and market towns. These circuits were sites of exchange of a wide range of commodities, which were either produced in the Thar itself or were imported from neighboring regions. The deficient areas of western Marwar along with Jaisalmer and parts of Bikaner were supplied food grains and other agrarian products from at least three directions. *Sayer* records of Marwar and Bikaner indicate movements of merchants and commodities in different directions. From Bali-Godwar belt a westward movement of grain was seen[143] while from Sindh the grain moved eastwards towards Jaisalmer, Sheo, Mallani etc.[144] The third span covered was from the fertile Punjab towards Bikaner.[145] A route for grain often mentioned was from Jaipur to Bikaner through Nagaur. The *mandis* of Rajgarh, Nohar and Churu in Bikaner are reported to have received wheat from Godwar,[146] while *kirana*, sugar, jaggery were reported to come into Marwar

---

[139] Enthoven, *Tribes and Castes of Bombay Presidency*, I, 273.

[140] *Report Mardumshumari*, 330–33.

[141] Tambs-Lyche, *Power, Profit and Poetry*, 190–91.

[142] *Report Mardumshumari*, 74.

[143] *Miti Sravan Sudi 6, SP Bahi No. 13*, VS 1830/1773 CE.

[144] *SP Bahi* No: 47, VS 1852/1795 CE, *Kagad Umarkot ro.*

[145] *Magre ri Khari Patti ri Zagat Bahi*, VS 1858/1801 CE.

[146] *Rajgarh ri Zagat Bahi*, VS 1858/1801 CE.

from Jaipur.[147] A good incoming trade existed with Gujarat in cloth including cotton, silk, and brocades.[148] The same record mentions import of ivory and saffron. Other imports from Gujarat were paper and coconut.[149] The southern frontiers of Marwar were the main inlet for the entry of opium, which came from Harauti, Kota and Udaipur mainly through Sirohi.[150] Opium covered a long route within Marwar to reach Pali, where was practiced an old trade of processing opium, its drying and packaging before further transit. Opium had a good consumption within the Thar too and an inferior quality opium was cultivated in Godwar area, mainly for local consumption. Around Sojhat and Jodhpur there were stone quarries that supplied building materials to areas around them. Jodhpur was also a centre for dying and printing of cloth from where such cloth was exported all over. Sojhat had lead mines that were a good source of income for Jodhpur state. Makrana near Parbatsar was famous for its marble that has been used widely in Mughal architecture.[151] Nagaur was famous for its brassware and utensils manufactured in Nagaur were quite in demand. The Jodhpur state too is reported to have bought brassware from the market of Nagaur many times.[152] The circulation of these commodities took place on networks of circulation described in the first chapter. The trade routes were sites of struggle and contestation between the state and the merchants, often passing through desolate stretches controlled by *thikanadars* or highwaymen. These routes were sources of considerable commercial income for the Rajput states. The *sayer thanas* were check posts to keep an account of traffic and realize the commercial duties. For the merchant they were mechanisms, both for extraction and redress. Networks of commercial circulation underwent drastic transformations in the nineteenth century with the construction of railway lines as will be discussed in the next chapter.

Itinerant Menial Artisanal Groups

Apart from groups involved in trading, there exist several 'service communities' that occupy the fringes of the rural society. They roam from village to village providing specialized services that only they can. In return, in past they were

---

[147] *Miti Magh Vadi 7*, Nagaur *Sayer, SP Bahi* No: 29, VS 1840/1783 CE.

[148] *Vyapariyon Kana Su Mal Utrayio Jin Ro Mel Ri Bahi*, Sanchor, VS 1857/1800 CE.

[149] *Jalor Sayer Bahi* VS 1861–63/1804–1806 CE.

[150] Ibid.

[151] *Majmui Haalaat wa Intijam*, 1883–84, 66.

[152] *Miti Sravan Vadi 4*, Nagaur *Sayer, SP Bahi* No 18, VS 1834/1777 CE.

either paid in cash or kind, but nowadays services are rendered invariably in return for cash. They set up camps outside the village and move to another village after a while, thus undertaking rounds of villages in this manner. Generally, these circuits of movements as well as patron-client relationships are old and repetitive. Joseph Berland identifies several groups as nomadic service communities in Pakistan under the category of *paryatan*, with overlapping sub-categories of artisans and entertainers. These include smiths of several kinds, baskets and broom makers, bangle and jewelry peddlers, tinkers, dancers, snake charmers, acrobats, prostitutes, jugglers, acrobats, magicians, bards and genealogists.[153] The peregrinations of such have been closely bound to the specific needs of communities that they service. For craftsmen like smiths, basket makers, sieve makers it could be preparatory time when the fields are being prepared for planting, or the time of harvest. For communities of entertainers, magicians, and jugglers it could be around the time of harvest and marriages. According to Joseph Berland while peripatetic artisanal groups prefer regularity of circuits to maintain regular relations with their client groups, communities of nomadic entertainers prefer the novelty factor.[154] Jaya Menon points out mobile groups have also sought peripatetic or sedentary artisanal service groups, particularly smiths, depending upon their needs on their own ambulatory circuits.[155] The mobility of itinerant service groups depends on location of their patrons, availability of raw materials, portability of tools etc. The mobile service groups are often bi or multilingual, depending upon the groups that they service. While most service groups claim to provide specialized niche services, they often need to evolve proficiency in more than one kind of service. While the nomadism of these service groups evokes mistrust, they are often also called in to provide ritual services, particularly for healing, both in case of humans as well as cattle.

In the Thar there are many such groups, though there exist no documented histories of these groups. The colonial ethnographic accounts that begin recording their presence in the nineteenth century invariably equate their mobility with criminality. Among the groups that were notified under the Criminal Tribes Act of 1871, the greatest number was of service groups like

---

[153] Joseph Berland, 'Servicing the ordinary folk: Peripatetic peoples and their niche in South Asia', in *Nomadism in South Asia*, 105–124.

[154] Ibid.

[155] Jaya Menon, 'Mobility and Craft Production' in *Mobile and Marginalized Peoples: Perspectives from the Past*, (Eds.) Rudolf C. Heredia and Shireen Ratnagar, Manohar, 2003, 89–121.

Nats, Kanjars and Kalbelias. The 1891 Marwar census records some of these groups as follows.

The Kalbeliyas, who are snake charmers are considered to have special powers and are regarded as the priests of snakes. They "carry their makeshift snake shrines to local neighbourhood, while offerings of milk and donation are made by the households".[156] In nineteenth century they were considered to be "*khanabadosh* (nomadic), who wander and live in the forests......they carry snakes in baskets and play the *pungi*.....some of them sell stone mills...they also know handicraft and only they can make the brushes for cleaning pit looms used by weavers...they go from village to village to repair these looms".[157] The Ganwarias were seen to make "ropes out of flax used by peasants and combs out of buffalo horns. They camp outside the villages and the women go to sell these combs".[158] The Sansis, "steal asses, oxen, goats and cows and sell them in far off markets....they also steal crops from the fields...but they make excellent ropes...better than the Ganwariyas".[159] The Nats always "keep always keep a rope and a drum with them...they climb on ropes and perform acrobatics for people.....they go from village to village.....and do not stay at one place...their houses are called *deras*".[160] The Ods "dig wells and tanks.... and the tale of Jasma Odan is a famous one".[161] They claim that they cannot drink water from a well twice. The *Gaduliya Luhars* are travelling blacksmiths. The name "Gaduliya Luhar is derived from their *gadis* referring to the open bullock carts in which they travel and live. To be a Luhar one must have been born one- when camped at the roadside – in or under the ox-cart".[162] They "always roam about in their carts ....and sometimes make huts of the most temporary nature outside a village".[163] They claim to be Rajputs and that they took a wandering life after the defeat of the Sisodiyas. They make and repair iron implements and "at times gelt the oxen, for which they are confused with the Santiyas. One speciality of this community is that the Gaduliya Luhar

---

156 Rustom Bharucha, *Oral history of Rajasthan: Conversations with Komal Kothari*, Penguin Books, New Delhi, 2003, 52.

157 *Report Mardumshumari*, 247–48.

158 Ibid.,581.

159 Ibid.,585.

160 Ibid.392.

161 Ibid., 502.

162 David T. Phillips, *Peoples on the Move, Introducing the Nomads of the World*, Piquant, UK, 2001, 381.

163 *Report Mardumshumari*, 466.

women too roam with the men, so in this way they are truly nomadic".[164]The Thories are a "clan of Bhils"…they "were Rajputs but were ostracized as they consumed a dead camel".[165]They were associated with Pabuji and the Thories who go about singing the *phad* of Pabuji are called *bhopas*.[166] The Baridars are the "carriers of news of Royal birth in Marwar….. they carry a red cloth imprinted with the feet marks of the newly born prince and go to the villages for which they are give a *neg*".[167]

Then there were several itinerant hill communities that lived in the Godwar hills. They traditionally had been warriors and had been pushed out of their areas towards the fringes like the Bhils, the Meenas or the Mers. The Bagris were employed for guarding the fields at night.[168] The Baories hunted with the help of nets called *bawars* but they were reported to be engaged in cattle theft and dacoity in the nineteenth century.[169]They were usually hunting and gathering communities and they sometimes supplied firewood to the villages and towns. Mostly they were itinerant labourers, receding to the hilly tracts where they were untraceable. These communities remained a matter of "concern" for administration and efforts were made to settle them as would be seen in the next chapter.

These artisanal communities were small in numbers and were ranked even below the pastoralists. They lived on the fringes of the settled society and as they could not be taxed they failed to be mentioned in any kind of state records until the nineteenth century. The services provided by these communities may appear insignificant but the rural society was composed of such small and marginal communities, whose services were considered irreplaceable. They had customary roles in the society and whether the state records mentioned them or not, they were a part of the popular understanding of the region and formed the complex systems as we would see in the next section. Tragically with the increased blurring of the specialized social roles, these communities of the fringe lost their specified roles and were criminalized by a system that placed a greater premium on sedentarism. So in the nineteenth century, efforts were made to settle these communities agriculturally.

---

[164] Ibid., 463–64.
[165] Ibid., 564.
[166] Ibid., 576.
[167] Ibid., 514.
[168] Ibid., 562.
[169] Ibid., 561.

Bardic and Genealogist Communities

So far we have observed that there existed in the Thar Desert several communities moved from one place to another in accordance with their occupational demand. We can see that the mobility of these communities makes them both accessible and inaccessible at different times. Thus, the task of remembering and transmitting their histories, experiences and narratives fell upon the various bardic communities of Rajasthan. These include the celebrated Charans, various kinds of Bhats and traditional folk singers of various sorts that were often associated with a single community.[170]

In the Thar region, traditionally every community had its own associate communities that would include its genealogists and singers. The task of these associated communities was to preserve the narratives of their patron communities, as memory was a valuable asset among mobile cultures. These communities operated at multiple levels providing the same services to a community that it received from another. This patron-client relationship existed on two levels, one where each community had an associated genealogist community that would periodically visit and memorize the births, marriages and deaths in that community since the last visit. The other patron client relation was between a community and their singer or the Manganiyar. These communities sang for their patrons at festive occasions. They would remember songs referring to those particular communities and transmit them orally to their next generation. Interestingly, even in the condition when the singers gave up their particular profession they were still entitled to their traditional share as decided upon.[171]

The Charans are the most well known bardic community of the Thar though, "they were found in Marwar, in greatest numbers". [172] There are two distinct lines of Charans based on regional differences, the Maru and the Kachchela. Due to their practice of demanding their traditional dues at the gates of a house they were also known as Barhats or Polpats. The genealogists for Charans were Brahmins from Ujjain who periodically inscribed their genealogies in their accounts. The bards of Charans were called Motisars, but

---

[170] The role of Charans in the Rajput social spectrum and the marginalization of their narratives as well as of Charans as a community in the nineteenth century would be discussed in the fifth chapter.

[171] Rustam Bharucha, *Rajasthan: An Oral History, 219*. He cites the example of *Langas* who gave up singing under the influence of Pir of Pagaro, but were still entitled to the *neg*.

[172] *Report Mardumshumari*, 346–47.

they were not genealogists. Instead they wrote poetry in honour of Charans as Charans did for Rajputs. The Motisars like Charans left their villages after Dussehra and returned only after four to six months.[173] Another client caste of the Charans were the Bhands, who sang and danced for the Charans on festive occasions. Their particular form of dance is still popular and called *rammat*.[174]

The Bhats were the true genealogical community. They claim to have originated from a Brahmin father and Shudra mother. In contemporary caste hierarchy they are placed lower than the Charans.[175] They kept accounts of all social groups with each group boasting of its own genealogist. Genealogies were memorized by the Bhat of each community, who would visit the village once in several years and recite the genealogy and add births, marriages and deaths to his phenomenal memory. The process of recitation of the genealogies attained an element of ceremonialism evident even today in the Thar. In return, the genealogist would be given grain, cattle or money. No community could dare to annoy a genealogist lest they refuse to keep genealogical accounts, in absence of which all memory of that family could be lost. There are two kinds of Bhats, the *pothi* or *bahi bancha*, who recite from a written account and the *mukh bancha* who recite from memory. If there was ever a contestation regarding generations, especially in case of marriages, the Bhats of the community were referred to. The Bhats and Charans received rent free lands that were called *sasan* grants.[176] There were several other communities of mixed origins that acted as genealogists of groups other than Rajputs, and were collectively called Bridhari Bhats.[177] The Rawuls, Sevags, Doms, Pancholis etc. were such communities that roamed around the villages singing and begging, and sometimes claiming to keep genealogical accounts.[178]

Then there were various musician castes in the Thar that are reported to be classified under the category of Dholis, Dhadis and Mirasis in Munshi Hardayal's report. The Dholis played drums, *nagaras* and *damamas* to announce special occasions.[179] The Dhadhis played *sarangi, rawanhatthas* and *rababs*.[180] The Mirasis are Muslims and they have traditionally sung a particular

---

[173] Ibid., 351.

[174] Ibid., 355.

[175] Rustom Bharucha, *Rajasthan: An Oral History*, 29.

[176] *Report Mardumshumari*, 334, 356.

[177] Russell, *Tribes Castes of the Central Provinces*, II, 165.

[178] *Report Mardumshumari*, 320, 252, 363.

[179] Ibid., 364.

[180] Ibid., 369.

type of song popularised by the name of *Khayal*.[181] They also serve as bards for
Muslim communities. The Manganiyars are also Muslims and they live around
Sheo, Mallani and Jaisalmer. They play a type of broad *shahnai* and *khartal*.[182]
Popular tales replete with instances with of love, war, fantasy, mysticism and
spiritualism were circulated through networks of social gathering like fairs by
groups like Manganiyars, Bhopas and Mirasis. Their music was evocative of
loneliness and separation. There was also a religious form of music popularized
by the groups like the Bhopas who are the priests of Pabuji. The Bhopas are
"itinerant, and they carry their temple about with them. The Rebarees are
semi-nomadic and not in a position to visit a temple in a fixed spot. So instead
the temple visits the worshippers."[183]The musicians of Kamad caste travel with
groups of dancers called the *teratalans*. They tie thirteen pairs of cymbals on
their bodies and dance in reverence to Ramdeoji.[184]

It is thus clear, that a significant section of the population of the Thar were
peripatetic communities. The groups mentioned were itinerant in nature and
their primary concern was with the circulation of several types on various
distributive networks. While these mobilities of pastoralists, traders, artisans
and bards appear distinct from each other, they were in fact very closely linked.
Pastoralists like Raikas were also cattle and wool traders, Banjaras were cattle,
grain and salt traders, Charans were cattle herders, traders and bards, Kalbelias,
Nats, Sansis and Kanjars were artisanal groups, peddlers as well as healers.
Thus all these groups were connected through the mobile circuits on which
they operated as well as through the commodities they supplied and services
they rendered. Mobility need not refer only to long distance movements. Groups
like Kalbelias or Ghattiwals performing seemingly insignificant services were
nevertheless very important for the rural society, as they performed specialized
tasks that were necessary for the society. Even if the distance that peddlers or
service providers covered was a small one, from their own village to villages
nearby they still were creating a network of circulation, whereby they exchanged
their particular skills for cash or commodities. They, in this manner were
not very different from the traders or carriers who brought distant parts of
the region together through their transactions. Then, there were the bardic
communities that created a network of lore that they exchanged from village to
village. On the other hand, despite complex and contested relationships with

---

[181] Ibid., 371.
[182] Rustom Bharucha, *Rajasthan: An Oral History*, 223.
[183] John D Smith, *The Epic of Pabuji*, Katha Books, New Delhi, 1991,15.
[184] Rustom Bharucha, *Rajasthan: An Oral History*, 181–182.

the sedentary peasants, the mobile pastoralists too were tied to them through networks of exchange. It was the sedentary communities that took over the task of maintaining the village pastures and *orans* when the pastoralists were gone in return for the pastoralists' herding their cattle and producing much required pastoral by-products. As Chetan Singh points out, "the relationship between the two was not simply one of outright confrontation. In many parts of Mughal India, the lives of peasants and pastoralists were inextricably entwined".[185] Also, while the discourse of settlement posited sedentary and mobile as two unconnected and rather oppositional regimes, the fact is that survival of one kind of regime was difficult without the other. Mobile groups linked settlements, sedentarism provided the basis for mobility to survive. As long as mobility and sedentarism were understood as two ends of a dynamic system, they did not necessarily appear as oppositional forces. As Khazanov argues, mobility and sedentarism are connected through a "continuum of specific and flexible economic strategies with an almost indefinite range of variations".[186]

Interestingly, most of these communities claim to have a Rajput past or Rajput association. The names of sub-clans within these groups are often similar to the Rajput groups. While in case of a community like Raika, marriage to Rajputs becomes the way to seek Rajputhood, for Charans inter-dining with Rajputs becomes a way of explaining loss of Brahmin caste. Communities like Bhils, Minas, Sansis, Kalbelias, invariably have a Rajput narrative in which loss of Rajput caste status is explained through either consumption of proscribed food, mostly cow meat in times of distress like famines, or through contact with dead cattle. As I have pointed out in the previous chapter that narratives of a shared past emerge not only from seeking a higher position from the present location of these groups but also from a belief in shared histories. The emergence of an endogamous Rajput caste as the core resulted in the marginalization of all other communities that shared a claim to Rajput *status* if not caste. Interestingly some of the Bania, and particularly Jain communities also claim a pre-Vaishya Rajput identity. For instance a Maheshwari origin myth traces its descent from a Chauhan Rajput who lost his kingship as he failed to honour Brahmins.[187]

---

[185] Chetan Singh, 'Forests Pastoralists and Agrarian society', in *Nature, Culture, Imperialism*, 34.

[186] Anatoly Khazanov, *Nomads and the Sedentary World*, Introduction to Second Edition, xxxii.

[187] Lawrence Babb, *Alchemies of Violence: Myths of Identity and Life of Trade in Western India*, Sage, New Delhi, 2004, 108–9.

The origin myth of the Oswals is also related to a Rajput king Upaldev and his conversion to Jainism, amidst a conflict between Jain monks and the goddess Chamunda who desired animal sacrifice.[188] Oswal *mutsaddis* continued to be in the service of Rajput kings till the nineteenth century. Munhata Nainsi, the *mutsaddi* whose work I have followed closely in this book, was an Oswal Jain. But he is known to have participated in several battles with Raja Jaswant Singh his employer. Jain *mutsaddis* like Nainsi mirrored what was understood to be the Rajput image in the seventeenth century. Rajput rulers of Jodhpur and Bikaner employed Jain *mutsaddis* like Singhvis and Bhandaris to counter Rajput *sardars*. Interestingly, while groups like Raikas, Charans, Nats, Bhils, Kanjars, Kalbelias constantly refer to a Rajput past and the loss of it, Banias have a very ambivalent attitude towards Rajputs. Lawrence Babb in his work *Alchemies of Violence* points out that Banias look down upon Rajputs for their "carnivorous diet, profligate ways and supposed lack of intelligence".[189] On the other hand, the Rajput derides the Bania for his greed and lack of courage. While violence came to be seen as an essential Rajput attribute, there appears to be an equally essentialist understanding of Bania attachment to non-violence. However, according to Babb, these binaries of violence and non-violence do not explain the ambivalence of Rajput-Bania relationship, as violence has been as much a part of Bania past. In fact, while Rajputs represent temporal control, with Charans, Bhats and to a lesser extent Brahmins as religious advisors, the trading castes represent the bridge between economy and polity. Therefore, it appears that for tribal communities like Bhils, Mers, Minas, Sansis, Kalbelias, Kanjars etc. denial of Rajput status is seen as the result of it being forcibly alienated from them through political, social and cultural subjugation. On the other hand, trading communities can see the loss of Rajput status as act of relinquishment, even renunciation. For tribal groups, loss of authority and political control due to subjugation by Rajputs resulted in social, cultural and political marginalization, and even notification under the Criminal Tribes Act. But for trading castes, control over economic resources guarantees continual control over political power, and thus a higher rank in the hierarchy of peripatetics, even if their claim over Rajput status was lost.

Therefore, I would argue that communities like Rajputs, Banias, Raikas, Gujars, Charans, Bhils, Mers, Minas, Kalbelias, Nats, Ghattiwals, Motisars, that appear to be fulfilling distinct roles in the Thari society and economy

---

[188] Babb, *Alchemies of Violence*, 167–169.
[189] Babb, *Alchemies of Violence*, 60.

were associated through deeper processes of emergence. The claims to a shared Rajput past were in a sense claims to the mobile space of the Thar. Peasants, pastoralists, traders, carriers, peddlers, and genealogists all had a role to play in the society and were not always juxtaposed against each other. In order to sustain the social and economic structures in a largely arid society it was important for sedentary and nomadic elements of society to supplement each other. The circulation networks in the Thar were created through the mutual interactions of diverse groups, where each group had a specific and significant role to play. In the nineteenth century however, the dynamics of these relations began to alter. I would discuss these changes, the resistance to changes and their outcomes in the next chapter.

Expanding State Contracting Space: The Thar in the Nineteenth Century

> Mobility confounds settled relationships. It raises uncomfortable questions about teleological theories of history, undermines attempts by states to territorialize and control their populations, and confounds accepted understandings of the relationships between property rights and efficiency, place and community.[1]

Mobility, as seen in the previous chapters, was an essential aspect of the social, cultural and economic structures in the Thar Desert. Mobile communities in the Thar had not always been placed in opposition to sedentary ones and interchangeability between mobile and sedentary ways of life had been a common phenomenon in the society. In fact, the opportune placing of each segment in the society was aimed towards optimum use of scant natural resources that were precariously balanced. Therefore, mobility in the context of the Thar refers to the ways in which people, cattle, merchandise, services and ideas have circulated over the region and not merely long range movements undertaken by the pastoralists or merchants.

However, by the late nineteenth century, as British officials began exercising greater influence in Rajput states, mobility and settlement were increasingly posited as two contrasting ideas in the administrative ideology of the states. As the chapter on Rajput polity discusses, Rajput states of the Thar region had been assimilated in the Mughal imperial structure by late sixteenth century

---

[1] Arun Agarwal, *Greener Pastures*, 21.

and had increasingly replicated Mughal administrative structures and practices, with emphasis on land as the basis of rule. The agro-centrality of these norms was also useful for the Rajputs as they progressively emphasized on sedentary agrarian administrative structures. Nevertheless, given the predominantly arid and mobile character of the Thar, even in the nineteenth century, a number of mobile groups regularly circulated across the Thar for socio-economic reasons. In the Mughal period, Rajput states like Jodhpur, Bikaner and Jaisalmer had attempted to regulate mobility by trying to monitor routes on which warriors and traders travelled. But there still remained large spaces, which remained outside the direct control of the rulers of these states, and thus in which the movements remained largely unchecked. However, as the present chapter discusses, the late nineteenth century in the Thar witnessed the extension of control over movements of a wider range of travellers including pastoralists and bards, as well as 'criminalisation' of certain kinds of mobilities in this region.

The signing of the subsidiary alliances between 1812 and 1818 opened the administrative policies of Rajput states to British influence. These treaties opened up several possibilities as, "Rajpootana being now redeemed from the grasp of the spoiler, it was of essential importance to its interests as well as those of British India, in order to prevent a renewal of the predatory system, and to interpose a barrier between our territories and the strong natural frontier of India, that newly settled states should be united in one grand confederation".[2] However, it was only towards 1870s that British officials were able to create a confederation and initiate major administrative changes in the region, which were to include revenue settlements, survey operations and boundary settlements, extension of rail and road networks and regularization of commercial revenues among others. While some of these changes can be viewed as part of general administrative restructuring as it was carried out in princely states in the late nineteenth century, I argue in this chapter that they were largely motivated by the idea of 'settlement' of what was experienced as a geographically 'hostile' region, swarming with dangerous itinerants.

The history of the Thar Desert is replete with cycles of migrations and settlement. The 17th century chronicler, Munhata Nainsi, contrasts *basti* with *khalidesh* to denote the difference between settled and unoccupied spaces. The arid nature of land and the sparseness of the settlements meant that a large amount of land was available for a number of groups that survived on pastoralism or trade, two major economic systems in the Thar. However, in

---

[2] Tod, *Travels in Western India*, xxxii.

the nineteenth century, unoccupied spaces began to acquire the connotation of 'waste'. The shift from *khalidesh* to 'waste' was reflective of the emphasis placed on settlement in the 19ᵗʰ century, whereby attempts were made to arrive at a more tangible connection between existence of 'waste' land and 'unsettled' groups that 'roamed' across the Thar. The presence of both was seen as unsettling for the stability and the economy of the state. Mobility increasingly came to be identified with criminality as well as inefficiency. The physiocratic view of land and production regarded settlement of mobile groups on unoccupied wastes a reasonable solution to both problems, of which the 'urge to move' was considered to be more dangerous.

The nineteenth century discourse of settlement, linked mobility of its travellers and the perceived hostility of the Thar with each other. The idea of an array of mobile groups on the move represented an unsettled state of affairs to the British administrators, who saw 'order', 'settlement' and 'improvement' as essential components of administration.[3] Therefore, a range of administrative policies and reforms were instituted in the Rajput states of the Thar region which were aimed at settling the region through agrarian expansion, forest conservancy, increased control on trading networks and increased surveillance of mobile groups. This chapter argues that, as a cumulative effect of these policies the networks of movements were radically transformed by the end of the nineteenth century. Attempts were made to 'open up' the region by creating more routes by way of roads and railways. These were primarily aimed at creating networks that could be monitored and managed, and thus resulted in enclosing a historically open region that had been united through the mobility of its travellers. With the formation of the Western Rajputana States Residency, the British involvement in administration in the Thar region was noticeable in its efforts to assimilate the region with the surrounding states by means of newer networks of roads, railways and canals, and thus create what appeared to be an uninterrupted flow of commodities. But in this region an uninterrupted flow of commodities had also meant an uninterrupted flow

---

[3] Ranajit Guha extensively explores the relationship between order, settlement and improvement in colonial India. See Ranajit Guha, *Dominance without Hegemony: History and Power in Colonial India,* Harvard University Press, Cambridge, 1997, 30–34. British antagonism to mobility has also been highlighted by several historians as discussed in the Introduction of this work, including Neeladri Bhattacharya, 'Pastoralists in a Colonial World', Mahesh Rangarajan, *Fencing the Forest* and Meena Radhakrishna, *Dishonoured by History.*

of people. To make sense of the region also meant making sense of the vast population that traversed through this region. Thus, the regulation of traffic of people and commodities came to form the most important aspect of the British administration in the Thar.

Just like the previous Mughal administrators of the region, the British too recognized the importance of the Thar as a corridor that connected Gujarat and Sindh with northern India. As discussed earlier, not only did important trade routes crisscross the Thar, it was also the route to Sindh, the frontier that British were struggling to control. Having created political alliances through the subsidiary treaties, the British administrators attempted to extend control in this region with the view that control of this space could provide them with a strong and much required foothold in north-western India. This could only have been achieved by both, settling sparsely populated tracts as well as regulating movements on networks of circulation.

As noted in the first chapter, most British travelers and administrators were struck by the desolation of the Thar region and by the fact that so little cultivation existed in so many parts. This was sometimes attributed to the indifference of the rulers and at others to the indolence of the people.[4] This vast tract of unfarmed land perplexed the British administrators who attempted to put it to use by either inducing cultivation or using it for controlled grazing and forest conservancy. That it had not been done so far reflected upon the administration of the region, whose, "only blight is bad rulers and therefore an unsettled people. To open up a through road would throw light into the land. Stimulate energy and enterprise; whilst its repressed evil doings are the part of the ruler and the ruled".[5]

The idea of 'settlement' ranked high in the British administrative psyche and in the context of the arid and semi-arid land of the Thar it presented both a complexity and an opportunity. In British understanding, a universal attachment to private property and the legitimization of the same was to prove the keystone to the edifice of the empire.[6] It meant that all forms of ownership in the Thar had to be brought under the ideal of private property that was attributable and taxable. All that could not be attributed to the individual then had to belong to state and aggressively harnessed to generate more revenue.

---

[4] Fagan, *Settlement Report of Bikanir*, 17.

[5] Civil Works, Transfer of Marwar Section of Agra Ahmedabad from PWD to Marwar Darbar 1874, File 4, RA, Jodhpur, 104, RSAB.

[6] Gilmartin, 'Cattle, Crime and Colonialism', 56.

The other aspect of this understanding was to attribute locatable identities to people in this region where they seemed to be constantly on the move. That, pastoralists from far off areas came and utilized the grasslands in villages as a matter of customary right did not appear to be a matter of tradition, but rather a transgression of the settled agrarian economy of the society. As Neeladri Bhattacharya observes in the context of pastoralists of nineteenth century Punjab,

> "nomads, vagabonds, wanderers were thus to be disciplined and settled. Their identities had to be fixed. They had to belong to a marked territory- a village, a district, a province. To exercise power the state *had* to know the identities of those over whom power had to be exercised, and confine them within controllable delimited spaces. Nomads appeared elusive, unknowable, anonymous beings whose identities were difficult to ascertain. Their mobility was, to an extent acceptable; their anonymity was not. Since the anonymity of the nomad threatened the very basis of power, their mobility had to be restricted and regulated".[7]

Therefore, towards the end of the nineteenth century the princely states of the Thar and the Western Rajputana States Residency came together to undertake a series of steps, which were aimed at the general process of settlement of princely states of Rajputana. This chapter attempts to study a few facets of this process to be able to understand what these steps meant for the largely mobile sections of society. I do so by examining the process of land revenue settlements in the princely states of the Thar, the formulation of fodder and forest policy in Marwar, the control of trading networks, canal expansion policy in Bikaner, and the question of control of salt in the Thar region. The primary purpose of this attempt is to explore how these policies, implemented in different parts of the Thar, over different time periods, cumulatively affected people who were occupational travellers in the Thar. I would also attempt to show how these interventions and also others like construction of roads and railways altered the circulatory and peripatetic networks in the Thar. Thus, in this chapter I argue that the settlement policies in the late nineteenth century were a result of the ways in which the British administrators tried to make sense of the essentially mobile character of the region. They were both awed and repulsed by the enormity of boundless space that the Thar represented and the idea of a predominantly mobile populace

---

[7] Bhattacharya, 'Pastoralists in a Colonial World', 84.

in a boundless region was a challenge to the notion of settlement that the British administrators propounded.

Sedentarisation and Settlement

An important aspect of the British administrative policy in the Thar region was the reframing and redefining of landed property rights. The absence of clear landed rights often meant fluid locational identities. This was allied with the classic European distrust towards mobility that British administrators essentially associated with purposelessness. Neeladri Bhattacharya notes, the "'lazy' pastoralist was inevitably defined in opposition to the 'sturdy industrious' Sikh peasant cultivating his field with care and yielding revenue to the British. Pastoralism was not a worthwhile enterprise, cultivation was".[8] The agrarian producer was perceived as the mainstay of the state, and not only the protection of the agriculturalists but also the settlement of more people in the agrarian structures was seen as a way of ensuring stability of the state as well as conquering the desolateness of the landscape. Besides, to produce from land not only added economically to the state, it also ascribed fixed identities.

The British settlement officers in the Thar attempted to understand and make sense of the variable system of land rights that existed in different parts. However, the framing of land rights was not an easy enterprise as the settlement officer of Bikaner P J Fagan realized. He was exasperated by the fact that, "revenue methods have varied in the State greatly, in fact it is now impossible to understand many of the methods in all their detail".[9] The process of assimilation of pre-existing agricultural groups or settlement of newer groups in uncultivated tracts by the early Rajput conquerors had created a variety of land rights. This included the overlapping rights of *bhomias*, rights of the first clearer, rights of the *sasan* grantees, rights of the cultivators etc., which appeared incomprehensible to the British settlement officers. Besides, as there existed several political units within Rajputana the nature of land rights also differed considerably from state to state.

Therefore, in Marwar and Bikaner states, it was decided to conduct detailed revenue surveys by 1880s. An important aspect of this exercise was to create means for increase in overall revenue extraction from the existing levels. In Marwar a new land revenue settlement process was started in 1878

---

[8] Ibid.,71.

[9] Fagan, *Settlement Report of Bikanir*, 41

and Colonel Loch was appointed the settlement officer. A boundary survey was conducted in January 1882 in order to fix the actual number of villages in Marwar. Boundary settlement was conducted and maps prepared for 3623 villages of which 579 villages were found to be *khalsa* villages.[10] In 1884–85, the older assessment systems were done away with and *ankbandi* was induced under which a rough estimate of the entire village was prepared. Eventually, a system of cash per *bigha* was introduced in 1894–95, with the rates for wetlands varying from Rs 2.35 to Rs 10 and for dry land from 8 paisa to 78 paisa. Two other taxes called *kharda* and *ghasmari* were charged, which were primarily non-agricultural taxes.[11] In Bikaner, it was decided to undertake a summary settlement of *khalsa* villages in 1884, "....the main object aimed at obtaining some idea of the revenue which such villages could be called upon to pay and to assess and collect the amount on some uniform system of assessment".[12] The settlement surveys were undertaken by Capt. P J Fagan and summary prepared by 1893. The state was divided into revenue circles called *chiras* where revenue collection was done through the *chirayat*, who periodically visited the villages and collected revenue. In certain *jagir* areas the revenue collection was done through the agency of the village *mahajan* who advanced the agreed revenue to the *tahsildar* and later collected it from the villagers.[13]

One of the major concerns that came up in settlement reports of both, Colonel Loch and Capt. Fagan, was the existence of large patches of common lands in Marwar and Bikaner. The village commons were composed of *orans* or sacred groves, *jods* or village grasslands and uncultivated land called *gair mumkin kasht* in administrative parlance and *partal* land in common interchange.[14] In Bikaner, Fagan lamented the fact that no tax was being realized from the cultivators on account of the 'waste' that they used.[15] His suggestions were to assess land whether occupied or waste and then to consider how to deal with considerable unoccupied waste.[16] Fagan thus proposed the enumeration of cattle and a taxation system based on calculation of averages per village.[17] This was in contravention of the convention under which so far

---

[10] P R Shah, *Raj Marwar under British Paramountcy,* 56.

[11] *Marwar Administration Report,* (1894–95), Chap IV, Sec IV, 28.

[12] Ibid., 42.

[13] Powlett, *Gazetteer,* 114.

[14] P R Shah, *Raj Marwar under British Paramountcy,* 56.

[15] Fagan, *Settlement Report of Bikanir,* 50.

[16] Ibid., 85.

[17] Ibid.

the resident cultivators paid *singhoti* (for grazing sheep) once every two years and did not pay any taxes for the use of village commons as regards the grazing of the cattle. Also, the migrant pastoralists, who paid grazing taxes based on the number of animals they grazed on the village commons, were to be taxed even higher.[18]

This was reflective of a larger colonial ideology in which the most desirable production was agricultural production and British administrators considered the expansion of agrarian revenue as their primary purpose. The preoccupation with the usage of 'culturable waste' indicated an attitude that implied that all possible land was to be put to good use by means of getting it under the plough. The possibility of large tracts of land lying unharnessed mirrored an administrative failure. But, to his dismay, Fagan found that, "Mirzawala has only 11 percent area cultivated, and even this small proportion falls to 6 percent in the southern portion of the *tahsil*, of the Suratgarh *tahsil* only 12 percent area is under cultivation, the Anupgarh *tahsil* is as a whole almost devoid of cultivation".[19] In fact, Fagan expressed disappointment that, "no attempts are made to increase area of irrigation by means of artificial cuts or *nalas* as is the case of neighbouring Sirsa"[20] In his view this reflected the lack of enterprise as he found that villages in Bikaner were peopled by Jats who were "far inferior as agriculturalists to the Deswali and Sikh Jats of the Punjab".[21] Nevertheless, the British settlement officers realized the importance of settling agricultural castes, preferably Jats, who could add to the productivity of the region, in ways that non-agricultural communities could not.

As discussed earlier, the British settlement officers in the nineteenth century experienced the physical geography of the Thar region as hostile to human settlement. Besides being unproductive, the vast desolate stretches were also seen as a barrier in the protection "of the western frontiers, from the incursions of plunder of thul or desert".[22] An example of this policy was the seizure and settlement of Mallani, on the western frontier of Marwar. Mallani was viewed as a country with "few scattered villages, which for ages had been a grazing ground for camels, kine, goats and sheep".[23] It was considered to be an anarchic region as it was totally devoid of cultivation and was inhabited

---

[18] Ibid.,88.

[19] Ibid.,10–11.

[20] Ibid.,16.

[21] Ibid., 8–9.

[22] FP, November 1832, 8, NAI.

[23] FP, October 1868, 69–80, NAI.

by the Khossa and Sodhas. According to Major Erskine, the geography of
the region rendered it suitable for providing refuge to freebooters who looted
caravans.[24] Thus, in order to set this right British forces seized Mallani in
1836 and then set about populating it in order to make it less hospitable for
the plunderers.[25] This meant introducing Jat peasantry "with no less labour
into these previously uncultivated tracts".[26] Jat peasantry was seen to be "fast
supplementing the roving class of graziers; and hamlets with their rain crops of
millet and pulse spring up to relieve these deserts".[27] In fact, between 1865 and
1868, seventy villages had been founded in Mallani and the British appeared
to complement themselves on "reclaiming people from their predatory pursuits
and encouraging cultivation, trade and peaceful sources of income".[28] Even till
the early twentieth century settlement and expansion of cultivation in western
stretches of the desert was being induced through incentives. The peasants
willing to settle and cultivate in drier areas like Sheo-Sankra on the western
frontiers were extended significant advantages in revenue assessment and also
in grazing regimes.[29] The settlement in this manner of what were considered
to be politically, economically and ecologically hostile areas, brings forth the

---

[24] Erskine, *The Western Rajputana States*, 200.

[25] FP, September 1836, 12, NAI. Mallani or Barmer was transferred to Bombay
Government in 1836, and later to AGG, Rajputana in 1839, and to Political Agent in
1843. It was restored to Marwar in 1898.

[26] FP, October 1868, 69–80, NAI.

[27] Ibid.

[28] Ibid.

[29] In 1920s in Paragana Sheo, which was one of the driest paraganas of Marwar, it was
offered that if anyone leased 300 *bighas* of land for cultivation, he would be charged at
a concessionary rate of Rs 10 per hundred bighas. He could cultivate any *gair mustakil*
land, graze as many cattle as he owned and cut fuel and fodder from the village commons.
He was to be considered a *pattadar* and this right was to be heritable. He also paid a
concessionary *ghasmari* at the rate of Rs 2 and 8 *annas* for camels, 12 *annas* for old stock,
10 *annas* for young stock and six pies each for sheep and goats. On the other hand anyone
wanting to lease less than 300 *bighas* had to pay a rate of 2 *annas* per *bigha* and he could
lease a number of *bighas* as per the number of animals he owned or desired to own. For
example he could lease 15 *bighas* per old stock, 5 *bighas* per new stock, 1 *bigha* per goat
and 1 ¼ *bigha* per sheep. Such a landholder was called *shartia pattadar*. If he raised any
more livestock he paid 2 *annas* and six *paise* per *bigha* required for the particular animal.
In case a person with no landholding in the village wanted to graze his animals he paid
the *ghasmari* as Rs 3, 10 *annas* and 8 *paise* for camels, Rs 2 and 8 *annas* for old stock,
12annas for young stock, 3 *annas* for sheep and 2 *annas* and 6 *paise* for goats. Forests,
Grazing and Fees General, 1920–26, Dated 28/10/1926, CR No 7, p 162, RSAB.

irony of land being viewed as a resource that could be tamed and domesticated and put to use. That, even seemingly desolate areas like Mallani and Sheo provided sustenance, particularly as pasture, was not considered of significance.

As it was observed by the settlement officers in the Thar that the fallows exceeded the cultivated land, various measures were considered to increase the land under cultivation from late nineteenth to early twentieth centuries. The expansion of cultivation through extension of irrigation was considered as a means of augmenting the agricultural revenues. In the survey undertaken by Fagan, he found that despite the fact that villages of Bikaner that bordered Punjab were very close to the Sirhind canal and the western Jumna canal ran westward across the Hissar border as far as Bahadra, there was practically no artificial irrigation. All cultivation, though sometimes benefitting from the floods in the Ghaggar, was dependent on the rains. Therefore, the construction of two canals for the distribution of water from river Ghaggar was undertaken between Bikaner and Amritsar Bund 6 miles into British territory. The two canals of the Ghaggar were, "constructed in order to utilize for irrigation, the water which collecting in pools here and there, not only submerged cultivable land but occasioned sickness and rendered the climate unwholesome".[30] The northern paraganas of Bikaner, which were also known as the *nali* lands due to the excess drainage of the Ghaggar, were actually excellent pastures. Of this region Powlett had noted that, "in spite of the difficulty about water, the grazing is excellent and the number of cattle possessed by people is great; and from Bahawalpur and British districts cattle are brought to graze, the Bikaner people being in return allowed, to take their cattle to the banks of the Satlaj. The Sotra occasionally flows. General Cunningham speaks of the water sometimes coming roaring down, but its waters are soon absorbed in the desert".[31]

However, soon after the canals were constructed it was found that there was "practically no demand for canal water in Bikanir during the rains when it was available".[32] Besides, the northern paraganas of Bikaner like Tibi and Suratgarh that were likely to benefit from the canals were sparsely populated. Also due to the uncertainty of supply of water from Ghaggar, cultivation was limited in this region so that the cultivators were ill prepared.[33] In fact, the Deputy

---

[30] Abstract of the Deputy Revenue Official's Note Dated 1/3/1899, Ghaggar Canal File, 1899, RSAB.

[31] Powlett, *Gazetteer*, 135.

[32] Inspection Note, Superintendent Engineer PWD, Ghaggar Canal File, A-602–706 1896–98, 23, RSAB.

[33] Ibid.

Revenue Official of the Bikaner state, in a strongly worded note, highlighted the sections of the agreement that actually benefitted the British territory instead of Bikaner. The agreement was so worded that Bikaner could get the water from the Ghaggar reservoir only in case of excessive rainfall when the land was irrigated by rainwater itself. This superfluous water was found to be less than the time the bund had not been constructed. It was found that by "bunding up the stream, every drop of water was utilized in irrigating the winter crop in British territory".[34]It was predicted that the flow in the canals would rarely reach the 55th mile and with passage of time as distribution channels come into use the flow in the canal would reduce further. Therefore, despite being aimed at providing artificial irrigation the scheme could not achieve much expansion. Marwar Darbar undertook similar schemes in the Luni Basin. Jaswant Sagar, Dipawas, Bagole and Chopra dams were built in Marwar. During the famine of 1899–1900, more reservoirs were commissioned at Sardarsamand, Kharda, Jogawas and Pali. These projects collectively irrigated 12,579 hectares of land.[35]These irrigation systems could not however, displace the traditional systems like *khadins*, stepwells and L shaped embankments.[36]

Nevertheless, these steps were reflective of how the manner in which land was viewed in the Thar region radically altered in the nineteenth century. It had to be increasingly harnessed and managed to be make it agriculturally productive. Any other kind of usage could only mean hostility to state and settlement, unless it could also manage to aid and augment the settled residents of the villages. The existence of land with indeterminate use had meant its availability to all possible kinds of users without much conflict. However, the straitjacketing of land use into categories inevitably led to increased conflict in the Thar region. In fact, increasingly bureaucratic rules and procedures were evolved for managing pastures and forested tracts, as we would see in subsequent sections.

Fodder, Fallows and Forests

As discussed in earlier chapters, the importance of pastoralism in the arid Thar region was attested by the presence of diligently managed grasslands and sacred

---

[34] Abstract of the Deputy Revenue Official's Note Dated 1/3/1899, Ghaggar Canal File, 1899, RSAB.

[35] P R Shah, *Raj Marwar under British Paramountcy*, 155.

[36] R. Thomas Rosin, 'The Tradition of Groundwater irrigation in North western India', *Human Ecology*, 21(1), 1993.

zones, whether state owned or community managed. Munhata Nainsi's late seventeenth century *Vigat* makes a careful note of grass *jods* in the *paraganas* he surveys. A number of *jods* are mentioned in the Marwar Administration Report of 1883–84 that were spread over various *paraganas* of Jodhpur.[37] Erskine mentions that these grass *jods* were used to produce and store fodder for lean years.[38] The Summary Settlements undertaken between 1881 and 1884 in Jodhpur and Bikaner underlined the need to manage the common resources in a more 'rational' manner. Though these resources were utilized on the basis of traditional rights of communities, yet a determined effort was made on behalf of the native states in the Thar region from late nineteenth century onwards to systematize the approach to the use of these resources. Population and cattle censuses had been undertaken in Western Rajputana from 1881 onwards.[39] By the year 1901, a separate cattle census was proposed in Jodhpur, with the view of assessing grazing dues.[40] The census in both *jagir* and *khalsa* areas was to be undertaken every five years to have a clear idea of the number as well as the kind of cattle grazing on common grasslands.[41] The other purpose of this census was to categorize the animals on the basis of their utility for cultivation and state carriage.

In the same period, an attempt was made to formulate a fodder policy that led to the proposal of establishment of fodder farms that could be the source of disbursal of fodder. The proposal of supplying fodder to graziers as compared to the graziers taking herds to grass lands was seriously considered. The fodder policy was also related to the idea of developing fallow lands into conserved forests. K D Erskine, who was the Resident of the Western Rajputana Agency, observed that, "forest conservancy though more or less in its primitive form, found favour, with people of Marwar, high or low, from very ancient times, as would appear from the relics of old institutions like *jods*, *orans*, *gochars* and *rakhats*".[42] He recommended that "fodder areas should be increased and on the other hand they shall not encroach on the existing cultivation, where it can be avoided, or interfere with future extension of cultivation where it can

---

[37] Hardayal Singh, *Majmui Halaat wa Intijam*, 1883–84, 13.

[38] K.D. Erskine, *Rajputana Gazetteers- The Western Rajputana States Residency and the Bikaner Agency*, Vol II, Allahabad, 1909, 119.

[39] Cattle Census, 1882, Sr 3 old 2, Vol 1, Bundle 1, RSAB.

[40] Cattle Census, Hawala Jodhpur, 1901, Official Note Revenue Department, Sr 3 old 2, Vol1, Bundle 1, RSAB.

[41] Ibid.

[42] Grass Jors and Fodder Reserves, File 134, 66, RSAB.

be profitably undertaken".[43] It was found that to a small extent, the practice
of cutting and preserving fodder for leaner times was already prevalent in
Marwar. But this was done mainly for stall-fed milch cattle and other animals
like sheep, goats and camels were usually grazed. In fact, Erskine enquired if
restrictions could be placed on the grazing of sheep and goats in a portion of,
or the entire *jod*. The absence of a widespread interest in the practice of cutting
and storing was blamed upon the "apathy and poverty of the agriculturalist".[44]
Erskine hoped that the encouragement on part of the state might lead to the
profitable extension of the practice so that a valuable reserve could be created
for the times when other sources became unreliable. The propagation of local
species over exotics was preferred for these reserves.

In 1888, a department of forests was set up that had started the work of
forest conservation in the hills of Godwar and Sojhat. Initially, a proposal was
also put forth to reserve about 15 square mile of the Jaswantpura forests. The
Department intended to bring about 360 square miles under reservation in
the year 1901, which amounted to about ninety percent of total hilly tract in
Marwar.[45] These hilly tracts were the perennial pasture grounds of the goat
and sheep herders from Jalor, Pali, Jaitaran and Merta. Further, it was proposed
to extend the work of reservation to the plains.[46] This extension work was to
be carried out in 700 *khalsa* villages with the aim of enclosing half the area
for sheep and goat and other half for cattle. As we can see, to begin with the
purpose of forest conservancy in Marwar was to provide fodder for the herds.
Therefore, it was proposed that "growth of Khejra tree in the reserved area will
be encouraged which furnishes an excellent fodder; the wood of the dried trees

---

[43] Grass Jors and Fodder Reserves 1901–22, File No 5 part 1, New File No 11, 3, RSAB.

[44] Ibid.

[45] Superintendent of Forests, Marwar to Secretary to Musahib Ala, 16/5/1901/Forest
Department, RSAB. Also Marwar Administrative Report, 1899–1900, 23.

[46] The forest reservation policy in this period was a reflection of similar attempts at forest
conservancy in thickly forested tracts in other parts of the country, as pointed out in
works like M Rangarajan, *Fencing the Forest: Conservation and Ecological Change in
India's Central Provinces 1860–1914*, OUP, Delhi, 1996, R. Guha, *The Unquiet Woods:
Ecological Change and Peasant Resistance in the Himalaya*, Delhi, 1989, and D D Dangwal,
'State, Forests and Graziers in the hills of Uttar Pradesh: Impact of Colonial Forestry
on Peasants, Gujars and Bhotiyas' *IESHR;* 34; 405–435, 1997. The forests in the hills of
Marwar were however distinct as they did not yield many products of commercial value
and were understood to be fodder reserves primarily for years of droughts. However, the
zeal of conservancy converted these into enclosed areas that were soon out of bounds
for the very people they were meant for.

would be used by the Raj or the Jagirdar".[47] It was claimed that the restriction on the grazing of goats and sheep would not cause any suffering to the herders and further it was also proposed to enclose 1/8 of *partal* land in every *Khalsa* village to create a fodder reserve. It was also claimed that a similar experiment had already been carried out in Sojhat, where fodder from these reserves had been sold to the villagers during the famine of 1899–1900. It was proposed that the "grass so stacked will be distributed to the villagers in proportion of number of *working ploughs* (emphasis added)".[48] The Forest department too proposed to let the cultivators of the respective villages cut the grass from the reserves free of charge.[49] So the fodder reserves while being created out of the land traditionally available to herders for grazing cattle, sheep, goats and camels, did not actually benefit them but rather the agriculturalists in the village in proportion to the land they farmed. It has been pointed out earlier that the herding communities were unlikely to be large landholders and thus there was not much possibility of their being able to benefit from such scheme. In fact, within another month it was proposed to increase the enclosed land to one fourth of the *partal* land, as against one eighth. The area thus enclosed was to "entirely and strictly be closed to cattle of all sort and hatchets of all metals and wherever practicable a thorn fence must be put around since orders by themselves do not go very far to ensure sufficient protection".[50] Along with fencing, digging of ditches was also proposed around the fodder reserves. Besides, the fodder reserves thus formed were to be administered by the Hawala department under the charge of the *hawaldar* or *patwari* and were to seek professional advice from the forest department. So the new fodder reserves while emanating from the old grass *jods* were to be more professional and bureaucratic, and for the cattle of the resident cultivators and not the transhumant herders.

By 1903, there was a renewed stress on improving the quality of the reserves, which it was claimed to be deteriorating due to fires and unrestricted grazing. It was maintained that the forests were deliberately set on fire, "with the object of making available for grazing the first shoots of young grass which appear in the hot weather and early rains. The fires remove dried stumps of previous

---

47 Secretary to Musahib Ala to the Resident WRS, Dated 22/8/1901, Mehkama Khas File No 322/24 (Forest) RSAB.

48 Ibid.

49 Supdtt. Forests to Secretary to Musahib Ala, 21/9/1901 Mehkama Khas File No 322/24 (Forest) RSAB.

50 Ibid.

season's grass which protect the young grass from the teeth of the cattle".[51] The resultant grass was not seen fit for the consumption of stall fed cattle. Therefore, the Hawala department was advised to "discourage promiscuous and unreasonable grazing … the restriction of grazing is in itself an important step in this direction".[52] It was anticipated that the improvement in quality of fodder would favourably affect the quality, health and stamina of the cattle and make them sturdy enough to last the ever-recurrent famines. Here, I would underline again that these measures were aimed at the stall-fed cattle within the village communities. Sheep, goats, camels that were not usually stall-fed are not mentioned in the proposed scheme of forest reforms. These animals grazed on wild grasses, shrubs and crop remnant and not on the grass thus grown and preserved in these reserves that due to small landownership and a high number of animals, would in any case not have been available to the graziers.

Around 1903, it was also proposed that a practice of supplying fodder from these farms be encouraged instead of the regular feature of the cattle migrating to the pastures seasonally. This was contested by the Hawala department claiming that it was not only "cheaper but more practical for the cattle to migrate to places where grass is to be had, rather than for the grass to be transported and distributed according to the wants in the affected parts, especially when we have to consider that large tracts of Marwar yet untraversed by the railways are not within easy reach".[53] A review of the scheme was undertaken in 1906, after the scarcity of 1905. It was declared that 94, 531 *maunds* of grass had been stored in Marwar. Yet it was noticed, that "sufficiently large quantity of grass was not stored and though the state had its requirements well satisfied, the public derived very little benefit from the arrangements".[54] Besides, the grass was stored at "places greatly out of the way and could not be easy of access when wanted".[55] It was thus proposed that large quantities of grass should be cut and pressed mechanically into bales and supplied to railway stations in the summer months where the peasants, who would be free of agricultural work

---

[51] KD Erskine, Resident, WRS to Mehkama Khas Raj Marwar, Jodhpur, 30/1/1903 File No 322/24 F-66 (Forest), fl 26, RSAB. Also mentioned by Neeladri Bhattacharya, 'Pastoralists in a Colonial World', 85, as a strategy for resistance.

[52] Ibid.

[53] Secretary, Musahib Ala to Resident, letter 6284, dated 11/10/04, File 6C, fl 44, Mehkama Khas, RSAB.

[54] Superintendent of Forests to Mehkama Khas, 26/4/1906/, no 100 of 1906, fl 39, Mehkama Khas, RSAB.

[55] Ibid.

at this time, could collect the bales. By 1907, it was proposed that the work of managing fodder reserve should be handled by a person suitably versed in forestry as work done was to be mostly of forest conservancy.[56] By 1910, the fodder reserves were proposed to be managed by a Forest Ranger with the help of 24 gangmen who would sow seeds and guard the reserved areas.[57]

Thus, as we can see that the scheme that was initiated to preserve and provide fodder for lean years was thoroughly bureaucratized within 10 years and instead of reserving areas for grazing in leaner times it actually went on to exclude the very people it was meant for. The grass and fodder reserves only served to supply the needs of the state departments that still owned large number of draught animals. The scheme of supplying fodder over long distance was hardly a realistic one, considering that the railways were mostly concentrated in the salt producing areas and that the railway stations were located miles from the villages. A scheme like this could hardly have mitigated the severity of the droughts and famines in Thar. So, while fodder reserves could hardly allay the scarcity of fodder in villages, they actually reduced the amount of common lands available to both the village community and the migrant pastoralists. The peasant proprietors already had legal rights to grazing in their own villages that were governed by social customs and rules. So the people actually affected were the migrant pastoralists who by virtue of smaller land holdings even in their own villages had minimal share and in other villages the open pastures available to them were reduced by areas ranging from one eighth to one fourth of the previously available grazing. This in effect would have increased the range of migration for the pastoralists who would have to move further to get enough grazing for their flocks. Even though it was desired, the movements of migrant pastoralists could not be restricted, though a process of reduction of their access to commons began which in future was to continue to cause conflict in rural areas. As far as natural resources were concerned, what needs to be in focus is not only the availability of resources but also affordability and access to them. Given the status of peasant proprietors in the society, they were in a better position to decide upon the mechanisms of management of common resources like the *jods* and *orans* that were used more intensively by the pastoralists for specific periods of time. Therefore, any change that occurred

---

[56] Superintendent of Forests to Mehkama Khas, 21/9/07, no 466 of 1907, fl 49 Mehkama Khas, RSAB.

[57] Superintendent of Forests to Mehkama Khas, 21/9/07 no 257 of 1910, Mehkama Khas, RSAB.

in the position of mobile communities can be attributed to the repositioning of the way the natural resources were shared in the society and the changes that were brought about in the way that resources could be accessed. [58]

It is clear that agrarian expansion and systematization of grass and fodder reserves actually led to the reduction of the share of migrant and transhumant pastoralists in the rural society. Since access to resources was proportional to share in landed property, there is little possibility that these could have been used by the migrant herders. Besides, when more areas were enclosed it led to the reduction in actual area available to the migrant pastoralists. Though considering the land to man ratio, the problem would not have been acute even in the nineteenth century, but the policies of the nineteenth century were the precursors of policies that led to agro-pastoral conflicts witnessed in later years. Besides, the focus on the agrarian expansion led to redefining of access norms in the society. As pointed earlier, the question was not only of availability but also one of accessibility. The colonial understanding of unused land presumed that access to it was not governed by any kind of rules. As we have seen in the above discussion, all resource use was governed by well-defined societal regulations. Any change in the availability and accessibility norms was bound to undermine any pre-existing regimes. The shift of control of common resources away from the village community into the hands of state apparatus redefined the norms of both availability and accessibility.

The increasingly preferential approach towards settled agriculture on part of the state led to settled cultivation and nomadic pastoralism being posited as competitive livelihood strategies. However, the social, cultural and economic need for mobility in the region could not really be ruled out. Categories like mobile and sedentary which were considered mutually exclusive were not so, and often complimented each other. In circumstances like dearth and famines, the sedentary peasant could abandon land and move away with his cattle in search of pasture. Resident peasants were also responsible for the maintenance of the village pasture while the pastoralists were away. The pastoralists in turn

---

[58] By 1920s there was a considerable difference between the grazing fees charged from rights holder and non-right holder herders in Marwar districts. The non-rights holders paid 12 *annas* per head for buffaloes, 6 *annas* per head for cows, bullocks horses and donkeys and 1 *anna* per head for sheep in especially demarcated *guzaras*. If allowed into forest jods they paid an additional of 8 *annas* 20 paise. In Jodhpur the grazing fees were Re 1 and 8 *annas* per buffalo and 12 *annas* per cow etc. On the other hand the rights holders paid 3 *annas* 10 *paise* per head of buffaloes while 1 *anna* and 6 *paise* per head of cows etc. Forests and Grazing Fees General, 1920–26,127, RSAB.

herded the livestock of resident cultivators, manured the fields on the way and returned annually to their homesteads. Thus, these were complimentary relationships that survived the interpretations as well as the interventions of the settlement policies.

However, the increased sense of responsibility towards the protection of agriculturally useful cattle and grain brought the conduct of what were labeled as mobile criminal tribes into focus. The multiple roles of these communities, as pointed out in the previous chapter, were subsumed in the discussion on their criminal intent as cattle lifters and grain thieves. The 19th century discussion on criminality failed to locate the behavior of these communities in the older practices of contested control and warfare. What the British understood to be the crime of cattle theft was in a sense a different kind of exchange that was meted out between political and cultural territories. However, cattle theft, and interestingly kidnapping of small children, appears to be an accusation surfacing repeatedly in the descriptions of mobile communities in the nineteenth century tribes and caste compendia. Criminality thus was inscribed onto mobile communities by positing them as disrupters of agrarian harmony. Thus in the nineteenth century, the policies regarding agrarian expansion, irrigation, pastures, forests and fodder created a discourse of criminality in which both pastoralists and mobile trading groups were implicated. However, while the pastoral nomadic herding groups managed to survive by employing varied resource use, it was the groups on exchange networks that were affected the most.

The Ordering of Trading Networks

As discussed in the previous chapter, the trading networks in the Thar region were old and apart from traders from the Thar had also been used by traders from Punjab, Multan, Gujarat and Central India. In the seventeenth and the eighteenth centuries the Rajput states controlled the trading networks by means of installing *sayer thanas* on important trade routes. The need to control the routes emerged from the fact that control over circulation of resources defined the nature and extent of authority in the Thar. As we saw in the previous chapter, the mechanisms of control exercised by the Rajput states in the Thar were negotiated ones. The state attempted to control trade routes and the merchants always found mechanisms to flout this control. The late eighteenth and early nineteenth centuries witnessed widespread extension of *ijaradari* to commercial taxes, with the *sayer thanas* being awarded to the highest bidder, often against loans raised to pay off the Marathas and Pindaris.[59] While the leasing of *sayer thanas* proved to be a lucrative source of income, it also meant weakening of state's control over trading networks, which as examples from areas like Pokhran, Mallani and Umarkot reveal, were largely controlled by the *thikanadars*.

The British involvement in the regulation of trade and transit through the region was justified by pointing towards the political and social unrest that the region, especially Marwar experienced in 1830s. The British officials' reports repeatedly pointed towards the "disorganized state of Marwar",[60] and the "plundering and predatory incursions carried out by the natives of Jodhpur ...on the frontiers of Marwar, a malaise slowly penetrating the interiors as well".[61] The Political Agent was alarmed by the "heightened state of plunder, looting and crime in Jodhpur state" which he claimed was causing a decline in trade and commerce with neighbouring states of Jaisalmer and Sirohi.[62] Concerns were also expressed about the safety of the routes that passed from Marwar into Gujarat and Bombay, the lack of which was claimed to be responsible for the traders' decision of "leaving Marwar to trade in some other place".[63] In fact, security of trade routes was cited as a major reason for the occupation of Mallani in 1843, which was supposed to check the spread of anarchy, protect the allies of British as well as ensure free communication and mercantile activitiy on the trade routes.[64] While the district of Mallani was restored to Marwar in 1898, the district of Umarkot, through which the crucial trade route to Sindh passed was never restored to Marwar citing concerns for mercantile safety.[65] These concerns, surprisingly, were in contradiction to the fact that armed guards employed by the state and traders themselves as well as bodies of Charans escorted most caravans. Apart from these measures, there

---

[59] Tanuja Kothiyal, 'The Market and the State in Late Eighteenth Century Marwar' Unpublished MPhil Dissertation, CHS, JNU, 1996.

[60] FP, 25th April 1838 (97–99), NAI. This perception was also a result of highly antagonistic relationship between the British and Maharaja Man Singh, in whose reign the Naths of Mahamandir became extremely assertive.

[61] FP, 25 December 1838, (10–24), NAI.

[62] Political Agent to Raja Man Singh, 30th August 1828, Khatut Maharajgan Series, No: 235, Jodhpur Records, RSAB.

[63] AGG to Raja Man Singh, 25th February 1837, KM Series, Jodhpur, RSAB.

[64] FP, 26th September 1838 64–66, NAI.

[65] Regent to Resident Western Rajputana States, January 4, 1919, Rajputana Agency Records, File no 19, RSAB.

existed a system of *bulawo*, which in essence meant a protection tax levied by *thikanadars*, who themselves in several cases were found to aid and abet the highwaymen. While taking stock of security on the highways, the system of *bulawo* was found necessary "for affording a ready system of local protection to travellers, merchandise and property in the cause of ensuring safe transit to person through tracts where both stand in need of armed escort".[66] These concerns about safety of trade routes will have to be seen in the context of the larger discourse of 'lawlessness' in the frontiers. As argued in the second chapter, networks of circulation, and particularly the trade routes in the frontiers of Thar, had remained the primary sites of contestations over authority between the ruling clans and the *thikanadars*. By controlling passage as well as revenue on these routes, the *thikanadars* forced the ruling clans to acknowledge their sphere of authority and negotiate with them. Even the assimilation of the Thar into the Mughal Empire had not affected these exchanges to a very significant degree, as is visible by the continuance of privileges held by the *thikanadars*. The imposition of indirect rule in the Rajput states however, fundamentally altered the relationship of the ruling Rajput clans with their frontiers as well as their cadet lines. The centralization of authority in the nineteenth century was achieved at the cost of privileges held by *thikanadars*, particularly in the context of control exercised on trade routes as well as local levies imposed by them. Trade routes in particular were seen as sites where the multiplicity of authority had to be curbed, by centralizing customs and closely guarding the routes, particularly ones through which certain high value commodities like opium were transported.

The British realized the importance of controlling circulation of certain commodities over others; in particular opium as its trade and circulation within the Thar was closely related to control exerted by powerful militarized groups on the frontiers. In western India, opium was produced largely in Malwa, and also to some extent in Kota and Godwar hills between Mewar and Marwar. Opium became important to this region because of two reasons. Firstly, opium was carried from Malwa to Sindh for export to China through overland routes through the Thar. As Portuguese controlled the trade route from Daman to Macao, the land route through the Thar became extremely important in securing the transit of Malwa opium to Karachi, from where it could be shipped by boats to Daman.

Secondly, Pali in Marwar had emerged as a centre of opium processing from where local trading firms controlled the circulation of opium. Pali was

---

[66] FP, 17th December 1838, 37–49, NAI.

the most suitable place to establish a headquarters for the opium trade. It was already a bustling emporium and a centre for sale and purchase of Godwar opium. Networks of Marwari, Gujarati and Sindhi merchants based their operations in Pali from where the overland part of the opium transport was controlled. Therefore, scales were established at Pali so that all opium going to Palanpur enroute Bombay or to Karachi had to pass through Pali.[67] Opium was carefully sorted, measured, packed and sealed in two layers of hides and a layer of coarse jute covering. It was packed in "forty paileas of ten pucca seers a piece which made a total of ten kutcha muns or about six and a half pucca muns exclusive of package".[68] Though opium was a restricted commodity, it was quite freely available in Marwar. It was highly priced due to the exorbitant duties that were levied on opium.[69] According to Boileau, the transit duties on opium that covered the long route from Malwa to Sind, via Marwar and Jaisalmer came to 1 rupee per *kos*.[70] From Karachi Bunder in Sind this opium was then shipped to China, "as the direct trade with China in cotton cease(d) for several months and the gap could be filled by opium".[71] Opium traffic through Marwar was strictly monitored and taxed. The duties were to be paid at definite entry points like Narsing Basni in Merta. Opium entering Sirohi had to go through the railway via Shivgunj.[72] The Ameers of Sindh too imposed heavy duties ranging to Rs 225 per camel load of 8 Surat *maunds*.[73]

The transit of Malwa opium through the Thar caused heavy losses in Bengal opium trade, by competing in the China opium market. However, as Sindh and the princely states of the Thar were not under British control, there was little that could be done to prevent the passage of Malwa opium through the Thar. The signing of treaties with Rajput states opened the possibility of controlling the transit of opium through the Thar. By 1826, all the Rajput

---

[67] FP 6th February 1871, 141–156, NAI.

[68] Boileau, *Personal Narrative*, 177.

[69] FP, 14th November 1830, 3–8, SC, NAI.

[70] Boileau, *Personal Narrative*, 178.

[71] FP 14th October 1864, 29–34, NAI.

[72] Ibid.

[73] In the year 1829–30, the Talpurs of Sindh received Rs 5, 40,00 Rupees in duties on opium transit. Claude Markovits, 'The Political Economy of Opium Smuggling in Early Nineteenth Century India: Leakage or Resistance?' in *Expanding Frontiers in South Asian and World History, Essays in Honour of John F Richards,* (Eds.) Richard M Eaton, Munis D Fauqui, David Gilmartin and Sunil Kumar, Cambridge University Press, Delhi, 2013.

states had signed opium treaties that led to the suppression of local opium production and trade as the East India Company became the sole dealer of opium in this region.[74] This boosted Company's opium trade as is visible in Alexander Burnes' observation in 1828, "for the last six years the exports (of opium) have never been less than fifteen hundred camel loads and more frequently two thousand".[75]

Despite these measures, trade in 'illegal' opium continued throughout the Thar region as neither its consumption nor its transit to Sindh could be controlled. Non-Company opium trade in Marwar was controlled by Marwari and Gujarati Banias as well as wandering Gosain ascetics. In the early part of the nineteenth century, as the controls on the frontiers of Rajputana were considered to be less stringent as compared to British territories, the Rajputana route to Sindh was preferred even though it was longer.[76] For instance, a significant amount of opium was found to be taking older routes like the one through Lathi, Bap and Bijolae to Jaisalmer, and further across the desert to Sind.[77] In this 'clandestine' trade, Pali became the centre of several routes that led to Karachi *bunder*. However, the company soon started pressing upon the native states to curb this clandestine trade, which was seen as smuggling. The British monopoly on opium trade tried to limit such movement through strict surveillance of trade routes. As Tod rued, "if the opium from Malwa and Harouti competed in China market with our Patna monopoly, we intervened not with high export duties which we were competent to impose, but by laying our shackles upon it at the fountainhead...and command that no opium shall leave these countries for the accustomed outlets, under pain of confiscation".[78] The strict imposition of terms of treaties disrupted the networks of local syndicates in this region, who reacted violently to the curbs and impositions. As Amar Farooqui demonstrates, in opium producing and trading zones of western India, several 'gangs' of 'banditti' mushroomed that included Pindarris, Bhils, Gonds, Ramoshis, Kolis and Minas, who had been

---

[74] Amar Farooqui, *Smuggling as Subversion: Colonialism, Indian Merchants, and the Politics of Opium 1790–1843*, Lexington Books, 2005, 123.

[75] FP, 14th November 1830 3–8 SC, NAI. Here it would be pertinent to point that opium was an important commodity for consumption within the Thar as well. It had great ceremonial value with no agreement being reached at without the practice of *amalpani*.

[76] Farooqui, *Smuggling as Subversion*, 143.

[77] FP, 14th July 1883, NAI.

[78] Tod, *AAR*, II, 128.

previously employed by the Marathas.[79] These groups, operating in the frontiers between Malwa, Gujarat and Rajputana were employed by indigenous opium traders to guard caravans carrying 'illiegal' opium. Besides, the *thikanadars* in the Thar whose interests were closely allied with transit of opium sheltered groups carrying such merchandise. While "the company unloaded a 'swarm of spies and opium seizers, whose hands were in every man's house and every man's cart', the smugglers 'armed themselves to oppose opium-seizers'".[80] By 1829, the company was forced to reverse its Malwa opium policy and issue passes against high transit duties to Gujarati and Marwari traders, who began supplying opium to the port of Daman. Malwa opium continued to reach Daman via Karachi, which was supplied by overland routes through the Thar. The volatility of opium market given the situation in China, affected the fortunes of most of the Gujarati traders supplying opium to Bombay that had emerged as the opium capital. But for the Rajputana opium routes, the blow was dealt by the annexation of Sindh in 1839, as it dried up the supply route from Karachi to Daman, and along with it the routes through the Thar. Pali, that had emerged as an important processing and transit point for this 'smuggled' opium declined with the routes.

Amar Farooqui regards the continuance of 'unofficial' opium trade and transit for at least four decades after the region came under British dominance as an outcome of "serious conflict between colonialism and Indian capitalists, wherein the indigenous merchants were able to contest at various levels including in that crucial arena, the market".[81] Markovits views it as 'residual leakage', which would not have been possible without the involvement of Portuguese Naval enterprise and British and Indian speculative capital.[82] In my understanding, the links between merchants, Rajput states and locally powerful potentates who controlled all kinds of mercantile circulation through the Thar including that in opium played extremely important role in the continuation of opium trade through the Thar. The Thar as a political-commercial frontier controlled partially by Rajput kings of Jodhpur, Bikaner and Jaisalmer, the Ameers of Sindh, a number of recalcitrant chiefs as well as various roaming bands, provided the possibility of all kinds of legitimate and illegitimate exchanges. These exchanges were made possible by careful negotiations between all the parties, which respected each other's zones of influence. Even in what was considered to be legitimate trade and transit, the

---

[79] Farooqui, *Smuggling as Subversion*, 187.

[80] Ibid., quoting *Imperial Gazetteer*, IX, 381.

[81] Farooqui, *Smuggling as Subversion*, 10.

[82] Markovits, 'The Political Economy of Opium Smuggling', 101.

spheres of influence were carefully managed, as I would demonstrate through the exploration of the changing relationships between merchants and state in Marwar in the nineteenth century.

In Marwar, commodities were subject to certain common transaction and transit duties under the heads of *pesar, nesar (nikal), behtiwan*, and *rahadari*. However, other local taxes like *dan, mapa, shriji mahajanan, darwaja ri lag* as well as a local protection tax called *bolawo* were realized both in *khalsa* and *jagir* areas by the state officials and the *thikanadars*.[83] In the seventeenth and eighteenth century, all transit of commodities, taxes paid and exemptions granted were recorded in the *sayer thanas*. The traders were expected to travel on the route listed in their *chitthis* and their movement was carefully charted. The *sayer thanas* were also the place for making petitions regarding any mishaps or undue exactions made by either the *thikanadars* or state officials. By the 1830s, the widespread network of *sayer thanas* had fallen in disuse and disrepair, and illegal exactions by thikanadars had become rather common. In order to counter the multiplicity of exactions, a major restructuring of customs department was undertaken in 1874, with the view of "securing comfort and convenience to traders...and removing all unnecessary obstacles which had hindered trade in Marwar".[84] Maharaja Jaswant Singh II issued a proclamation "consolidating and revising the tariff of duties leviable on goods imported into, exported from or passing through Marwar".[85] The reason cited was the fact that import duties were being levied at multiple junctures and at varied rates. This subjected the traders and merchandise to "unnecessary detention at several places enroute before they could finally be reached to their destined markets".[86]Therefore, the *Darbar* fixed a uniform rate of import duties, which were the mean of the amounts realized as import duties at important *sayer thanas* in Marwar. This tariff was to be realized once at the frontier *sayer thanas*, irrespective of the distance covered by the merchandise within Marwar before being sold. Apart from this *rahadari* or transit duty as well as all duties levied by *bhomias, thikanadars* and *jagirdars*, with exception of *mapa*, were abolished. The privileges of communities like Charans, Bhats, Brahmins as well as Rajputs were curbed and they were now required to pay duties as applicable to all other communities.[87]

---

[83] Report on Marwar Customs, 31st March 1892, Customs, Jodhpur, RSAB.

[84] FP, B, 23 October 1874, 4–8, NAI (Proclamation of the Maharaja).

[85] Ibid.

[86] Ibid.

[87] Hewson, Report on Marwar Customs, 1884, para 7, Customs Jodhpur, RSAB.

In 1882, all duties except import, export and transit were abolished, and in the new scheme the tariffs on the luxury merchandise were increased.[88] The entire administration of the *sayer* department was taken over by the Jodhpur Darbar and in place of multiple cesses a single rate for all merchandise was introduced.[89] Permit system was introduced, under which the traders obtained numbered and stamped *rawannas* from the *thanas*. The person paying duties received a counterfoil that he could show at any custom post. Another free permit called *khali chitthi* was issued to traders wishing to conduct trade within Marwar.[90] In 1883, another major reduction in duties was undertaken.[91] In 1884, the *sayer* department adopted the English calendar in record keeping.[92] Efforts were also made to improve the salaries and working conditions of the employees of the *sayer* department with the view of keeping a check on pillage and corruption.[93] Despite abolition of several duties, due to the reorganization and centralization of custom duties, the realization from *sayer* increased from Rs 4,61,442 Rupees in 1863–1882 to Rs 9, 96, 180 for the years 1883–91.[94]

In order to achieve this reorganization, Marwar state had to enter separate agreements with local *thikanadars*. In 1883, the Thakur of Kuchman was awarded an annual fixed sum in return for abolishing all local duties on grain, *mapa, chungi, parlai, kalali,* duties realized on camel loads and Banjaras within the town of Kuchman which lay on the trade route to Bikaner.[95] Some of these changes also became necessary as the extension of railways altered the flow of routes and commodities, particularly salt through the Thar. For instance, the reason for agreement on part of the Thakur of Kuchaman was that since a railway depot had been constructed at Kuchman Road (Nawa), the traders instead of paying duties at two places shifted their trade to Nawa, which was within the boundaries of Marwar.[96] The Kuchaman Thakur thus entered an agreement according to which, "his share in Darbar duties would be increased

---

[88] Loch, Report on Marwar Customs, 31st March 1892, 1, Customs, Jodhpur, RSAB.

[89] Hewson, Report on Marwar Customs, 1884, para 13,Customs Jodhpur, RSAB.

[90] Ibid., para 21–24.

[91] As a result of overshooting the income from customs by 2,00,000 Rupees, a reorganization of duties was undertaken as follows, all duties on 41 items, export duties on 210 items, import on 215 items, and transit duties were reduced on, 7 items were abolished. P R Shah, *Raj Marwar under British Paramountcy*, 54–55.

[92] Hewson, Report on Marwar Customs, 1884, para 29,Customs Jodhpur, RSAB.

[93] Ibid.

[94] Loch, Report on Marwar Customs, 31st March 1892, 9, Customs, Jodhpur, RSAB.

[95] Assistant Resident to Musahib Ala, 21st April 1883, Customs, Jodhpur Records, RSAB.

[96] Ibid.

along with an yearly compensation for the losses they sustained".[97] On the other hand the Pokhran *thikana* continued to levy *behtiban* despite repeated requests by the Jodhpur Darbar.[98] Similar agreements had to be drawn with the neighbouring states of Sirohi and Kuchaman. When the Rao of Sirohi requested that the transit duties on goods from Sirohi being charged at Erinpura be abolished, Jodhpur agreed on the condition that Sirohi state too abolish duties on goods from Marwar going into Mewar.[99] Similarly, Marwar also exempted the duties on charcoal being taken from Todagarh forests through Marwar to Ajmer, in return for similar exemptions.[100]

The above discussion on the trading networks in the Thar region attempts to demonstrate how the states in the Thar region attempted to organize the trading networks in the nineteenth century. The difference between the pre-nineteenth and nineteenth century trade was manner in which the idea of a political space had been applied over a wider commercial space. The attempt to impose restrictions on goods moving across the boundaries of several political states altered the nature of manner in which commodities had circulated in the Thar region. The employment of monopolies especially on salt and opium also transformed the circulatory space with increasing use of categories of 'legal' and 'contraband'.

Besides, the British attempted to reframe old trading networks according to their own stakes in long distance trade of various commodities. Their interest in local exchange was largely overridden by their policy of treating the Thar region largely as a transit zone. This led to breakdown of local trading networks and to the decline of thriving markets in the region, which no longer fell on the emerging railway network, which were designed to suit British commercial interests. By the end of the nineteenth century market towns of Pali, Bhinmal and Sanchor that are celebrated emporia in Tod's account were no more than small *qasbas*. The restructuring of the market networks did not result in the securing of older trade networks of the region. Certain markets like wool *mandis* in Bikaner and Nagaur survived by allying with the larger wool circuit in Punjab. But there was a general decline in the status of market towns in late nineteenth century, which was also a result of the newer networks that emerged with the railways.

---

[97] Hewson to Musahib Ala, 2nd July 1883, Customs, Jodhpur Records, RSAB.

[98] Pokhran Thikana in reply to a *purcha* issued by Mehkama Khas, 13th December, 1892, Transit Duties and Passes, Customs, Jodhpur Records, RSAB.

[99] Secretary Musahib Ala to Resident 26th March 1886, Customs, Jodhpur Records, RSAB.

[100] Commissioner Ajmere-Merwara to Resident WRSR, 20th June 1899, Customs, Jodhpur Records, RSAB.

Roads and Railways: The New Channels

As discussed in the first chapter, in the nineteenth century there developed a discourse of hostility, which viewed the Thar increasingly as a closed and difficult land to traverse. It was increasingly felt that it was important to open up the region by creating new, efficient and effective means and modes of travel in the Thar, particularly for the commercial traffic through this region. In absence of such travel networks that were comprehensible from the British point of view, the traffic on the routes had to depend on native guides, as well as an old system of *bolawo*, or protection taxes. Besides, the expediency of connecting the salt works of Rajputana with British controlled salt networks and markets required that these salt works be connected with the rapidly developing railway network in British India. In such circumstances, creating a new network of railways and roads appeared to be a way of making the region more legible, and imposing the states' authority in a more direct and evident manner. As Major Norton, the Superintendent Engineer, Rajputana circle noted in 1862, "there is not, a canal opened, a road laid down, or a railway which pushes its iron horns through into the heart of the province, which does not assist us in stimulating the energies and in developing the half formed ideas of community, and in creating those demands which it is afterwards necessary to supply".[101] Thus, while on the one hand, the road and rail networks were intended to 'open' the region to larger commercial networks, on the other they were also expected to enable the state to implement its policies, like new land settlements in areas considered inaccessible. These networks, particularly the railways were seen as necessary in the light of repeated famines in the region, particularly for transporting food and fodder.

In 1863, three high roads were proposed through Rajputana, one from Agra through Jaipur, Ajmer, Pali and Sirohi towards Ahmedabad, which largely was along the old Mughal highway. The second road proposed was from Ajmer through Naseerabad and Neemach towards Mhow and the third, from Ahmedabad through Neemach and Jhalrapatan towards Guna.[102] Initially, the commercial importance of these roads was questioned and it was enquired if these routes served any other purpose than that of connecting the military cantonments. It was however found that the "road between Erinpoorah and Beawar serves as an outlet for the export of wool and cotton from the great

---

[101] Transfer of Marwar Section of Agra Ahmedabad Road from PWD to Marwar Darbar, Civil Works, 1874, Jodhpur Records, 104, RSAB.

[102] Circular Rajputana Agency, Mount Aboo, 16th September 1863, File no 13 of 1862, WRS Jodhpur, fl 8–9, RSAB.

mart of Pallee to Bombay".[103] These roads were to form a "connected system of high roads through Rajputana and Central India".[104] The road through Pali was also considered to be of utmost importance as, "there can be little doubt that the road between Erinpoorah and Beawar is of Imperial importance as it is the link in the route between Guzerat and north western provinces".[105] Thus, the roads were to connect military outposts as well as become means of connecting commercial flows. The construction of these roads took place between 1869 and 1875. The most important was the Imperial highway connecting Agra to Ahmedabad that covered 155 kilometers between Sendra on Ajmer border to Erinpura on the Sirohi border, passing through Bali, Pali, Sojhat and Jaitaran. This road was transferred to Marwar state in 1874, and one Jahangeer Khan was appointed as the in-charge of the road by the Darbar.[106] This network of roads was considered independent of the impending Railways, as, "the rail will create its own traffic and with a single line will be unable to accommodate, that the rail moreover is a matter of some 6 to 8 years to come, and lacs and lacs of produce with troops and mails must pass over this important line".[107]

However, it was the introduction of the railways in the Thar region that was the most decisive step in determining the future of travel networks. In the British understanding, while on the one hand the importance of Thar was largely as a transit zone, on the other it was also a vast unexplored frontier, bordering Sindh. The construction of railway lines connecting Gujarat, Rajputana and Sindh to northern India could serve both purposes. It could expedite the traverse of commodities through the region while also helping in controlling the western frontier, particularly Sindh better. Another purpose of extension of railways in Rajputana was to facilitate the practice of the "Through Traffic" system discussed later in the chapter.[108] The British

---

[103] Political agent Marwar to Capt. Griffith, 26/5/1862, Civil Works, Jodhpur, Marwar Section, Agra Ahmedabad Road, 2, RSAB.

[104] Circular Rajputana Agency, Mount Aboo, 16th September 1863, File no 13 of 1862, Jodhpur Records, fl 8–9, RSAB.

[105] Political Agent, Marwar to General Lawrence, 3/7/1863, Civil Works Jodhpur, Marwar Section Agra Ahmedabad Road, fl 26, RSAB.

[106] Mehta Bijay Singh, Diwan Marwar to C K M Walters, Pol. Agent, Tranfer of Marwar Section of Agra Ahmedabad Road from PWD to Marwar Darbar, 4/1874, Civil Works, Jodhpur, RSAB.

[107] Note from Superintendent Engineer Rajputana Circle February 1869, Civil Works Jodhpur, Marwar Section Agra Ahmedabad Road, p 103, RSAB.

[108] FP, September 1874, 140–147, NAI. In the Through Traffic system the price and duties on merchandise could be paid at any Government treasury in India, thus doing away with the need of employing local agents.

imperial design found it necessary to "connect Ajmere with Ahmedabad...
thus completing the direct line of railway communication with North West
Provinces through Rajputana and Bombay".[109] The railways through the
Thar region were considered necessary in order to carry grain and heavy
merchandise through Marwar as well as an "advantage in fighting a famine
which every line penetrating the interior of Marwar will be".[110] This railway
was also expected to "facilitate the export of numerous cattle, sheep, and horses
of Marwar".[111] There is no doubt, that the laying of these railway lines was
considered highly profitable from the point of view of trade and commerce
as well as governance. These plans were met with tremendous support from
several sectors. Bombay Chamber of Commerce repeatedly urged for the
construction of the line from Rajputana to Bombay.[112] Jodhpur state agreed
to cede lands for railway construction as well as to relinquish duty on transit
of goods through railways in 1866.[113]

Another important reason for attempting to connect trading zones in this
region was the British interest and subsequent monopoly in salt. Given the
existing British monopolies in Bengal salt, controlling trade in Rajputana
salt was vital. The easy availability of Rajputana salt had to be curbed and the
trade in Rajputana salt had to be made profitable for the British. Therefore,
it was considered important to control major salt works and link them to
depots and markets outside Rajputana. The first line to be constructed was
the Malwa-Rajputana Railway, which ran 24 kilometers from Sambhar Lake
to Kuchaman Road.[114] This line connected the salt works of Sambhar with
Malwa, Central India and United Provinces. The other was the Jodhpur
Bikaner line that connected the Thar region with Punjab and Sind. The
towns through which the railway lines were to pass were Didwana, Marot,
Sankra, Sheo, Jalor, Jaswantpura and Sanchor in Marwar and Hanumangarh
and Suratgarh in Bikaner.[115]

However, the Malwa-Rajputana line was constructed on without taking the
important town of Pali under consideration as it meant a detour of 40 miles.[116]

---

[109] FP, September 1874, 08–110, NAI.
[110] Resident to Musahib Ala, 22nd July 1884, Civil Works, Jodhpur, RSAB.
[111] FP, September 1874, 08–110, NAI.
[112] FP September 1874 108–110
[113] Erskine, *Rajputana Gazetteers*, Vol III A, 122–32.
[114] Ibid.,120.
[115] Ibid.
[116] *Majmui Haalaat wa Intijam*, 1883–84, 642.

The "detour to Palee.... about 40 miles though once a centre of considerable commerce ... was considered unnecessary...on account of so *unimportant* a town" (emphasis added).[117] But Maharaja Jaswant Singh decided to undertake the construction of a branch line, on state means, which would join Pali, and ultimately Jodhpur with the Malwa-Rajputana line at Marwar Junction.[118] This line was planned to pass through Pali, as well as other important towns. A cheap line was constructed with second hand rails and hired engines and coaches were used to operate the first Jodhpur State Railway in July, 1882, though on a slow speed of 20 Kms per hour.[119] This line was also supported by the Assistant commissioner of PWD who argued in favour of the line with respect to "local claims and rights and for political, commercial and local convenience".[120] However the arrangement of booking and interchange of merchandise remained primitive. The stationmaster had to make all bookings for Through Traffic, pay freight, take delivery and rebook the consignments on the other lines.[121]

Subsequently, the construction of the Pali-Luni-Pachpadra section was undertaken as it was to join the Pachpadra and Luni salt works with the larger railway network. In 1880, the Prime Minister Marwar, reiterated the need of a "line direct from Pali to Pachpadra with the object of facilitating the transport of government salt from the Pachbadra salt mart".[122]

This section of the Jodhpur State Railway became operative in 1884. Pachpadra salt was edible but it was located far from the railway lines. Bringing Pachpadra into the railway net could yield upto 6,00,000 *maunds* of salt per year.[123] The Pachpadra line was also expected to serve Jaisalmer and Mallani. It was also hoped that, "a considerable export of cotton, grains and seeds from the fertile lands all along the banks of Luni.... and a large quantity of wool would be carried by the Railway".[124] The added advantage was the fact that Marwar could exercise a better control on Mallani. But there still remained a considerable distance between the railway depot at Balotra and the salt works

---

[117] FP, December 1873 (4) Gen A, NAI.

[118] FP, May 1882, (22–25) Gen A, NAI.

[119] FP, July 1882, (14–16), NAI and *Majmui Haalaat wa Intijam*, 1883–84, 643.

[120] FP, September 1874(108–110), Gen B, NAI.

[121] *Majmui Haalaat wa Intijam*, 1883–84, 643.

[122] Prime Minister, Marwar, to Political Agent, 23 December, 1880, Railways, Jodhpur Records, RSAB.

[123] Memorandum on Financial Prospects of Branch line of Rajputana Railway, 1880, Railways, Jodhpur Records, RSAB.

[124] Resident to Musahib Ala, 7th November 1883, Railways, Jodhpur Records, RSAB.

at Pachpadra, which led to higher portage rates and greater dependence on local carriage and labourers. Sambhar salt could be delivered at the Jodhpur junction at 3.75 annas while Pachpadra salt freight and portage came to 6 annas. Even at Erinpura, Sambhar salt could be delivered for two and a half annas less than Pachpadra salt.[125] This led to the steady decline in the importance of the Pachpadra salt that I would discuss in the next section.

In 1882, Jodhpur State Railway was merged into the Rajputana-Malwa Railway and in 1885, Jodhpur was connected to this network from Marwar Junction with a meter gauge track and later became part of the Jodhpur Bikaner Railway. In the subsequent years, two more lines of considerable commercial and strategic importance were constructed, the Jodhpur-Bikaner line and the Jodhpur-Hyderabad Railway. The Jodhpur-Bikaner line that was built on the request of the Bikaner state and passed through Nagaur was expected to tap sugar traffic from Bikaner, wool from Bikaner and Nagaur, salt traffic from Sambhar and Pachpadra, and marble from Makrana that was so far untapped.[126] The Bikaner line was connected to Punjab thus giving an important strategic access to the desert. Similarly, the extension into Sindh was to be carried further from Balotra on Pachpadra line to Shadipalli in Sindh that was expected to make the connection between Jodhpur and Sindh through the desert. This line, that was extended upto Mirpur Khas in Sindh, was opened in 1900 and was considered of importance as it gave direct access to the port of Karachi.[127]

Table 4.1. Construction of Jodhpur-Bikaner Railway

Branch	Length in miles	Opened for traffic
Marwar Jn to Pali	19	July 1882
Marwar to Luni	25	June 1884
Marwar to Jodhpur City	20	March 1885
Luni to Pachpadra and Balotra	60	March 1887
Jodhpur to Merta Road	64	April 1891
Merta Road to Nagore	35	October 1891
Nagaur to Bikaner Border	24	December 1891
Merta Road to Kuchaman Road	73	March 1893
Balotra to Shadipalli	135	December 1900

[125] Major Murray to Resident, 12th October 1883, WRSR, Jodhpur Records, RSAB.
[126] Note by Home on JB Railway, FP, 18th September 1888, (78–91) Int A, NAI
[127] Marwar Administration Report, 1900–01, 11.

The expansion of Railways in The Thar region effectively altered the older networks of travel through the desert. The construction of the railway lines also necessitated the construction of feeder roads at right angles to the Railways as "the old dak roads were on lines parallel with the railways, and, on the construction of the railways they would be become of less importance".[128] Only with the construction of such roads could the railway line be of any use, as all merchandise would have to travel on these roads to arrive at the railway depots.

In this manner, older networks of travel came into disuse and newer emerged. The railways were laid according to the British commercial needs. For instance, the Ajmer-Ahmedabad line bypassed Pali, which was the most important market town in Marwar, as the emphasis was on the Thar as a transit zone. Pali could only be joined by a cheap branch line on which the train could not travel at a speed more than 20 kms per hour.[129] The railway lines also evaded important market towns like Sojhat, Jaitaran, Jalor, Bilara, Bhinmal, Sanchor, Shergarh, Phalodi, Bisalpur, Kaparda, Khandap, Tinwari, Pokhran etc. where commerce gradually declined. By the time railways were extended in early twentieth century, these marts had lost their importance as the railways had already directed the commerce away from these towns. Towns like Jalor, Sanchor, Bhinmal, Barmer, Sojhat, Nadol, Mandor, Pali etc had been ancient market towns. The older routes in the Thar region allied to these old market towns. But the railway networks ignored these towns that led to their ultimate decline. On the other hand, towns like Balotra, Mathania, Gotan and Makrana that were on the railway lines flourished. So we can say that a new travel network emerged in the nineteenth century, which was expected to open far flung regions and join them with larger trading systems. Paradoxically, not only did the new travel network ended up isolating older market towns, the movement on these networks was far more restrictive. Older norms of circulation were no longer operative as new rules and regulations emerged. One of the major consequences of this was that a number of communities that had been operating on older travel networks could no longer function with the railways. Nowhere was this change felt as drastically as on the salt trading networks in the Thar region.

The Question of Salt

No salt would be brought into Marwar.... nor salt from the daribas of Government at Pachpadra, Nawa, Sambhar, Didwana and Phalodi shall be

---

[128] Superintendent Engineer, PWD to Musahib Ala, 16th January 1867, PWD, 34–39, RSAB.

[129] FP, July 1882 (14–16) Gen A, NAI.

taken out.... any one found doing so will be fined Rs 250 and or jailed for upto 3 months and all salt found in possession confiscated.[130]

Salt was a major produce of the Thar, as well as one of the largest exports of the region. In 1877, A. Adam, the Inland Commissioner, observed that, "the great peculiarity of Jodhpur which gives the state so much interest in our eyes at present, consists in its numerous salt sources."[131] Archibald Adams in his study, *The Western Rajputana States*, observes that, "this part of the country is rich in salt which is obtained in large quantities from natural salt lakes of Sambhar and Didwana, but also from artificial pits of Phalodi, Pokharan, Bhasti and many other places. The salt jhils of Sargot and Kuchavan possess unknown capacities of manufacture although it is not now manufactured at these places".[132] Salt was produced at the Sambhar lake, Didwana located northwest of the Sambhar lake, Pachpadra which was located about fifty miles southwest of Jodhpur, the salt marsh at Phalodi and another one at Pokhran.[133] Besides, salt was also manufactured in *agars*, or salt pans "working factories that turned out good edible salt".[134] Salt was also manufactured in the dry bed of Luni during the summer season. In Bikaner state, salt was produced from the salt lakes of Chhapar and Lunkaransar. While salt from sources like Phalodhi and Pokhran was often sent to neighbouring *mandis*, salt from the Sambhar Lake was the one that was preferred for export. In the Mughal period, Agra was a favoured destination from where annually 10,000 tons of Sambhar salt is reported to have been exported further by boats through the river route down to Bengal.[135] James Tod notes that the, "salt is exported to every region of Hindustan, from Indus to Ganges, and is universally sold under the title of *Sambhur Loon*, or the 'salt of Sambhur', notwithstanding the quality of different lakes varies, that of Pachbhadra beyond the Looni, being most esteemed."[136] By late nineteenth century it was reported that apart from about "2,00,000 *maunds* of salt, of which 50,000 *maunds* was supplied from Pachbadra, 15,000 *maunds* from Luni; 5000 *maunds* from Pohakaran, 5000 *maunds* from Deedwana, 10,000 *maunds* from Phallodee and the remaining

---

[130] *Kayda Sazaa Mukadmat Mutalke Namak Raj Marwar*, May 1910, 1.

[131] General Report on Salt Producing Capabilities of the Jodhpur State: Superintendent of Government Printing Calcutta, 1871, 1

[132] Adams *The Western Rajputana States*, 2–3.

[133] Ibid.

[134] Ibid.

[135] Irfan Habib, *Agrarian System of Mughal India*, 64.

[136] Tod, *AAR*, II, *133.*

1,25,000 *maunds* being supplied from local salt works" was exported from this region.[137] Of this Pachpadra salt went into "Meywar, Malwa, Gwalior, Harowtee, Indore...all over central India right upto Nerbudda".[138] The Luni salt was sent to Sirohi, Gujarat, Jaisalmer and Sindh, while the Sambhar-Didwana salt, "crossed through Shekhawuttee and Bickanir into British territory... leaving enroute about 30,000 *maunds* for local consumption in these states".[139] On the other hand, Phalodi and Pokhran salt was consumed in Bikaner and Bahawalpur.[140] Salt produced at Lunkaransar and Chhapar in Bikaner was used for tanning and dyeing. Besides these sources, several small sources were located in the *jagir* areas.

Table 4.2. Salt Agars and Pans in Marwar in the Seventeenth Century [141]

Paragana	Village	No. of Agars	No. of Houses of Kharauls
Jodhpur	Kapardo	25	30
	Kankani	1	1
	Mori Baman ri	1	1
	Chavadha Dhiyan	40	40
	Bilara	8	20
	Dessor	3	0
	Kaluano	7	7
	Kharla	2	2
	Kharo Bero	1	1
	Bhakri	2	1
	Waul	6	6
	Bahlo	4	6
	Ghaghani	13	15
	Panchiyak	3	4
	Sanvalto Bado	15	20
	Hamavas	9	12
Merta	Punalto	4	4
	Javo Sisodiyan	3	3
	Labardar	2	2
	Kherwo	4	4

Contd.

---

[137] General Report on Salt Producing Capabilities of the Jodhpur State: Superintendent of Government Printing Calcutta, 1871, 1.

[138] Ibid.

[139] Ibid.

[140] Ibid.

[141] Nainsi, *Vigat*, II, 34–36.

Contd.

Paragana	Village	No. of Agars	No. of Houses of Kharauls
Sojat	Badiyalo	4	50
	Godhelav	1	1
	Hasalpur	4	4
	Dhavlera	3	3
	Khakhro	3	4
	Moklavasni	2	2
	Mahev	3	4
	Hasiya Hedo	13	12
	Chopro	6	6
	Sobhrawas	2	2
	Popla	2	2
	Dhudhiyavasni	6	6
	Sinlo	2	2
Siwana	Pachpadra	300–325	50
Phalodi	Godhanila	100	60–70
Pokhran			20

The manufacture of salt was a significant industry in the Thar and chiefly involved the community of Kharauls or Lunias, who held leases for its manufacture. Adams described "the method of obtaining the salt at Sambhar, Didwana and Phalodi as "simple, viz by evaporation after the rainy season".[142] In Pachpadra a different process was followed where as Adams observed, "oblong pits of various sizes are dug, a supply of brine percolates through the pit bed, and when that has become sufficiently concentrated, so as to show signs of crystallization around the pit edge, branches of a thorny shrub, called *morali,* are sunk in it. On these branches salt crystals form and continue to grow for two, or sometimes three years".[143] This salt was extracted manually by men who "enter the pit, and with an iron chisel, through the thorny branches, break up the salt which is caked on the bottom. By shaking the branches the crystals are detached. The salt thus broken up is drawn to the sides by a broad iron hoe and is removed in baskets to the top of the pit". [144] Further, "it is then gathered and heaped up into immense masses, on whose summit they burnt a variety of Alkaline plants, such as saji, by which it becomes

---

[142] Adams, *The Western Rajputana States,* 3
[143] Ibid.
[144] Ibid.

impervious to the weather".[145] The salt thus acquired were used not only for domestic consumption and trade; it was also needed by pastoralists and for various industries like dyeing and tanning. Thomas Hendley in his *General Medical History of Rajputana*, observed that the "abundance of cheap salt has been necessarily of value to all in the past times, both for the food of man and beast, and for the preservation of skin of the latter".[146]

The salt pits were customarily the property of the state that were leased out to the Kharauls, who were entitled to deal with the manufacture and sale of salt. The states of the Thar region recognized the rights of the Kharauls and leased out salt pans and mines to them as *ijaras*. The leases of the salt pits or mines were largely hereditary and were treated as properties against which the Kharauls secured loans and mortgages, though only with the approval of the state. When the mines were given on *ijara*, a preference was given to the Kharauls, as an *ijara* document of 1785 regarding the *ijara* of salt pits in Pachpadra shows, "Kharauls of the village have been manufacturing salt traditionally, hence the *ijara* of the *agar* be granted to them".[147] The salt sources that were located in *jagir* areas were similarly leased out by the *jagirdars* who paid a share to the state from the income thus derived. The presence of official leases to the salt manufactures did not, at any point of time mean that salt was manufactured only at these sources. There existed small and seemingly insignificant sources where salt was obtained perhaps for local consumption, which often escaped the revenue net of the state. For example, in 1788 the *sayer* officials in Merta reported that Kharauls of village Kherwo were selling salt and not paying any duties.[148]

A number of communities were engaged in salt trade within and outside the Thar. The community of Banjaras or Baldiyas that was involved in the carriage of grain was also traded and transported salt across the Thar. In late eighteenth century, Pema Nayak, a Banjara seems to have been a prominent trader in salt, and received yearly *chitthis* to procure salt from Sambhar and Didwana as well as numerous exemptions enroute. For example, in 1765 a *sanad* was issued in his name instructing the Kharauls at Sambhar to weigh him seventy *maunds* of salt and the *Sayer daroga* to exempt him from the payment of *rahadari*.[149] Between 1765 and 1773, the Banjara is mentioned several times in the *sanads*

---

[145] Tod, *AAR*, II, 133.

[146] Thomas Hendley, *General Medical History of the Rajputana*, 4.

[147] *Miti Margshish sudi 3, Dariba Pachpadra, SP Bahi* No.33, VS1842/1785 CE.

[148] *Miti Paush sudi 4, Merta Kacheri, SP Bahi* No. 39,VS 1845/1788 CE.

[149] *Miti Vaishakh Sudi 6, Dariba Sambhar, SP Bahi* No.4, VS 1823/1766 CE.

pertaining to *dariba* of Sambhar. Other communities engaged in the salt trade and often exempted from the payment of custom duties were Charans, Bhils as well as Pushkarna Brahmins, particularly in Pachpadra.[150] The whole process involved thousands of people and pack animals like oxen and camels. It was a well established and recognized system with clearly defined places where carriers could stop to graze or water their animals. Besides, salt trade was intricately linked with exchange of other commodities like grain, sugar and livestock. While the Kharauls held leases for salt production, theoretically, the salt traders bought the salt from the state and paid duties for the further sale of salt. In order to procure salt from various sources the dealers received a *chitthi* to the effect from the respective *sayer thanas* that enabled them to buy the quantity of salt mentioned in the document.[151] In return, the states realized a considerable amount as taxes and dues from manufacturers as well as taxes like *dan, mapa, rahadari, kayali* etc on the sale of salt. Whole sale merchants controlled the bulk trade in salt with established business in the market towns, operating through their agents at the salt works.[152] Merchants with trading houses in Jodhpur, Pali, Merta and Nagaur were found to purchase salt from Pachpadra, Sambhar, Nawa and Didwana, presumably for export.[153] These merchant firms were granted various degrees of exemptions in transit duties by various states in the Thar.

The subsidiary treaties of 1818 opened the Thar salt trade to the British who turned it into a monopoly, leading to the ruin of flourishing salt trade in the Thar as had happened in Bengal, Orissa and Berar among other regions. The British already controlled salt trade in Bengal by the time the subsidiary alliances were signed in the Thar. The British involvement with salt trade in Rajputana began in 1830s with near takeover of Bharatpur salt works and increasing the duty on salt entering British territory at Agra and Allahabad by four to twelve *annas* per *maund.*[154] This made Bharatpur salt unsaleable in British territories. The salt returns were so profitable that by 1834 it was no longer necessary to concentrate the energies of the customs department on other

---

[150] *Miti Ashadh sudi 7, Dariba Sambhar, SP Bahi*, No.7, VS1824/1767 CE.

[151] *Miti Magh sudi 13, Budhvar, SP Bahi* No.4, VS 1823/1766, The *sanad* is in form of *chitthi* to Kharauls in Nawa to give salt worth rupees one thousand.

[152] *Miti Sravan vadi 4, Dariba Sambhar, SP Bahi* No.6, VS 1824/1767 CE.

[153] Ibid. The order mentions that merchants from Jodhpur coming to Sambhar to buy salt should be granted one fourth exemption in *rahadari*.

[154] M S Jain, *Concise History of Modern Rajasthan*, Vishwa Prakashan, New Delhi, 1993, 80–85.

commodities. Between 1835 and 1843, the British took over the Sambhar salt works from the Jaipur state in order to set off the expenses on the Shekhawati Brigade.[155] For some years after the lease of the lake, the Government officers there posted merely supervised the manufacture and sale of salt and duty was collected on the salt when it crossed the customs line.[156] By the 1850s, commerce related salt treaties with the princely states began to be discussed. By the year 1869, the need to control free trade of salt in India had already led to the creation of a 2,472 miles long customs hedge dedicated entirely to preventing the circulation of cheaper Rajputana salt. It ran from "Torbela to a point on the Mahanadi,... and was guarded by an army of 12,911 officers and men. Consignments of salt and sugar were liable to examination when they entered the jurisdiction of the Line, which was a zone 10 to 15 miles in width, and were examined at the posts open for trade and traffic, all of which were under European supervision. Wherever it was possible, a hedge was cultivated so as to form a natural barrier; and in places the hedge thus formed was impenetrable".[157] Despite the customs hedge it was also felt that in order to control the salt trade it was important to gain control of the salt works in Rajputana, as cheap salt produced there was the source of vast quantities of legal or smuggled salt that entered the Bengal Presidency. Besides, the salt trade in Rajputana was thought to be a lucrative enough venture that could be rendered very profitable through "skilled supervision of the sources and improving communications".[158] Though, there were some taxes on salt trade in the Thar, none of them were as high as the British salt tax, or as diligently realized. Therefore, the British entered a series of salt treaties with the princely states of Rajputana, with Jaipur State in 1869–70 and Jodhpur state in 1870.[159] The events preceding the signing of the salt treaties led to the appointment of the Agent of Governor General (AGG) Colonel Keating in Jodhpur in 1868. In Bikaner state, the agency was established the same year in Sujangarh against the wishes of Maharaja Sardar Singh, with Capt. Powlett as the AGG. In 1868

[155] Ibid.

[156] Richard M Dane, 'The Manufacture of Salt in India', *Journal of the Royal Society of Arts*, Vol. 72, No. 3729 (MAY 9, 1924), pp. 402–418. Published by: Royal Society for the Encouragement of Arts, Manufactures and Commerce Stable URL: http://www. jstor.org/stable/41356565, 404.

[157] Ibid., 405. Also Roy Moxham, *The Great Hedge of India*, New Delhi, 2001.

[158] Manual of Northern India Salt Revenue, Vol I, 15.

[159] Report on Salt Royalty Dispute between Jaipur and Jodhpur, 15th December 1883, Salt, Jodhpur Records, RSAB.

and 1869 respectively, Jodhpur and Bikaner signed extradition treaties with the British. In 1869, Jodhpur also agreed to the construction of the Imperial road from Agra to Ahmedabad through the Jodhpur territory. These treaties in a real sense were preliminaries to the control of salt trade.

In 1870, the first Salt Treaty was signed between Jodhpur state and the British. Sambhar Salt Lake was leased by the Government from the Native States of Jodhpur and Jaipur, to which it jointly belonged. According to this treaty, the lease of the Jodhpur side of the Sambhar lake was given to the British on the payment of 1.25 lakhs annually and a royalty of 40 percent in case there was a sale in excess of 8.25 lakh *maunds*. The Jodhpur state was to receive 7000 *maunds* annually.[160] Similarly, the salt works at Gudha and Kuchaman were rented out for a sum of Rs 3 lakhs annually and a promise of 40 percent royalty in case of sales exceeding 9 lakh *maunds* per annum.[161] By 1878, it was decided to "assume the management of Didwana at once".[162] However, at this point the British were content to get Didwana and A C Lyall, the AGG, advised that, "no duty will be imposed at first and existing Durbar arrangements should be as little interfered with as possible. The great thing is to get the source in our hands, to learn all about its capacities and local politics, and above all get the people accustomed to our management and all our arrangements".[163] The very next year the Jodhpur Darbar sought a loan of Rs 24 lakhs and against this loan the salt works of Didwana, Pachpadra, Phalodi and Luni basin were leased to the British.[164] Thus by 1879, the British had "assumed the management of the Pachbadra, Didwana, Phallodee and the Luni salt tract, the major salt sources in the Jodhpur state".[165] According to the provisions of the treaties, the British were also to have the sole right of manufacturing salt in Jaipur, Bikaner and Jodhpur and production in all other salt works was to be stopped. They negotiated for "the closure on payment of unimportant sources for the restriction in the out turn of works left open under the management of the Darbars".[166] The treaties also provided for the

[160] Report the Administration of Northern India Salt Revenue Department, 1883–84, Mehkama Khas, Jodhpur Files No 25, 15–16, RSAB.

[161] Ibid.

[162] Salt Demarcation, File 9 of 1878, 9–16, A C Lyall Esqr, AGG Rajputana to W S Halsley C S, Commisioner of Customs Agra, RSAB.

[163] Ibid.

[164] The Didwana salt source that lay north of Sambhar and bordered Jaipur state was also shared between the states.

[165] Manual of Northern India Salt Revenue, Vol I, 1.

[166] Ibid., 18.

"suppression of manufacture of earth salt, and for abolition of transit duties and of all restrictions upon the trade in salt on which duty had been paid to Government of India".[167] Also, the only salt that could transit through Rajputana was one on which British duties had been realized, all other salt being regarded as contraband.[168] In return for this the Jodhpur Darbar was to be paid Rupees 1.25 lakhs annually and the *jagirdars* in whose territories any salt mines should happen to be, were to be given Rs 19,600. British were to provide 2.25 lakh *maunds* of salt at the rate of 8 *annas* per *maund* for domestic consumption in Marwar and 10,000 *maunds* for the personal consumption of the Darbar. The Jaipur Darbar was offered identical conditions, considering the revenue sharing arrangements between the Jaipur and Jodhpur states with regards to salt revenues from Sambhar.[169] A similar agreement was drawn with the Bikaner state regarding the salt lakes at Chhapar in Churu and Lunkaransar near Bikaner. Not more than 30,000 *maunds* of salt could be extracted from these lakes annually, and 20,000 *maunds* had to be purchased from Phalodi and Didwana. All other salt making in Bikaner was declared illegal. Bikaner too could not impose any duties on salt and no salt except the British salt was to pass through Bikaner. In order to ensure the imposition of the salt treaties, the inspection of all big and small salt works was undertaken by the Northern India Salt Revenue Department. One of the objectives of these inspections was to check the "irregularities in the way of removal of earth salt and material salt".[170] The *hakims* of Pali, Parbatsar and Bilara were strictly ordered to, "exercise a close and rigid supervision of the tracts- to put an end to the manufacture of salt" in the prohibited sources.[171] There were also pronounced measures for suppressing the cases of "smuggling" of salt. Measures for suppression of cases of "smuggling" of salt were also announced. [172]

As a result salt became an expensive commodity even within the salt producing areas of Marwar. By 1883, salt was being sold in Marwar at a price that varied between Rs 5 to 10 per *maund*, which had more than doubled over

---

[167] Dane, 'The Manufacture of Salt', 405.

[168] *Kayda Sazaa Mukadmat Mutalke Namak*, 2.

[169] Administration of Northern India Salt Revenue Department, 1883–84, Mehkama Khas, Jodhpur Files No 25, 18.

[170] Commissioner, NISR to Resident, 30th July 1886, Salt, Jodhpur, No 37, RSAB.

[171] Secretary to Musahib Ala to Resident, 9th September 1886, Salt, Jodhpur, RSAB.

[172] Hakim Pachpadra to Secretary to Musahib Ala, 4th May 1885, Salt, Jodhpur, RSAB. These included diligent checking of all caravans entering the state, as well as arrest of people found to be trading in untaxed salt.

the past fifty years.[173] This price was attributed to the heavy royalties that had to be paid to the native states, and thus every effort was made to make the enterprise of salt profitable. The salt administration of the British was based on the logic of maximum extraction and for this they attempted to rationalize the process of salt extraction and distribution. They were concerned about the wastage at the salt pits that was caused due to the negligence and complicity of the salt manufacturers and dealers. By 1882, in Sambhar it had been found that "very large deficits (were) found on clearance of heaps......which not unfrequently amounted to from 20 to 30 percent on the entire contents of a heap and were far too large to be accounted for by fair wastage and that they were due to over-estimation or to unauthorised removal or to both causes combined. The first of these causes implied that the extractors received payment in excess of what was due to them while the second involved far more serious loss".[174] To counter this, the British adopted a system of cubic measurement, in which they used an estimating rod that was "so constructed to shew 25 percent less than the actual quantity".[175] Thereafter, payment was made for, "a quantity less by 25 percent than the amount extracted and of the excess 12 ½ is credited in the stock books and 12 ½ percent is left out of the stock account altogether to allow for wastage and loss".[176] The salt administration at Sambhar also experienced difficulty due to the old practice of spreading the salt to dry before weighing. The "practice does not prevail at other works either here or in Bombay and it gives an undue advantage of some 3 or 4 percent to purchasers at Sambhar".[177] The prohibiting of this practice brought down the losses due to wastage to 10.24 percent.[178]A uniform system of weighing was also introduced. Earlier, the salt dealers, who held *rawannas* or passes for salt, could use bags of any shape or size and ten to twenty percent of their consignment was reweighed by the inspecting authorities to estimate the gross weight of salt being carried out. The local salt dealers also used irregular stone weights. This situation was remedied by the use of uniform bags with capacity of storing 2 ½ *maunds* of

---

[173] Report of Salt Admin 1883, Salt, Jodhpur Records, RSAB, 16. In 1820s Tod reports the price of salt to be between 2 to 4 Rupees per *maund*, with price in Jodhpur being Rs 2 per *maund*, while in Didwana, Sambhar, Pachpadra and Phalodi as being Rs 4 per *maund*. Tod, *AAR*, II, 133.

[174] Report of Salt Administration, 1883, Salt, Jodhpur Records, RSAB, 19.

[175] Ibid.

[176] Ibid.

[177] Ibid.

[178] Ibid.

salt per bag and by introduction of new weighing machines, which displayed a difference of 5 to 10 percent in comparison with local scales of the dealers.[179] Thus, while in 1881–82, the total production of salt in Sambhar was 32,90,057 *maunds*, in 1883–84 it rose to 71,11,353 *maunds*.[180] A record production of salt was observed in year 1910–11, being 1,00,05,412 *maunds* from the Sambhar source alone.[181] The increase in output from the Sambhar salt works corresponded with the expansion of the Rajputana-Malwa Railway in 1875, which ran 24 kilometers from Sambhar Lake to Kuchaman Road connecting Sambhar with Malwa, Central India and United Provinces.[182] However, not all salt works experienced similar increase in production, as is visible with the case of another important salt source, that in Pachpadra.

After the Salt Treaty of 1879, the British sought to evaluate the salt sources at Pachpadra, Phalodi and Luni Basin. It was found that the Pachpadra salt source was at a "grievous disadvantage as compared to other sources owing to the remoteness of its position".[183] The Pachpadra salt mines lay south-west of Jodhpur and were the third important source of salt in Marwar state. Nainsi described Pachpadra as a *rann* located 12 *kos* from Siwana, with 300 to 325 salt pits containing brine out of which a crop of salt could be obtained every four months. In late seventeenth century, as estimated by Nainsi, Pachpadra housed 50 Kharaul families who paid half of the salt extracted as tax.[184] The *hasil* from Pachpadra was 10000 *duganis*, which largely was from the salt works.[185] There were also smaller salt works in Luni Basin scattered around villages of Khed, Talwara, Mahewa and Jasol. These were in *jagir* areas while Pachpadra was in *khalsa*. In 1881, Pachpadra was reported to have about 1000 pits surrounded by *partal* land out of which 298 were in working condition. At this time, it was claimed that a crop of salt could be produced every 18

---

[179] Ibid., 19–20.

[180] Between 1869–69 and 1877–78, the production at Sambhar had risen from 20, 700 tons to 29, 800 tons, while it fell at Bharatpur from 43,000 tons to 37,000 tons. Dane, 'The Manufacture of Salt', 405. Richard Dane considers imposition of uniform salt duty irrespective of quality as an important reason for the decline of less valuable salt sources.

[181] Extracts from Administration Reports regarding the Disadvantages of the Pachpadra Source, Appendix (1).

[182] Erskine, *Rajputana Gazetteers*, Vol III A, 120.

[183] Report of Salt Admin, 1883–1884, Salt, Jodhpur Records, RSAB, 25.

[184] Nainsi *Vigat* II, 36.

[185] Nainsi *Vigat* II, 247.

months. The production in the said year was 5,53,544 *maunds*.[186] The average production in Pachpadra varied between 5 to 6 lakh *maunds* per year with a record high of 12,99,862 *maunds* in 1920–21 and low of 56,825 in 1926–27. In 1881, it was reported that though Kharauls claimed hereditary proprietary rights in the salt pits of Pachpadra there were actually no *sanads* to the effect of it. By the early twentieth century, Pachpadra salt source was reported to be declining in the light of reduced demand from the source. By 1926, the production had fallen to a record low and it was considered that "it could ever be worked except at a loss".[187] The salt source was located in barren country and was inaccessible in summer. Banjaras could reach Pachpadra only in monsoon when sweet water and fodder was available. Banjaras who did come to Pachpadra were forced to water their oxen at Balotra. In the worst case, for instance, in the year 1921–22, it was reported that water had to be carted by road or rail. Besides, in drought years the lack of rains rendered the salt pits useless.[188] In order to deliver a consignment to the railway yard, Banjaras were forced to make a journey of 60 miles.[189] It was hoped that the opening of a new railway at Pachpadra itself, will have the effect of stimulating the trade in Pachpadra salt.[190] Since the pits were scattered over a wide region, even after sidings were constructed pack animals were still required to carry salt to the railway wagons.

It had been anticipated that with the construction of the railway, the Pachpadra source would be in a better position to compensate for the shortfalls in Sambhar salt. It was soon realized that owing to the cost of portage, the Pachpadra salt was pitted highly disadvantageously against the Sambhar salt. Even at junctions like Jodhpur and at Erinpura, which were farther off Sambhar, the cost of Sambhar salt worked out to be lower than the Pachpadra salt. At Jodhpur Junction, the freight on Sambhar salt worked out to be 3.75 *annas* per *maund* as against 6 *annas* per *maund* for Pachpadra salt. At Erinpura junction too Sambhar salt could be delivered at 5 *annas* per maund as against Pachpadra salt for 7 ½ *annas* per *maund*. [191]

Another problem that besieged Pachpadra was shortage of local labour. While on the one hand the scarcity of water discouraged the immigration of

---

[186] Report of Salt Admin, 1882–1883, Salt, Jodhpur Records, RSAB, 59.

[187] Report of Salt Admin, 1883–1884, Salt, Jodhpur Records, RSAB, 69.

[188] Report of Salt Admin, 1918–19, Salt, Jodhpur Records, RSAB, 8.

[189] Report of Salt Admin, 1885–86, Salt, Jodhpur Records, RSAB, 32, 1886–87, 20.

[190] Report of Salt Admin, 1886–87, Salt, Jodhpur Records, RSAB, 20.

[191] Major Murray to Resident, 12th October 1883, Salt, Jodhpur Records, RSAB.

required labour from neighbouring areas, as often happened around salt mines, on the other, local labour too often migrated to Sindh thus creating shortfalls at Pachpadra. In 1893, it was reported that a load of 3000 *maunds* could not be cleared for lack of portage and carriage.[192] In the same year, the Hakim of Pachpadra reported that he had to assist the contractors in finding portage and carriage.[193] It should be pointed out however that 1891–92 were years of a *trikaal*, a severe famine with acute shortages of food, water and fodder, and the population around Pachpadra would have migrated to Sindh in this period. Also, it was reportedly difficult for the contractors to find labourers at the time of the *Chaitri* fair, held closeby annually. In any case, the availability of carriage also depended upon the availability of good grazing for camels, asses and oxen, which could not always be assured.[194] Increasingly, the local Kharauls appeared to abandon the business of salt extraction at Pachpadra causing great concern allegedly because of their "laziness and improvidence".[195] If viewed from the perspective of the Kharauls who after losing manufacturing and trading rights had been reduced to being mere labourers, there was little interest in reviving a trade already lost.[196] They either migrated to thriving salt works or continued to work as labourers. However, by 1925–26, the Salt Administration appeared to accept of the decay of the Pachpadra salt source "with its lack of water, its disabilities of position, the impossibility of its competing with a fully developed Sambhar and the apathy and indolence of its workers though to be deplored, appears inevitable".[197]

It is clear from the above discussion that the decline of Pachpadra as a salt source occurred within fifty years of the Salt Treaty of 1879. It is difficult to believe that ecological conditions could have altered so drastically within a span of fifty years to cause such acute water shortage and render the pits dry. The fact remained that the British did not see Pachpadra as an independent salt source, but rather an adjunct to the supply from Sambhar. The urge to exploit each and every source to its fullest and associate it with the larger salt networks led to the destruction of local salt networks to which sources like

---

[192] Number 615 of 27th August 1893, Pachpadra Salt Papers, Salt, Jodhpur Records, RSAB.

[193] Musahib Ala to Assistant Commissioner, 26th September 1893, Pachpadra Salt papers, Salt, Jodhpur, RSAB.

[194] Musahib Ala to Resident WRS, 25th February 1898, Pachpadra Salt papers, Salt, Jodhpur, RSAB.

[195] Report of Salt Admin 1910–11, Salt, Jodhpur Records, 7, RSAB.

[196] Ibid., 6.

[197] Ibid.

Pachpadra supplied. The denial of traditional rights of local Kharauls and their distancing from the process of salt production in Pachpadra also affected the supply and sale on the smaller local networks that they supplied to.

These changes were not accepted without protest at all places. An area where a resistance was offered to these changes was the Luni basin that had some small salt mines. These mines were in *jagir* areas of Mallani and there were several discussions over their transfer to Raj Marwar. They were a source of perennial income to the *jagirdars*, who understandably were not willing to part with their traditional rights. The transfer of the rights of mining salt around the Luni basin, which included mines at Jasol, Khed Talwara and Mahewa had taken place in 1841. In evaluation of mines in Luni tract, it was found that there were 23 salt mines in the Jasol area of which there was no salt in 15 mines. Rest of the mines was destroyed by rain so only 4 mines were workable.[198] These mines were in the jurisdiction of various *jagirdars* of Jasol, Khed and Mahewa. In return for the mining rights the *jagirdars* were to be paid 1 *anna* per rupee from the income from salt. They were also to get Rupees 1000 for grazing and watering of oxen that came to carry the salt from 1843 onwards. In case the animals despoiled the crops the *panchayat* was to be called to decide upon the fine. [199] This agreement was renewed in 1875, where by the Raj Marwar sought to expand the production of salt in the *partal* lands of the Luni Basin. It was stipulated that the *jagirdars* of Jasol would be paid the said amount even if there was no salt to be extracted from the mines.[200] Some mines in Khed Talwara were shut down and in lieu the *jagirdars* were to get fixed amounts of Rupees 105 annually.[201] Besides, since the railway had already been planned so that whenever the railways came and pack animals were no longer required the dues for grazing and watering would stop.[202] The *jagirdars* were also to receive 125 *maunds* of salt annually for personal consumption.

The earlier agreement of 1841 denied the right of the local Kharauls to sell salt anywhere outside Mallani. This was contested time and again by the small *jagirdars* of Mallani who put forth various objections. Rawal Sivdan Singh of Jasol complained that, "our Kharauls cannot sell salt anywhere outside Mallani

---

[198] Statement of the Hakim on the Salt works near Jasol, 25/7/1841, Mallani Salt Cases, 8, RSAB.

[199] Agreement between Raj Marwar and Jaigirdars Jasol, Mallani Salt Cases, 303–309, RSAB.

[200] Ibid.

[201] Agreement between Raj Marwar and Jaigirdars Jasol, Mallani Salt Cases, 307, RSAB.

[202] Ibid.

and if they do so they are arrested with salt, even though we have the right to mine the salt since the time of Mallinathji. But the Charans from Sindhari, Desi Bhats from Jalor and Gadit from Jodhpur come to take salt from Talwara and sell. This is an injustice to our Kharauls. Besides the traders graze and water their animals in our villages. The Hakim Pachpadra is unjust to our Kharauls...so they should also be given letters to sell salt".[203] In December 1842, the *jagirdars* and residents of village Mahewa of Mallani complained that merchants from Pachpadra while collecting salt halt at the village and "their oxen graze in the gram fields ...there is very little water in the village... earlier there was no halt in the village... the *hakim* is posted here now to count loads of oxen earlier there was no *chowki* here. ...please discontinue the halt". [204]In 1845, it was complained "the *jod* of our villages is near Pachpadra and the traders who come to take salt halt at the *jod* ...they despoil our grass".[205] A similar complaint was lodged by *jagirdars* of Mahewa regarding the despoiling of the gram crop in the village by the pack oxen of the traders coming to Pachpadra.[206] In the same year the *jagirdars* protested that while the traders from Jodhpur could came and take salt from Khed- Mallani if the Kharauls from Mallani went to Jodhpur they paid fines. They also claimed that the volume of salt extraction in Mallani had increased and so the villages could no longer support the number of traders coming.[207] In 1859, Sivdan Singh and others the *jagirdars* of Jasol complained that the Hakim of Pachpadra had arrested the Kharauls of Jasol and attached their salt. He refused to return the salt and he "harasses the Kharauls of Jasol....they sell salt and have always been doing so...we request that the Kharauls and Bhils of Talwara should not be bothered".[208]

However, by the agreement of 1875 these objections lost their meaning. While the *jagirdars* were financially compensated for their loss, there is little information on the way the Kharauls and Bhils of Mallani, were compensated. In all likelihood they took to extracting salt for the British in Luni basin and Pachpadra, losing their status as proprietary salt extractors.[209] While Phalodi and the Luni were worked for a time, but without railway communication it

---

[203] Administration of Mallani Salt Cases, 1840–1873, Part I, 9/5, 205, RSAB.

[204] Ibid., 20–21, 23 December 1842.

[205] Administration of Mallani Salt Cases, December 1845, 36, RSAB.

[206] Ibid.

[207] Ibid. 16.

[208] Administration of Mallani Salt Cases, 1840–1873, Part I, 9/5, 251, RSAB.

[209] *Report Mardumshumari*, 510.

was found to be impossible to work them profitably.[210] The Luni basin salt works, therefore, were closed in 1887–88, and Phalodi in 1897–98. The salt sources of Luni basin appear to have entirely lost their commerce by the early twentieth century.

As we saw in the last section on the Railways that the laying of the Railways was planned in accordance with the circulation of salt trade in The Thar. By 1904, 80 percent of salt transport in Marwar was being carried out by the railway.[211] By this time, 50 percent of Didwana salt was being sent to Punjab and 46 percent of Pachpadra salt was being sent to United Provinces.[212] Major Erskine, Resident at Jodhpur, noted that there were, "three major railway stations on the (Sambhar) lake-at Sambhar, Gudha, Kuchaman Road or Nawa- and the line runs into all the principal manufacturing works or walled enclosures; the salt is stored close to the line and loaded directly into the wagons; it is largely consumed in United Provinces, Central India and Punjab south of Karnal, and it also finds its way into the Central Provinces, Bihar and Nepal".[213] He further noted that earlier, "the trade was in the hands of the Banjaras but with the extension of the Jodhpur-Bikaner Railway to Balotra and the continuation of the branch line to the works, very few of these wanderers visit the place and practically all salt is removed by rail."[214]

In order to control and facilitate the long distance British trade in salt, the British, along with extending the Railways, also brought in use the 'Through Traffic' system in 1876.[215] This system enabled the, "traders throughout... United Provinces and Central Provinces...to obtain salt direct without employing local agents by depositing the price, duty and a small amount for bagging and clearing their salt into the Government treasury".[216] By 1875, the Sambhar works had been connected to Delhi through the Rajputana-Malwa railway and the Jodhpur line had been extended to Pachpadra. This meant that the traders, by employing 'Through Traffic' system, could purchase salt from anywhere in Marwar without ever setting foot in the region. In Sambhar,

---

[210] Dane, 'The Manufacture of Salt', 406.

[211] Report on Administration of Northern India salt Revenue, 1904–05, Salt, Jodhpur Records, RSAB.

[212] Ibid.

[213] Erskine, *Rajputana Gazetteers*, 239.

[214] Ibid.

[215] Ibid.

[216] Manual of Northern India Salt Revenue, Vol I, 1905, 22–23.

about 36 percent salt trade and in Pachpadra practically all rail-borne salt trade was 'Through Traffic'.[217]

The measures to control the salt trade in the Thar transformed it completely and irreversibly. As we saw in the previous chapter, salt was one of the most important commodities on the trading networks in the Thar. The community of Kharauls and Lunias was entirely dependent on the extraction and trade of salt. Besides communities like Banjaras, Charans and Bhils were also engaged in salt trade. In the eighteenth century Marwar, salt mines and salt pans were often granted on *ijaras* annually, preferably to the Lunias or Kharauls who then proceeded to sell the salt in open markets on payment of appropriate duties. The British monopoly on salt reduced the community of Kharauls from independent salt manufacturers to labourers in salt mines and pans.[218] Kharauls are reported to have taken to petty labour as "because as the *ijara* of salt and saltpeter has been granted to the British, the mines have shut down in several places."[219]The community of Banjaras lost not only the trade in salt to the British, but the expansion of Railways affected the carriage of other commodities on trade routes as well. These communities were rendered 'criminal by intent' for doing the task that they had traditionally been doing. The salt laws gave the administration the right to search houses of Kharauls for salt and, in case illegal salt was found in a Kharaul's house, he could be fined Rs 100 and or imprisoned for a month and have his salt making equipment confiscated. In case a Thori or Baori was found making illegal salt, he could be caned and imprisoned.[220] Besides, communities like Kharauls, Bhils, Thories, Baories and Banjaras were not landowning communities. They had no resources apart from the salt that they manufactured, transported and traded in. The loss of livelihood caused them to take to either petty agricultural labour or at times, crime. It is not surprising that the communities mentioned as salt traders in the seventeenth and eighteenth century, like the Banjaras and Bhils, were referred to as cattle thieves in the late nineteenth century. The 1891 census of Marwar clearly states that the status of these communities had deteriorated due to the fall in the carriage of salt.

---

[217] Ibid.

[218] *Report Mardumshumari* mentions that Banjaras and Lowanas lost the salt trade due to Railways, 443–446.

[219] *Report Mardumshumari*, 510.

[220] *Kayda Saza Mukadmat Mutalke Namak, Raj Marwar*, 1910, rule 9.

Table 4.3. Comparison of Population of Salt Workers in1660 CE and 1891 CE[221]

S.no.	Name of the Paragana	Number of Houses	House-Man ratio of 1891	Population (c 1660)	Population of salt workers (1891)
1.	Jodhpur	208	5.39	1,121	0
2.	Merta	13	3.72	48	5
3.	Sojhat	98	4.05	397	0
4.	Siwana (Pachpadra)	250	4.39	1,098	823
5.	Phalodi	60–70	4.30	258–301	0
Total				2,922–2,965	828

The calculation provided by Bhadani shows that in paraganas of Jodhpur, Merta, Sojhat and Phalodi the population of salt workers had sharply declined. The villages where salt had been manufactured locally on petty basis for local trade could no longer manufacture salt. The control on salt would also have affected the dyeing and leather industry, though the extent of it is not known.[222] The effect of salt control led to the excessive exploitation of salt sources of the Thar at the cost of local communities and industry. The eventual popularization of cheaper sea salt effectively rendered most of the salt sources in the Thar redundant by the early twentieth century.

The repercussions of the British salt policy in the Thar were severe and led to the disruption of long established trading patterns and practices. The British monopolies turned a commodity into contraband and trade into smuggling. They turned what had been a perfectly legal enterprise, criminal; leading to criminalisation of communities conventionally involved with salt trade. Communities involved with production, trade and carriage of salt in the Thar were also involved in trade of a wide variety of commodities. For example, Banjaras, Bhils, Charans and Lowanas traded in salt, grain, sugar and cattle. This was a well integrated network, functioning with exchange of one commodity for another in rural and urban markets. The control on production and circulation of salt irreversibly altered the trading networks in the Thar region. The absence of salt on the traditional networks of exchange affected the exchange in other commodities as well. By late nineteenth century, the carriage of salt was shifted to the Railways, which reduced the movement of the Banjaras, Bhils, Charans and Lowanas in the region thus affecting overall trading patterns. The control on salt trade in the Thar not only affected the

---

[221] Bhadani, *Peasants, Artisans and Entrepreneurs*, 102.
[222] Bikaner salt Agreement 5/8/1877, fl 176, RSAB.

salt as commodity, but a whole culture that was associated with it. On the one hand, the manner in which salt monopolies were signed away affected the prestige of the Rajput states among their subjects, on the other, intricate patterns of trading were disrupted and vibrant trading networks faded forever altering the position of the communities operating on these networks.

The Outlaws

The changes in the networks of circulation in the nineteenth century severely impacted the movement of itinerant communities that were considered detrimental for the emerging settlement plans in the Rajput states. The mutual relationships of settled and mobile groups had been dynamic, but also fraught with tensions. In a society where political power was increasingly identified with territorial control and had come to rest with settled groups, mobile groups maintained their control on mobile resources like cattle and circulatory networks. However, in the nineteenth century the mobility of certain communities was increasingly identified with criminality, and was sought to be controlled by restricting such groups. The imposed restrictions were located within discourses of heredity and criminality, whereby the customary occupations of several groups were labeled illegitimate. Besides, in the emerging compendia of knowledge, ethnographic documentation and census records mobile communities like Banjaras were recorded as habitual criminals with the descriptions often laden heavily with allegations.[223] Land settlement and expansion of agriculture not only brought about with them identification of property rights but also the need to protect resources of settled agricultural communities. Thus, the charge most often cited against the mobile communities like Banjaras, Baories, Bhils, Minas and Sansis in the region was of cattle and grain theft and also of abduction of children.[224] In fact, it was claimed that with the advent of railways and decline of caravan trade, the crime of child theft had also declined.[225] The relationship of criminalization of Banjaras with the changing norms of salt trade has already been observed studied in the previous section. The very merchandise that the Banjaras carried was declared contraband and they were often arrested for carrying salt stealthily. Based on these considerations and following the implementation of the Criminal Tribes Act, 1871, in Punjab, its provisions were extended to Rajputana thereafter.

---

[223] *Report Mardumshumari*, 111,112–113,127–128,444–445.

[224] *Report Mardumshumari*, 445.

[225] *Report Mardumshumari*, 445.

In Marwar, a separate department called *Mehkama Baorian wa Jurayam Pesha* was established in 1882 to undertake close surveillance of this community.[226] The British labeled these tribes to be habitual and hereditary criminals, who were expected to remain so "until his death as crime was his trade, his caste almost his religion".[227] The 1891 census of Marwar reiterated that the Baories, Bhils and Minas were criminal tribes and thus had to be engaged in cultivation and watched in every way possible.[228] Steps were taken to ensure that the movements of the 'Criminal Tribes' were restricted. In the "proposed measures for the suppression of the lawless Baorees" it had been suggested that "the whole race of Baorees might be at once brought under surveillance... in an effort to reclaim them to peaceful pursuits".[229] These injunctions were implemented in Rajputana, including a restriction on keeping horses and camels as well as on sale of the same to the members of 'Criminal Tribes'.[230] They were not allowed to leave their villages without passes.[231] The above department also undertook the task of allotting landholdings to the members of these communities on concessionary rates, relocating their children to capital and other towns with the intention of training them in crafts etc.[232] The members of these communities were divided into classes according to the degree of their sedentarisation, with class 'A' being awarded to the least sedentary people. These classifications were interchangeable, dependent on the conduct of the person, who was always kept under surveillance. Efforts were continuously made to settle these communities by providing land, agricultural implements, livestock etc.[233] In Bikaner, expansion of canal irrigation was seen as an opportunity to engage and settle what were labeled as the criminal tribes. It was proposed that the Ghaggar scheme be utilised as a famine relief

---

[226] Erskine, *Rajputana Gazetteers*, Vol III, 168.

[227] Meena Radhakrishna, *Dishonoured by History*, 35.

[228] *Report Mardumshumari*, 560.

[229] FP, 14th September 1873, 40–47.

[230] Ibid. Also, Manual for the Guidance of Native States in Rajputana and Central India for the Control and Reclamation of Criminal Tribes, 1896, Rules 4 and 5, RSAB.

[231] Manual for the Guidance of Native States in Rajputana and Central India for the Control and Reclamation of Criminal Tribes, 1896, Rules 7 and 8. Not more than two members of the tribe could receive the *ijazati chitthi* for a maximum of three days at the same time, except for marriages and funerals, when the number of members could not exceed five, RSAB.

[232] *Majmui Halaat wa intijam*, 1883–84, 338–40.

[233] Report on Settlement of Criminal Tribes in Marwar, 1890–91, 1–6 and Marwar Administration Report, 1890–91, 48, RSAB.

measure.[234] Special orders were given to employ Baories in these works and police was made responsible to see that "no opportunity they find to commit crimes or emigrate".[235] In 1896, some Baories sought permission to emigrate to British territory to seek employment in Canal works but this was denied citing the availability of similar work in Bikaner territory.[236]

Another community that was viciously labeled criminal, though not notified, in the nineteenth century was the Charans. They were accused of demanding excessive amounts as *tyag* at the occasion of marriages of the daughters of Rajputs leading to the practice of female infanticide.[237] In the British understanding, the Charans and the Bhats always looked out for occasions when they could extort from the Rajputs. In 1844, Major Thornesby called for, "some restraint upon the exactions of Bhats, Charans, Dholees and Mirasis... who assembled in crowds from all quarters ....both from the foreign as well as the native paraganas....and according to their degree make their demand of money, jewels and horses".[238] The Bhats and Charans who in this manner argued for *tyag* on the gates were traditionally called *Barhaths* or *Polpaats*. Among other reforms proposed by Rajput social reform bodies like the Desh Hiteshini and Walterkrit Hitkarini Sabha, was the fixing of the amount of *tyag* that could be given by the Rajputs to these groups. The state also made efforts to keep a check on the movements of the Charans and Bhats. In 1897, on the occasion of the marriage of Maharaja Kishengarh's daughter, the Resident of Jaipur asked the Resident of Bikaner to ensure that the Bhats, Charans from Bikaner do not come to Kishengarh as it was "exceedingly undesirable that there should be the usual assembly of Bhats, Charans etc.".[239] The princely

---

[234] Famine and Famine Relief work, Bikaner, 1896–97, 239, RSAB.

[235] Ibid., 224.

[236] VP, Regency Council Bikaner to Political Agent Bikaner, File 624, Political Deptt, Basta No 8, 6–9, RSAB.

[237] *Tyag* was a customary due given to Bhats and Charans at the time of a Rajput wedding. Though no clear antecedents of this practice are known, it can be seen in the context of *birat/jajmani* relationships in the Thar. Charans and Bhats kept genealogical accounts, settled alliances and disputes and at the occasion of the wedding recited verses glorifying the two families. *Tyag* appears to be remuneration for the services thus rendered. Enthoven provides the Charan explanation for demand of Tyag as being the tax fixed by Raja Prithu to be granted to Charans by all Khatris in return for their agreeing to accompany them to the mythical Telang. Enthoven, *Tribes and Castes*, I, 273.

[238] FP, 30th November 1844, 157–160.

[239] Resident Jaipur to Resident Bikaner, 366 of 27th Nov 1897, Walter Krit Rajput Hitkarini Sabha file 146/B Political Deptt., 1896–97, Regency Council Bikaner, RSAB.

states of Jaipur, Jodhpur, Bikaner, Udaipur, Kishengarh, and Sirohi etc. passed stringent laws and prosecuted bards. As a result, not only were regulations passed, fixing the amounts to be paid as *tyag* to the Charans and Bhats it was also decided to restrict their movements between states on the occasions of marriages, quite reminiscent of the behavior meted out to the 'criminal tribes'. The restrictions on movements were particularly applied as it was believed that the mobility of the Charans helped them avoid prosecution. In fact, in the mid nineteenth century, as Charans began resisting the restrictions through the traditional mode of *chandi* and *dharna*, the horrified Marwar state declared self immolation and human sacrifice acts of murder.[240]

Both, conspicuous display of wealth at marriages and female infanticide had been the feature of the aristocratic Rajput communities. The Rajputs and Charans shared a patron-client relationship following which, Charans composed poems describing the glory of the patron at the occasion of the marriage, in return for which they would be lavishly rewarded. Rashmi Dube-Bhatnagar has interpreted the charges against the Charans to be a result of the collaboration between the British and the Rajput states, that attempted to lay the blame of female infanticide on the shoulders of what were branded as extortionist communities.[241] The allegations against Charans were based on the pragmatics of the British-Rajput polity, rather than the actual demands posited by the Charans. Dube-Bhatnagar points out the fallacy of Rajput patrons being subordinate to the Charans as the latter had already been reduced to the position of a service group. Besides, the practice of female infanticide was more prevalent in rich landed households who were conducting ostentatious ceremonies, rather than the poor ones.[242] She thus argues that Charans in fact were the watchdogs of daughter's rights, as by the act of praising and memorizing the items of dower they underlined the daughter's inheritance. As I have already pointed out, Charans' role as social and political arbitrators emerged from the kind of control that they had exercised on frontiers, both territorial as well as social. Treaties with the British excluded the Charans

---

[240] Ibid.,257.

[241] Rashmi Dube-Bhatnagar, 'A Poetics of Resistance; Investigating the Rhetoric of Bardic Historians of Rajpuana' in *Muslims, Dalits and the Fabrications of History, Subaltern Studies XII*, (Eds.) Shail Mayaram, MSS Pandian and Ajay Skaria, Permanaent Black, New Delhi, 2005, 253.

[242] The Krishna Kumari episode discussed in the previous chapter points out that the Rajput notions of clan superiority and hypergamous marriage may have played a greater role in the 'unmarriageability' of Rajput girls, rather than the fear of ridicule by the Charans.

from arbitration over boundaries. This would have further led to the decline in polygamy as marriages were no longer seen as a way of resolving disputes. Thus, in the nineteenth century as the "daughters became disposable, the Charans became inconvenient". [243] Besides, with the decline in the institution of polygamy, the role of marriage alliances in intra-Rajput relationships diminished. Daughters, who in the past were given away in marriages to settle political and social scores became dispensable, giving way to practice of female infanticide among the Rajputs. The attribution of blame to Charans alone, was a process that involved both, the British as well as Rajputs. [244]

I have pointed out earlier that Charans were not just bards but also traders, carriers, agriculturalist and moneylenders. But the nineteenth century portrayal of Charans represents them as beggars. This transformation was perpetuated by the British, who viewed the mobility of the Charans with suspicion and were perplexed by the social and spatial mobility of the Charans. This multifaceted community confounded them as they could not come to terms with a community that could herd and compose at the same time. That is why they constantly defined the Charan-Rajput relationship in the context of extractive patron-client relationships. Besides, for the British the Charan narratives were worthless as histories, as the nineteenth century British histories of Rajput clans, like that of James Tod, focused on chronological accounts. Thus, the role of Charanic tradition of remembering and writing histories was also being taken over by British chroniclers, who were increasingly taking over negotiatory positions as Political Agents and Residents that traditionally had been occupied by Charans. However, what was most disturbing about the Charans, was their unrestricted access to different Rajput states, and their ability to move across difficult desert terrain.

---

[243] Rashmi Dube Bhatnagar, 'Poetics of Resiatance', 234. The decline of the Mughal Empire and the power and prestige of most Rajput principalities resulted in the undermining of the advantages that could be gained from marrying off daughters to powerful kings. By this period there was a noted decline in trend of polygamy, rendering daughters worthless in socio-political exchange systems of Rajputs.

[244] As evident in the rhyme composed by Trevor, in G H Trevor, *Rhymes of Rajputana*, 1894, 112.

> "The Rajput may not marry in his clan
> A daughter's dower has ruined many a man
> *Charans*, or bards who came to bless or bane
> At every marriage feast , than locusts worse
> Beggared the simpleton who feared the curse."

Here, it would be pertinent to raise a question regarding the nature of communities that were branded criminal. While the straightforward conclusion to be drawn would be that these communities were criminalized because of their mobility, I would like to stretch this argument a bit further. It has generally been argued that the purpose of such classification was to restrict the movement of these groups and it emerged from a general fear and distrust towards mobility. However, I would argue that the inherent purpose was to minimize the kind of control that mobile communities exercised on circulatory networks in the Thar. As I have argued in the second chapter that despite the emergence of sedentary Rajput states, the larger frontier spaces in the Thar continued to be controlled by a number of mobile groups. These groups had always held deep seated knowledge of geographies that were labeled hostile in the nineteenth century. As argued earlier, that Charans' ability to guide caravans emerged from their knowledge of the region as well as from their ability to control traffic on hazardous desert routes. Similarly, Minas, Bhils and Mers were used as *pagis* or trackers, precisely because they understood the hilly terrain between Marwar and Mewar better than any other community.[245] I have argued earlier in the book that most of these groups shared the Thar as mobile warriors, and emergence of Rajputs as landed aristocrats had involved either assimilation or subjugation of such warrior groups. The continued dominance of Rajputs from the fifteenth century onwards was based on several 'power sharing' arrangements with such groups, in which fringes were often controlled by what were seen as recalcitrant groups, like Bhils, Minas, Khosas, Bidawats, Sodhas among others. This gave rise to multiple nodes of authority that were as incomprehensible for the British as was the land itself. Not only did the British seek to diminish the power exercised by Rajput *thikanadars*, they also sought to extend the singularity of control on circulatory networks. Thus, the mobile groups were labeled criminal because their existence and their ability to control mobile networks challenged the authority of indirect rule in the Thar, which had been attempting to render the frontiers of Thar legible and comprehensible. The nature of desired political and administrative control demanded clear and sharp boundaries and not grey frontiers, as challenges to sedentary polity in the Thar had always arisen from the frontiers. These

---

[245] Also, Manual for the Guidance of Native States in Rajputana and Central India for the Control and Reclamation of Criminal Tribes, 1896, Rule 16, prohibits the employment of members of Criminal Tribes as *chowkidars* or informers in return for informer's fee or *mehr khai*, as had been the tradition earlier.

groups were distrusted because they belonged to the frontier, geographically as well as ideologically.

In this chapter, I have argued that in the nineteenth century the networks of circulation were altered in the Thar region in response to policies that were adopted to promote the larger interests of the colonial state. While we can say that transformative changes occurred in the nineteenth century, it would not be correct to assume that the pre-colonial state systems were altogether devoid of disharmony between various interacting elements of society. Social institutions like caste controlled access to resources and in an arid region where subsistence is dependent on prudent use of scant natural resources, it was access to these resources that was decisive in the survival of communities and livelihood practices. Thus, for instance pastoralists while being an important segment of rural society, had little say in the way the resource use was determined and in the way patterns of resource use changed overtime. Yet, there is no doubt that an important feature of the rural social structure in the Thar region was the mutual dependence of sedentary peasant and mobile communities. The reason for this was that sedentary and mobile were not two mutually exclusive categories in the Thar region. It was not only possible, but natural for sedentary communities to become mobile in times of distress. The fluidity between the categories was indicative of the internal dynamic of the region, which neither Rajput states nor British could intervene in. Besides, sedentary and mobile communities were socially, economically and ritually dependent on each other. Therefore, it was pastoral communities that appear to be least affected by the policies of the nineteenth century colonial state. Despite attempts at controlling grazing and movements, the pastoral nomadic communities already on the fringes of the society, managed to survive by forging close alliances with sedentary communities like creating large herds by including herds of sedentary peasants or utilizing increasing crop stubble. These adaptive techniques helped nomadic pastoralism not only survive but also respond creatively to changing environmental situations, an image very far from being the 'silent spectators' of their own decline.

On the other hand, the policies of nineteenth century had disastrous consequences for carrier and small mercantile communities. As we saw, the salt laws destroyed the communities of Kharauls, Banjaras, Charans and Bhils that were actively involved in the salt trade through The Thar region. Railways and monopolies devastated the flourishing trading networks and reduced large emporia into small *mandis*. The infringement of salt networks affected trade in other commodities and a general decline in commercial economy initiated the exodus of the Marwari mercantile community of The Thar region. The state in the nineteenth century was willing to promote certain kinds of regulated circulation of both goods and people on the peripatetic networks of the Thar Desert, provided that its interests were protected and promoted. Those who did not fall into its scheme were dubbed as 'illegal' and hence criminal. Loss of occupation in some circumstances, severe dislocations caused by natural factors like famine and draughts in others, altered circulatory networks. The control imposed upon circulatory patterns by the colonial state worsened the situation for certain groups and classes, who at times turned to robbery and other criminal acts. Newer identities were generated through the processes that turned hitherto legal activities like grazing, or trade in salt into crime which is reflective of the fact that crime was closely related to control. Thus, within a century circulatory spaces as well as community identities were drastically altered in the Thar region. The Thar, rather than the vast open circulatory space that it had been for centuries, gradually became a closely monitored and regulated arena, circumscribed by newer boundaries. While earlier circulatory networks could have existed independent of political boundaries, the new networks were confined within the political space. Thus while the state expanded its reach in the nineteenth century, the space for mobility actually contracted.

5

Narratives of Mobility and Mobility of Narratives

"There are no entrails left in my belly
So I cannot take opium or food any more
Accept the respects of your warrior Dhembo
For we shall now meet in the court of God Rama!"

With these words Dhembo gave up his life
And became a dweller in heaven
Pabuji and his men reflected on this with great sadness
But Dhembo played a trick on the knights.[1]

Medieval Rajput courts in the Thar were home to some of the richest poetic and prose traditions in Sanskrit, Dingal and Braj Bhasha. However, as discussed in earlier chapters, these traditions were instrumental in reiterating Rajput norms of kingship and fostering the idea of Rajput exclusivity. Even as Rajput claims to kingship were contested by groups exercising control on the frontiers, Rajput literary traditions focusing on Rajput heroism and sacrifice, completely glossed over claims of other groups. Therefore, though a range of mobile groups continuously traversed the Thar, they often failed to find space in the written accounts and histories of the region, which based themselves on Rajput narrative traditions. However, the existence of a rich body of oral folk traditions in Rajasthan, ranging from lore dedicated to folk deities to songs of

---

[1] Smith, *The Epic of Pabuji*, 125.

separation, preserve narratives that are lost in written traditions. The mobility of such groups often gives an impression of narratives of lost in movements, but in my understanding they are narratives of movement over time and space.

The networks of pastoral and commercial circulation in the Thar region were also the very networks on which the oral traditions of the region traveled. Social groups continually on the move in the Thar preserved their history through bards, not just the legendary Charans but also genealogists, storytellers and musicians of various kinds like Bhats, Motisars, Nayaks, Bhopas, Langas, Mirasis and Manganiyars. These bards put genealogies together, created the corpus of oral accounts that were transmitted across the wide space of the Thar and also across generations as well as nurtured rich musical traditions in the region. The transmission of these traditions recreated them over space and time at multiple levels and led to several interpretations of these traditions by the communities themselves. Ambulatory bards traveled around the vast Thar Desert connecting hamlets and populations spatially and temporally. The more significant link was the temporal one, where by the bards acted as the transmitters of narratives preserved in time and memory, especially the ones that did not find space in the hegemonic historical traditions that centre around Rajputs. Oral narratives dedicated to folk deities like Gogade, Pabuji, Ramdeoji, Devnarayanji, Tejaji etc. while appearing to extol Rajput heroism, actually bring forth very complex processes of community identity formation among low caste mobile groups. These are reflective of how communities as social groups perceive their past. They also reflect how past is represented in narratives that are continuously shaped by marginal positions of these communities in the society and their understanding of their own marginality. These narratives thus take shape of subaltern histories, and emerge as counter-narratives to the dominant Rajput narratives.

The dominance of Rajput-centric perspective in the historiography of the Thar has resulted in the marginalization of other histories, as the processes through which histories were composed, recited, circulated and received continued to be controlled by Rajputs themselves. This resulted in the marginalization of histories of mobile communities like Bhils, Gujars, Mers, Jats, Rabaris, Charans, Bhats, Kamads, Kanjars, Luhars, Sansis, Kalbelias etc. Besides, the focus on the notion of 'Rajput' as the centre constantly forced these communities to posit themselves as the 'other' to all that was understood as Rajput. The process of creation of Rajput identity has constantly reformulated the identities of all other communities in the region, to the extent that most communities allude to a 'Rajput' past that was lost due to ritual pollution, consumption of proscribed food or taking up of non-martial occupations.

Bardic communities of Bhats and Charans, who composed the corpus of panegyric accounts addressed to Rajputs played a pivotal role in making of the 'Rajput'. An exploration of medieval literary traditions like the *vat, vachanica, sakha, geet, kavitt, raso, jhulano, pravado, pidhivali, vamsavalli, khyat* etc. helps in mapping out how Rajput social and political identities emerged in the first place. Most of these forms, some poetic and others in prose were composed by Bhats and Charans and were patronized by Rajputs. However, several oral poetic and prose narratives patronized by a number of other communities like Jats, Gujars, Rabaris etc., that not only re-orient the 'Rajput' versions but bring forth their own perspectives on Rajput narratives, also continue to circulate in the Thar. For instance, the tradition of *phad vachan* which involves singing a narrative while displaying painted visual narratives called *phad* and has led to the appropriation of Charan compositions by communities on the margins like Gujars, Kamads, Bhils and Raikas, thus significantly altering these compositions to underline the transformations in their own community identities. An exploration of these traditions allows me to see how expression and reformulation of community identities can be traced and to ask some questions about circulation of traditions and identity formation in the Thar.

This corpus of oral traditions in the Thar can be understood through several inter-related contexts. The first is that of Dingal heroic poetry composed, usually but not always by Charans and patronized by Rajputs. These poetic traditions while extolling the exploits of Rajput heroes emphasized the sacrificial spirit of Rajputs who laid their lives defending forts, cattle or women. Poetic works like *Kanhaddev Prabandh, Achaldas Khichi ri Vacanica, Rao Jaitsi ro Chhand* can be seen as examples of these traditions. Over time these poetic traditions also came to include genealogies, which associated Rajput clans with illustrious Kshatriya lines or celestial figures. However, these compositions also circulated all over the Thar as part of *bats*, which can be understood to mean accounts, tellings, tales, fables or myths. Understood as inspirational biographical narratives *bats* were ceremonially recited in Rajput gatherings and became the media for celebration and perpetuation of the idea of Rajput sacrificial spirit. Similar narrations exist for other social groups as well, whereby origin myths, genealogical accounts and heroic tales were woven together by bards of groups like Jats, Gujars and Bhils. Often appropriating high bardic tradition of Charans, a narration specific to each of these communities emerged. The narration of the Pabuji epic by the Nayak Bhopas in Raika gatherings, the Devnarayan tradition of the Gujars, the tradition of Ramdeoji among the Meghwals or Tejaji among the Jats reveal the processes through which groups other than Rajputs have interpreted historical events and processes in the Thar. Another context is of the popular tales of love, war, fantasy, mysticism and spiritualism that circulate through networks of social gatherings like fairs by groups like Bhopas, Manganiyars, and Mirasis. The retelling of these tales is indicative of the manner in which popular tales of love and separation like Dhola-Maru, Moomal-Mahindro, Umar-Marvi etc. also comment on processes of identity formation as well as their relationships with mobility in the Thar.

In this chapter I explore all these contexts with the purpose of unraveling connections between them. This exploration allows me to ask questions about circulation of yet another kind in the Thar, that of narratives. I am particularly interested in understanding if the kinds of dynamics that I explored in earlier chapters between sedentarism and mobility also operate between written and oral narrative traditions in this region. Is the position lost by mobile groups in the political-culture of the region reclaimed in the oral narrative traditions? Does the exploration of relationships between the oral traditions and mobility tell us something about the circulation of alternative historical perspectives in the Thar desert. In this chapter I examine the bardic traditions of Charans, traditions of *phad vachan* of Pabuji and Devnarayan among Bhils, Raikas and Gujars, and love-ballads, particularly of Dhola Maru to ask these questions.

Bardic Narrations: Rajput-Charan Exchanges

"In these golden times of Rajput life, when the swords were never allowed to rust, nor steed to rest....... bard was always wanted at the side of the warrior as a witness of his deeds and a singer of his praises."[2]

The bardic literature in the Thar region was composed over centuries by specialist bardic community of Charans and patronized by Rajput courts. Charans, a composite community that performed multiple tasks of poets, genealogists, arbiters, religious functionaries, graziers, cattle rearers, farmers, caravan guides, messengers, traders, and money lenders have a complex history of origin as discussed in chapter three. The process of transformation of Charans from graziers to genealogist-poets and traders is as complex as the emergence of the Rajputs. While it is usually assumed that Charans turned into traders in order to advantageously use the sanctions that they had received as poets, it is worthwhile to examine these assumptions. Sigrid Westphal-

---

[2] L P Tessitori, 'A Progress Report on the Preliminary work done during the year 1916 in connection with the Bardic and Historical Survey of Rajputana', *JPASB, (New Series)* Vol. XIII, 1917, No: 4, 250.

Hellbusch points out Charans had undergone the same kind of processes in their emergence as groups like Jat-Baluch, Rebari, Banjara and Bharvad, as reflected in the clan names that were common not only between various Charan grazier clans, but also with other grazier communities.[3] Charan, like Rajput can also be thought of as an umbrella term under which communities with different backgrounds that traversed the medieval Thar came together. In reference to 'Rajputisation', the term used for emergence of Rajputs from tribal pastoral groups, Janet Kamphorst, uses the term 'Charanization', which refers to emergence of Charans as a composite group.[4] By the end of fifteenth century, however, just as 'Rajput' became a closed category, the origins of Charans too had been eclipsed by their role in the emerging Rajput culture.

The emergence of Rajput identity was dependent not only on the creation of genealogies linking them to older Kshatriya clans by Brahmin Bhats but also on the perpetuation of the newly constructed imagery through stylized Dingal poetry composed by Charans that emphasized on bravery, valour, chivalry and generosity as Rajput attributes. It is through this imagery that Rajputs could uphold their status as a distinct caste group with specialized rites, practices and more importantly, values. This bardic literature in fact was the basis of the colonial 'medievalist' interpretation of 'Rajput culture' that furthered the idea of Rajputs as distinct group identified by these values. Jos Gommans argues that the creation of genealogies suggesting strong ascendant lines as well as of narratives underlining values like bravery, chivalry and loyalty was undertaken largely keeping the Mughals particularly, Akbar in view.[5] David Henige holds that advent of literacy and adaptation to foreign rule had the greatest impact on the chronological content of oral societies, which were altered to suit the position of the emerging clans. The newly acquired literacy was used to elaborate and synthesize older oral traditions.[6] In case of medieval Thar, the historical traditions that had emerged focused on heroism, a value that was believed to be transmitted through lineage. The process of reciting and

---

[3] Sigrid Westphal-Hellbusch, 'Changes in the Meaning of Ethnic names as exemplified by the Jat, Rabari, Bharvad and Charan in North-western India' in *Pastoralists and nomads in South Asia* (Eds.) L S Leshnik and G D Sontheimer, Otto Harrassowitz, Wiesbaden, 1975, 117–138.

[4] The origin stories of Charans refer to loss of Brahmin caste because of interdining with Rajputs in times of their distress. As they could no longer claim a Brahmin status they were accorded special status by Rajputs. Kamphorst, *In Praise of Death*, 259–260.

[5] Gommans, *Mughal Warfare*, 83.

[6] Henige, *The Chronology of Oral Traditions*, 95.

writing of episodes of valor legitimized continuity with an illustrious heroic past as well as authority in the present state. As Ramya Sreenivasan puts it, "when descendants of particular lineages remembered the exploits of their ancestors, they claimed the heroic essence of the lineage as an instrument for legitimising their authority over inherited resources-both material and moral, involving both territory and character."[7]

An interesting form witnessed in medieval Rajasthani historical literature is the *vat* or *bat* which, emanating from Sanskrit *varta*, can be interpreted variously as an account, telling, tale, fable or a myth. But as Norman Zeigler argues, as a genre of literature, *bat*, developed its own form and content. He views it as, " 'inspirational biographical narrative' which deals with either the life history of an important individual, such as the leader of a particular Rajput clan (*kul*), or with particular episodes in his life, which are seen to be significant".[8] They could refer to hostilities and settlements as well as conquests or marriages. Sometimes genealogies were included in and reiterated through *bats*. Munhata Nainsi's *Khyat*, makes extensive use of *bats* as a way of recollecting events that brought Rajput warriors in contact with each other. Most of the *bats* collated by Nainsi are anonymous, unlike the *duhas*, *sorathas* or *kavitts* that are attributed to Charan poets. Unlike the stylized language of the interspersed verse, the language of the *bat* is in the vein of relating an incident, of reportage, almost testimonial, thus primarily oral. The *bat* recalls events and conversations, while incorporating myths, legends, poetry and hearsay to make an interesting tale to be told. The *bats* scripted by Nainsi appear to be collectively memorized accounts, perhaps passed over generations, practiced as recitals, altered by incorporations and deletions over time. Shrouded in anonymity and rather prosaic in character, the *bats* were nevertheless conscious acts of remembering and forgetting that crafted a certain kind of Rajput past.

Norman Ziegler in his study of Marwari historical chronicles asserts that *bats* were to be ceremonially recited by Charans in courtly gatherings or in the houses of patron Rajputs with whom hereditary patron-client relations were maintained. These occasions were marked by the consumption of opium, which was also seen as a Rajput and Charan attribute. Here the Charans, particularly the Maru Charans, were shown great respect as these occasions could not be organized without the presence of a Charan.[9] The recitation of *bats*

---

[7] Sreenivasan, 'The 'Marriage' of 'Hindu' and 'Turak', 96.

[8] Zeigler, 'Marvari Historical Chronicles', 221.

[9] Jhaverchand Meghani describes one such gathering in where a Charan Mandan, addressing a Rajput gathering, intoxicated by opium and his heroic verses forgot about

was preceded with the invocation of the family deity usually a goddess. This was followed by a short recitation of genealogy and then the bats that usually referred to the ancestors of the family, but could also narrate accounts of valor of any Rajput hero.[10] What did these gatherings signify for both Rajputs and Charans and how did they impact the process of remembering and writing of Rajput pasts? In order to explore the dynamics of Rajput-Charan relationship we would need to understand how Charans were placed in medieval Thar.

Charans in the medieval social spectrum were placed below Brahmins but above Rajputs. While lacking the sanskritic sanctions available to Brahmins, they were still accorded a semi-priestly status. The fable of origin of Charans claims that they were created to herd the celestial bull Nandi, after the Bhats could not graze him as they were scared of the lion of the goddess Parvati. Pleased by the devotion of the Charan the goddess granted him a unique position encompassing all three qualities of the Charan, that is, the priestly feature of devotion, the Rajput fearlessness and the dexterity with herds.[11] Of the Rajput-Charan relationship, another fable of their origin recorded by Russell in *Tribes and Castes of Central India* states, "when Rajputs migrated from the banks of Ganges to Rajputana and their Brahmin priests did not accompany them, and hence the Charans... equally devout and of bolder spirit arose and took charge of the animals".[12] In another version available in Enthoven's *Tribes and Castes of Bombay Presidency*, Charans are said to have "accompanied the Kshatriyas in their southward flight and took charge of carrying of supplies".[13] However, in the description provided in the Marwar census, 1891, they claim a divine origin along with the Brahmins.[14] All these claims are indicative of the shifting self perceptions of Charans vis-à-vis Rajputs, Brahmins and Bhats. In a sense, Rajput value of bravery and Brahminic relationship to knowledge made Charan an entity that was not only distinct from both, but also perhaps superior, to the extent of replacing Brahmins in the medieval Rajput world. The amalgamation of these qualities granted them the special position of guards and guarantors of caravans as their person was believed to be as sacred as that of the Brahmins but their fearlessness inculcated in them a blatant disregard

---

his wife, who was swallowed by the rising current of the river. 'Elegiac "Chhand" and "Duha" in Charani Lore' *Asian Folklore Studies*, Vol 59, No.1, (2000), 41–58, 49.

[10] Zeigler, 'Marvari Historical Chronicles', 221.

[11] *Report Mardumshumari*, 330.

[12] Russell, *Tribes and Castes of Central Provinces*, I, 162.

[13] Enthoven, *Tribes and Castes of Bombay Presidency*, I, 273.

[14] *Report Mardumshumari*, 330–33.

for their own body and life and paved the way for rites of self mutilation as well as self immolation. This sacrificial spirit placed Charans close to Rajputs while their control over the 'word' attributed an inviolable sacred status to their person. So sacred was the person of the Charans that it was considered unthinkable to let them resort to *dharna* or *chandi*.[15]

Janet Kamphorst argues that the fear invoked by the destruction of the self of the Charan originated from the fact that Charans were poets and priests, worshippers and mediators of the Charani *sagatis*, living goddesses of Charan origin.[16] Charans regard themselves as devotees of a goddess named Hinglaj, a *mahashakti*, who herself was a Charani born to Charan Haridas of Gaviya lineage in Nagar Thatta.[17] Several Charani goddesses like Avad, Karni, Nagnechi, Sangviyaan, Barbadi, among others are revered by Rajputs as patron deities. They revered them through construction of temples dedicated to the *sagatis*, like the Hinglaj temple in Ludrova, the Karni temple in Deshnok, the Nagnechi temple in Bikaner etc. These goddesses were "thought of as historical women recognized as living goddesses in their lives and deified after their deaths".[18] They were usually unmarried goddesses with the exception of Karni, who became a part of the Rathor world by marriage. Their status as inviolable virgins confers on them a standing far superior to any other celestial category. In the Charanic accounts, the Charani *sagatis* were accorded an important place in the Rajput worldview as influences that directed the Rajputs towards the correct behavior and were protectors of their lives and territories. Their position both, as terrestrial and heavenly beings made them fearless conscience keepers of Rajputs, who could be cursed for their rapacious behavior.[19] In face of such power it is no wonder that the Rajputs accepted

---

[15] *Dharna* usually meant the rites of self mortification that Charans are supposed to have undergone in order to get their demands fulfilled. Since Charans were guarantors to agreements between Rajputs, or of the caravans on trade routes, this guaranty was based on the self-sacrificial sprit of the Charan, according to which the Charans could mutilate or annihilate their selves by either cutting their own bodies (*dhage*), fasting unto death (*tyag*), self immolation by pouring hot oil and burning (*teli*), self immolation, entombment, cremation or drowning (*samadhi*). The party guilty for flouting and agreement or despoiling a caravan was held responsible for the death of the Charan. Kamphorst, *In Praise of Death*, 230.

[16] Ibid., 221.

[17] Ibid., 236.

[18] Ibid.

[19] The tale of Umar Soomro and Sangviyan can be seen as an example. Umar proposed marriage to Charani Sangviyan who cursed his land to become destitute of water and

the cult of Charani goddesses and awarded them special status in their socio-religious practices. However, the guidance provided by Charans and Charani goddesses was far more than just spiritual guidance. The location of Charani goddess temples on trade routes in the heart of the desert, was also indicative of their deep knowledge of the geography of the region as well as the control that they extended, as discussed in the second chapter. The role of Charans as arbiters in mutual conflicts of Rajputs is clear from the manner in which Karni, facilitated the Bika Rathor's marriage to Sekho Bhati's daughter, and thus provided Rathors with a toehold in the northwestern Thar.[20] The presence of Charans as negotiators and witnesses in events of wars, negotiations as well as marriages, is indicative of the fact that all three were closely related in emerging Rajput polity.

The position of the Charani goddesses also led to the emergence of a filial bond between Rajputs and Charans, not shared with any other community, as Charanis were considered to be sisters by Rajputs. This bond awarded a high-ranking socio-political status to the Charans as poets, historians, ministers, advisors and counsels, mediators, as well as protectors of caravans, forts and palaces. The warrior-poet status was also desired by the Rajputs for themselves, with several Rajput kings excelling in the art of poetry. Charans were not only responsible for preserving and transmitting the Rajput past, but also for introducing the younger generation to their own past. The recitation of heroic verses and *bats*, in Rajput courts and gatherings emphasizing valour, honour and loyalty were believed to reinforce these values in the next generation. It was through "this medium that they were brought into the history of their families, lineages and clans, were schooled in the moral values of their fathers, and were tutored in their future roles in the society".[21] So in this sense, Rajputs themselves entrusted the creation and reiteration of their pasts to the Charans, thus creating sacrosanct bonds between the communities.

The Charans, thus, became responsible for the writing of their histories of the royal as well as of cadet lines. For this they were rewarded through gifts of wealth, cattle as well as revenue free land grants called *sasan* grants, apart from other ceremonial gifts. In order to maintain this order, young Charans

---

knowledge. This resulted in the shifting of the course of river Sutlej to north and left Marwar a desert. Nainsi, *Vigat*, III, 88. A number of versions of these tales are popular in Rajasthan and Gujarat, involving a Charani goddess and a Rajput chief, usually Hamir/Umar Soomro.

[20] Tambs-Lyche, 'Marriage and Affinity among Virgin Goddesses', 63–87.

[21] Zeigler, 'Marvari Historical Chronicles', 222.

were trained in the art of composition and recitation early in life. Memorization of older compositions was stressed upon, and the younger Charans were encouraged to hone their skills in order to continue receiving patronage.[22] This indicates that the growth of Charanic literary-historical tradition depended upon memorization and reworking of narratives already available to the Charan poets. Thus, while a large body of Charanic composition appears in written form, a significant amount would also have circulated as oral compositions.

This aspect of the Rajput-Charan relationship often suggests that the Charans often managed to extract gifts and grants from the Rajputs in what Tod terms as the barter of 'solid pudding against empty praise' or the 'sale of fame'.[23] Nainsi, relating the *Vat Jam Unand Sanwal Sudh Kavi Rohadia nu Aauthkor Saamai di*, describes how Jam Unand gave away his kingdom to match the generosity of Lakha Phoolani.[24] Thus Nainsi acknowledges that the relationship between the Charans and the Rajputs, as one based on exchange, which involved both, the Charan's 'greed' and the Rajput`s 'desire for fame'.[25] Tessitori observes that, "the lavishness of the chiefs to the bards had known no limits".[26] It has also been suggested that Charan ridicule or *bhumd* could be more damaging than praise could ever be valuable. Tod writes, "many a resolution has sunk under the lash of their satire, which has condemned to eternal ridicule names that might otherwise have escaped notoriety. The *vis* or poison of the bard, is more dreaded by the Rajpoot than the steel of foe".[27] Yet, Tod points out, "these chroniclers utter truths, sometimes unpalatable to their masters. When offended, or actuated by a virtuous indignation against immorality, they are fearless of consequences; and woe to individual who provokes them".[28] This does not appear to be surprising as the Charans were

---

[22] Ibid.

[23] Tod, *AAR*, I, xv.

[24] Nainsi, *Khyat*, II, 236–37.

[25] While discussing the dynamics of Charan-Rajput relationships, it becomes worthwhile to mention the *dharna* of Ahuwa in VS 1643 (Kajesar Mahadev, according to Nainsi) organised to protest against the revoking of *sasan* grants of some Charans by Motaraja Udai Singh. Nainsi, *Vigat*, I, 78. Unlike the imagery projected of the hold of Charans over Rajputs, Motaraja Udai Singh refused to be forced by the *dharna* and Munshi Hardayal even writes that, "since Charans were killing themselves, there was no need to send an army to kill them". *Report Mardumshumari*, 343–345.

[26] Tessitori, *Report on the Progress of the Bardic Survey*, 1917, 250.

[27] Tod, *AAR*, I, xvi.

[28] Ibid.

the possessors and preservers of the power of "word, the corpus of sounds by which the moral order of the society is maintained and altered".[29]

Here, we need to examine this contradiction in the context of conception of history writing that emerged in the nineteenth century. During the nineteenth century, an increased interest was witnessed among the Rajput princedoms to explore their own history, which they sought to be written in annalistic tradition. James Tod's *Annals and Antiquities of Rajast'han* not only was the first of such histories to be written, it also became a template that continued to be followed even by Indian historians well into the twentieth century. A number of other Europeans devoted themselves to Indology during this time, associating with the cause of 'discovering' Indian civilization. A phenomenal contribution was made by Dr L P Tessitori, who was engaged by the Royal Asiatic Society of Bengal and the Maharaja of Bikaner to launch a search for the manuscripts and epigraphs pertaining to the history of the Rathors of Bikaner scattered throughout the state. Tessitori not only collected epigraphs and manuscripts but also wrote a series of articles on the process of collection and the nature of the material collected. Above all, his descriptions provided valuable insights in to the manner of retention of historical knowledge in Rajputana over a long period of time. Tessitori was basically a linguist and a grammarian, but one who thought that language and form held the keys to understanding past in this region. His interest, like Grierson, was in the old western Rajasthani language that he painstakingly learnt. To him it was "... obvious that without a clear knowledge of local conditions in Rajputana, and especially of the languages no serious attempt can be made to prepare a scheme that will work and bring results".[30] He realised that changes in language, forms and styles indicated deep societal changes and lamented, quoting Grierson, that, "Marwari has an old literature about which hardly anything is known".[31] He indicated a marked change in the style and composition from the old western Rajasthani, Dingal, whose identity was denied by contemporary bards confusing it with the language preserved in Jain texts or *Jatiya ri boli* by early twentieth century.[32] This, in his opinion, was also reflective of change in language of bardic compositions as a result of inclusion of lower caste bardic groups that composed in Brajbhasha, as well as increased patronage to

---

[29] Zeigler, 'Marvari Historical Chronicles', 226.

[30] Tessitori, *A Scheme for the Bardic and Historical Survey of Rajputana*, 1914, 374.

[31] George Grierson, *Linguistic Survey of India*, Vol. IX, Part II, 1908, 19.

[32] Ibid.

Braj bhasha poets by Rajput courts.[33] Tessitori was highly critical of the selective approach adopted by James Tod in his Annals as well as Alexander Forbes, who used Gujarati Charanic compositions in his *Rasmala*, a history of Rajput clans in Gujarat. Tod had depended mainly on *Khumana Raso* and *Suraj Prakash*, both poetical works and had thus conveyed an impression that the bardic chronicles were "chiefly poetical works in which the plain facts are mixed with legends, altered by love of poetry, and distorted by poetical exaggeration".[34] On the other hand, Tessitori had uncovered a large body of prose literature distinct from the poetic compositions in Dingal, that he identified as Pingal or "Braja bhasa, (which) more or less vitiated against the influence of the former (Dingal)".[35] Jhaverchand Meghani, one of the earliest folklorists and ethnographers in Gujarat, attests that Dingal, the bardic Charani tongue developed from Prakrit and Apbhramsha, flowed freely between Rajasthan and Suarashtra and conformed to the contours of other phonetic tongues like Sindhi and Kutchi. According to him, Dingal is "a poetical medium based on sonorous flourish. It is neither a language nor a dialect. It is mode of rendering poetry, a mode with distinct flair of its own, a mode that has sprouted from the original Rajasthani and been shaped by historical events".[36] Janet Kamphorst holds the opinion that Dingal could best be seen as a poetic style employed by speakers of different tongues including Gujarati, Marwari, Jaisalmeri and Mewari. She points out that in the eighteenth century a distinction was made between Dingal and Pingal, as forms used distinctively by Charans and brahminical Bhats respectively.[37] However, by nineteenth century while Bhats were found largely to be genealogists, it was Charans who were composing both in poetic and prose forms, works that no doubt had significant literary as well as historical value. Considering that the very popular form of *bats* consisted of both poetic verses and prose narratives, it is reasonable to propose that both these forms had been composed simultaneously and had been in circulation in Rajput courts as well as assemblies. Moreover, there is no reason to question the merit of Charanic verses as historical literature, because of their form. In the Rajput-Charan world, verse was the form in which historical literature

---

[33] Rashmi Dube Bhatnagar, 'A Poetics of Resistance', 269. Allison Busch, *Poetry of Kings: The Classical Hindi Literature of Mughal India*, OUP, NY, 2011, Reprint, 2012, 166–199.

[34] Tessitori, *A Progress Report on the Proposed Bardic and Historical Survey of Rajputana*, 1919, 18.

[35] Tessitori, *A Scheme for the Bardic and Historical Survey of Rajputana*, 1914, 375.

[36] Meghani, 'Elegiac "Chhand" and "Duha", 43–45.

[37] Kamphorst, *In Praise of Death*, 34.

was composed, before annalistic prose forms came to be regarded as history. Charanic verses formed the bridge between the oral and the written ways of imagining the historical in the Thar.

However, Tessitori posited poetic Charanic work against the *Khyat* and *Vigat* composed in the seventeenth century by Munhata Nainsi, the Oswal Jain Diwan of Jaswant Singh of Marwar. He claims that both *pidhivalis* and *khyats* were composed after the sixteenth century keeping the Mughal court in view. *Khyats* were, "remarkable for their accuracy, sobriety and dispassionateness. They contained no legends, no quotations of bardic verses, no flatteries, no lies, but merely plain statements of facts teeming with names and dates, these facts were contemporary and many of them witnessed by the writer with his own eyes".[38] Posited against this, referring to the Charanic verses, he writes, "there is probably no bardic literature in the world, in which truth is so marked by fiction, so disfigured by hyperboles; as in bardic literature of Rajputana. In the magniloquent strains of a Carana, everything takes a gigantic form, as if he was seeing the world through a magnifying glass; every skirmish becomes a Mahabharata, every little hamlet a little Lanka, every warrior a giant who with his arms upholds the sky".[39] However, Tessitori too acknowledged the "kernel of truth" that could be unearthed in the Charanic literature if they were studied in context by "reducing things to their natural size, at the same time denude the facts of all fiction with which they are coated".[40] He acknowledged that Charans were, "no doggerel verse- makers, nor mere repeaters of oral songs, they are lettered poets and their works have not only an historical and ethnological value, but also a literary one".[41] Besides, the distinction made by Tessitori is quite inaccurate, as it is only the later *Khyats* like *Maharaja Man Singhji ri Khyat* and *Maharaja Takhat Singhji ri Khyat* that are annalistic in style and follow dispassionate chronological order. *Nainsi ri Khyat* bases itself on older Charanic narratives, which recapitulate recent incidents as well as legendary events.[42] In fact, Nainsi uses Charanic poetics to substantiate claims

---

[38] Tessitori, *A Progress Report on the Proposed Bardic and Historical Survey of Rajputana,* 1919, 26–27.

[39] Tessitori, *A Progress Report on the Proposed Bardic and Historical Survey of Rajputana,* 1916, 228.

[40] Ibid.

[41] Tessitori, *A Scheme for the Bardic and Historical Survey of Rajputana,* 1914, 379.

[42] The other work by Munhata Nainsi, *Marwar ra Paraganan ri Vigat,* also incorporates *bats* but it attempts to compile a rather chronological narrative along with a gazetteer like description of the paraganas under Jaswant Singh. The *Vigat,* which is closer to

that he puts forth through his narratives. In a sense, Nainsi acknowledges Charanic verses as historical accounts on which he bases his own narratives.

Besides, the methodology of composition of a court chronicle like *Nainsi ri Khyat* was very different from Charanic verses as the intent involved in composition was different. *Nainsi ri Khyat* is a collection of *bats*, interspersed with *duhas, kavitts* and *pidhivalis*. Nainsi sometimes attributes the authorship of *duhas* and *kavitts* to Charans but *bats* often appear to be anonymous. Reading through the corpus of *bats*, it is evident that all these *bats* had been composed over a very long period of time, and thus representative of acts of collective remembering. Nainsi bases his work on what he claims to have *heard*. What he claims to have *heard*, is not an individual act of hearing, but an awareness of knowledge in circulation. However, with newer forms of history writing emerging, both as Rajput states were incorporated in the Mughal Empire and as they later came under indirect British rule, the Charan narratives increasingly became obsolete. Posited against the histories based on dispassionate facts, the *bats* of the Charans were quickly discredited. They were no longer regarded as history but were to be studied in the realm of poetics and linguistics. But we need to question if Charans ever claimed that their work be regarded as 'history'. The desire to see an 'objective' history in the works of the Charans was pointless as the role of the Charan was not of an, "'objective' historian, but that of a seer, a guardian of legend and a conserver of tradition".[43] He did not consciously manipulate the truth, but represented the truth that he saw, as he himself was a part of that 'truth'. This 'truth' did not associate so much with the objective facts, as it did with the social order and ideal that the Charan endeavored to preserve. This discrediting was again related to the attribution of motive of financial gain that supposedly led the Charans to heap praises on the Rajputs. The 'sale of fame' had apparently impaired the unconfined flow of the 'free pen'.[44] Therefore for the European Indologists, like all other Indians Charans too were unreliable, and thus their accounts had to be dismissed, to be replaced by new objective histories. In the writing of these histories Bhats and Charans were to be employed as mere information gatherers, without having the privilege of sifting through the contents. From the outlook of the nineteenth century Rajput princely states, the idioms that were acceptable in the Rajput and Mughal courts were no longer required in the offices of the

---

narrative 'history', appears to be written in the method of the Mughal Chronicle *Ain-i-Akbari*, dependent on use of older sources as well as official informants.

[43] Zeigler, 'Marwari Historical Chronicles', 225.

[44] Tod, *AAR*, I, xvi.

Residents. Even the internal dynamics of the Rajputs had been transformed and the task of valorizing Rajputs had been taken over by the British authors and administrators. Charans were no longer required and could be dispensed with. As I pointed out in the previous chapter the discrediting of the Charans as bards and historians pushed them towards social obscurity and further into the realm of 'criminality' by expressly blaming Charans for the practice of female infanticide among Rajputs. The need for 'pure information' negated the context in which the Charan narratives had been composed. Charan narratives were largely meant to be oral narratives, as they were not to be read as chronicles, but were to be recited with particular emotion and listened to with awe and reverence in socio-religious gatherings that were marked by rituals and etiquette. The recitation of Charan narratives was a two way process in which the listener's participation was as important as that of the Charan. Therefore, Charan narratives were meaningless without Charans themselves.

But the European method and practice had space for the information and not for the informant. It had to be pure information that was verifiable, free of native emotion. The Charan narratives certainly could not pass this test and thus Charans were discredited on both counts as historians and as custodians of social order. In fact, even a well known Charan historian of the Mewar court like Kaviraja Shyamaldas Dhadhavadia, wrote his historical account *Vir Vinod* in an annalistic style reminiscent of Tod's *Annals*. Therefore, within two hundred years, the Charan identity was completely transformed and with it the indigenous tradition of *bat*, narratives that were constantly created and recreated through performance were lost.

Narratives from below: Re-appropriating Pabuji

As the narratives of Charans were getting lost in obscurity, other narratives flourished in the rural social space patronized by tribal, pastoral and artisanal groups in The Thar. These narratives were sometimes religious in nature like the epics of Pabuji, Devnarayanji, Ramdeoji, Gogaji, Tejaji, Harbhuji etc. At other times they centered on long lost tales of love and separation like Dhola-Maru, Moomal-Mahindro, Umar-Marvi etc.[45] These were often in the form

---

[45] Even though there appears to be some distinction between the two kinds of narratives mentioned above, they often deal with similar kinds of emotions of mysticism, fantasy, battles, love and separation. It is difficult to categorise one kind of narration as religious and the other as secular. For example, Umar Marvi jo Kisso was rendered in a poetic form by the Sufi, Shah Abdul Latif Bhitai, which explains the elements of Sufi mysticism in this epic poem of love and separation.

of epic poems that were sung in particular gatherings and had been recreated
several times over in these gatherings. Like the narration of *bats*, these 'folk'
narrations were also marked by observance of strict rituals and etiquette, with
clear demarcations of the prescribed and proscribed. It was these narrative forms
that appear to have remained free of the Rajput and colonial domination and
continue to represent the spirit of the communities that were marginalized and
criminalized. The subsequent discussion about, the oral epic traditions popular
in the Thar, will raise some question about mobility, identity and circulation
of traditions in this region. This attempt would explore the manner in which
social and political identities were represented in these epics and what they
meant for the mobile communities of the Thar.

The epic of Pabuji, a narrative centering around a Rajput martyr, who is
believed to have sacrificed his life while rescuing cattle, appears to have been in
circulation from about the sixteenth century. The earliest attributable version
appears to be *Pabuji ra chhand* composed by Meha Vithu, a Charan, following
which several versions of the epic seem to have been composed with titles like
Pabuji ra Duha, Pabu Prakas, Pabuji ra Kavitt etc. All these versions of the epic
underline the mythical status of Pabuji as one born of a celestial nymph and
thus exercising a semi religious power. The epic of Pabuji is part of the warrior-
cattle protector tradition, in which Rajput heroes have emerged as the protector
deities of pastoral castes. For instance, Gogaji a Chauhan hero is believed to
protect cattle against snakebites. He, along with Harbhuji, a Sankhla Rajput,
Mallinathji, a Rathor Rajput and Ramdeoji a Tanwar Rajput is seen as part
of the *panj pir* tradition in Rajasthan, though the names of the five *pirs* may
vary. All of them are worshipped as folk deities in the Thar, in some cases
by particular communities. Ramdeoji is revered by the Meghwal community
with his *thans* being constructed prominently in the villages. Similarly Pabuji
is considered a patron deity by the camel and sheep herding Raika community.

There are several variations in the different versions of the Pabuji narrative,
but a general lay of the epic is as follows:

> Pabuji, who is believed to have lived sometime in the fourteenth or the
> fifteenth century CE was the son of Dhandhal Rathor of Kolu. He was born
> of a celestial nymph, who extracted a promise from Dhandhal that he would
> never spy on her. She gave birth to two of Dhandhal's children Pabuji and
> Sonabai, who had two other children from another wife Kamalade, called
> Buroji and Pemabai. Dhandhal failing to overcome his curiosity decided to
> visit her unannounced and found her in the form of a tigress suckling the infant
> Pabuji. As Dhandhal had spied on her, she left him and Pabuji was raised by

Kamalade, the first wife. Pema Bai was married to Jindrao Khichi of Jayal, in order to settle a feud that had resulted in the death of Sarangde the father of Jindrao, while Sona Bai was married to the Devra ruler of Sirohi. Buroji's marriage was contracted with a Guhilot clan of Dadreva. Pabuji resisted his own marriage, but later was convinced to marry the Sodha princess Phulwanti of Umarkot, a marriage that was never consummated.

Pabuji displayed his miraculous powers even as a child and though all lands and authority rested in Buroji, as he was the elder son, Pabuji continued with his exploits around Kolu. Seven Bhil brothers, Chando, Dhembho, Khapu, Pemalo, Khalmal, Khangro and Chasal, who had become involved in a feud with Ano Vaghela of Gujarat, and thus sought refuge in Marwar, approached him. The Bhil brothers became his companions and followers as they realized that he was no ordinary mortal. Along with the Bhils, a Raika named Harmal and Salji Solanki, an augerer, also accompanied Pabuji in all his pursuits. Meanwhile, one day Pabuji's sister Sonabai was whipped by her husband when she entered an argument with a co-wife, who was the daughter of Ano Vaghela over Pabuji's association with the Bhils. She appealed to Pabuji who decided to revenge the insult meted out to his sister and settle the feud of the Bhils with Ano Vaghela.

For this he needed a mount, which he got in the form of the black mare Kesar Kalami, given to him by Charani Deval, who had moved to Kolu with her cattle and clan. She had earlier refused to give the mare to Buroji as well as to Jindrao Khichi. The reason for this was that the mare had in previous birth been the mother of Pabuji. In return Pabuji promised to protect the cattle of Deval, particularly against Jindrao Khichi, who had set his heart upon the mare. He also granted her the pastures of Jujaliyo and the Nibali Tank. As promised he defeated Jindrao and extended his protection to the cattle wealth of the Deval.

At the occasion of the marriage of niece Kelam with Gogade Chauhan of Sambhar, Pabuji promised Kelam the gift of *rati bhuri* (red-brown) she-camels from Lanka. On being teased about the she-camels by her sisters-in-law Kelam appealed to Pabuji, who decided to send Harmal Raika to spy on the she-camels in Lanka. Harmal reluctantly went to Lanka in the garb of a Nath ascetic and won over the trust of the camel herders, with the help of the powers that his guru Gorakhnath had given him. While coming back he branded the she-camels in the name of Pabuji. When Harmal Raika returned with the news of camels, Pabuji set off with his men and magically crossed the seas to reach Lanka. After a fierce battle with Ravan, the camels were obtained. Half of these were given to Kelam, who left them to graze in the gardens of her sisters-in-law in Sambhar, while the other half was entrusted to Harmal Raika.

On his way back with camels, Pabuji passed Umarkot where the Sodhi princess Phulwanti beseeched her parents to marry her to Pabuji Rathor. After the camels had been delivered a date was set for Pabuji's wedding but Pabuji still prevaricated and desired to plunder the saffron fields of Lakhu Pathan to dye the turbans of the wedding party. Finally after the saffron was obtained and the turbans dyed in the fashion of Rajputs going for climactic battles, Pabuji agreed to go to Umarkot to marry Phulwanti. Chando, who was marrying off his seven daughters around the same time, was left behind in Kolu. As the wedding procession was setting off, the Charani Deval approached it and demanded to know who would protect her cattle while Pabuji would be gone. He promised to return as soon as she called even from the wedding pavilion. The wedding party encountered a number of ill omens and was advised to return by the *sagani* Salji Solanki, but Pabuji refused as he had made a commitment to the Sodhas.

As Pabuji was getting married, Jindrao Khichi stole the cattle of Deval. She appeared as a bird and appealed to Pabuji. He immediately cut the nuptial knot and rode out to fight the Khichi. The seven Bhil bridal parties of twenty Bhil archers each, who had arrived to marry the seven daughters of Chando also joined the battle. Dhembo and Pabuji freed the cattle of Deval. Pabuji made Dhembo spare the life of Khichi as he was his sister's husband. In this battle his fiercest companion Dhembo was killed, as he had fed his entrails to the vultures on the way to the battle.

As they were returning the cattle to Deval, she insisted that Pabuji water the cattle from a well that she had already caused to be dried up through the agency of Susiyo *Pir*. Pabuji defeated the *Pir*, and watered the animals but Deval found a one eyed calf missing who was finally located in the opium box of Dhembo.

In the meantime, Jindrao Khichi managed to enlist the support of his uncle Jaisingh Bhati and attacked Pabuji, who at the climactic moment ascended to heavens in a *palanquin* while his brother Buroji was slain in battlefield. The Sodhi princess and Dod-Gehali, the wife of Buroji both committed sati. Buroji's wife was pregnant at the time and she cut her belly and gave away the child who was called Jharadoji, to be raised by her mother. On being finally told about his father's and uncle's death, again through the agency of Deval, he killed Jindrao Khichi and sought his revenge. He later became the Nath sage Rupnath, who is worshipped along with his uncle Pabuji.[46]

Interestingly Buroji, despite the fact that he died in battlefield while defending cows, is neither deified nor worshipped as he was initially reluctant

---

[46] Smith, *The Epic of Pabuji*, 47.

to extend his protection to Charans and their flocks. On the other hand Dhembo, Chando, Harmal and Salji find a place in the Pabuji icons and the epic performance, despite their lower social origins. In some versions of this narration a connection is made with the pan-Indian epic tradition of Ramayana. In this narration, Pabuji is believed to be an incarnation of Lakshman, while Jindrao is believed to be the incarnation of Ravan and Phulwanti of Surpnakha, the sister of Ravan. The whole episode of Pabuji was thus played out to fulfill Laksman's obligation to Ravan and Surpanakha. As Lakshman in his previous birth, he not only had been the responsible for ferreting out the secret of Ravan's death but he had refused to marry Surpanakha and had disfigured her. It is for this reason that to Pabuji has to die at the hands of Jindrao. It is also for this reason that though Pabuji marries Phulwanti, yet the marriage is not consummated.[47] The Charani Deval is believed to be an incarnation of goddess who is there to ensure that all these obligations are fulfilled and thus she creates conditions in which Pabuji and Jindrao Khichi repeatedly face each other. Yet before Khichi can kill Pabuji, he ascends to heaven. Though, Pabuji is born in order to be killed by Khichi, he cannot die a mortal human death. The entire epic thus is deliberated towards the climactic moment of the death/disappearance of Pabuji, all events pointing towards that predestined moment.

A prose version of this epic was cited by Munhata Nainsi in his *Khayt* as *Vat Pabuji ri*. In this *bat*, which evidently was literized much later, the actual geography is adhered to with the camels being brought from Sindh, from the herds of Dodo Soomro. Charani Deval is replaced by a group of Kachela Charans, who appeal to Pabuji for protection. Lengthy descriptions of mutual display of mystical powers between Gogaji and Pabuji are played out. The marriage is actually consummated and it is on his way back from Kolu that Jindrao Khichi steals the cattle of the Charans.[48] In Nainsi's version, Pabuji is represented to belong to a tradition of Rajasthani folk deities, the *junjhars* and the *pirs*. No connection to previous births is made and the story is played out in the temporal dimension of medieval Rajasthan. Both these versions would have been composed by Charans at different times. The epic in either version is fraught with possibilities of several kinds of interpretations. One of the most important questions that it raises is regarding manner in which relations of the Rajputs with Charans, Bhils and Raikas have been depicted in the epic.

---

[47] Pabuji's narrative follows the classic patterns of narrative development drawn by Stuart Blackburn. Pabuji's death is followed by his deification, and identification with Lakshman. Blackburn, 'Patterns of Development of Indian Oral Epics', *Oral Epics in India*, 15–32.

[48] Nainsi, *Vigat*, II, 57–79.

The main temple of Pabuji is located in his village Kolu, but small shrines are put together in Raika hamlets and orans, which can contain small icons depicting Pabuji on his mare, Kesar Kalami, and sometimes his Bhil archers. Another popular form of worship is through the recitation of the epic of Pabuji, with or without a visual narrative called *phad*. The recitation without *phad* is carried out in the Kolu temple and is called *mata* performance. It involves singing of the epic of Pabuji by two Bhil singers alternately accompanied by earthenware drums called *mata*.[49] The unaccompanied group singing of the Pabuji epic by Raikas themselves is called *jhurava* and is usually carried out without the ceremonialism evident in *mata* or *phad* performances.

The other and more popular form, which is the *phad* vachan, is carried out by Bhil Bhopas. The *phad*, a visual narrative painted on cloth is about fifteen feet in length and contains episodes from Pabuji's life arranged in a formation where Pabuji and his court form the focal point of all events. It is painted by a particular community of *chiteras* called Josis, who live around Bhilwara. The *phad* is usually owned by the *bhopa*, who carries it from village to village, a case of temple visiting the devotees.[50] If a new *phad* is required it may be commissioned by the Raikas and given to the *bhopa*.

The pictures appear to be chaotic, but they are actually arranged so with reference to the manner and sequence of the story as it is recited by the Bhopas. The places mentioned in the epic like Jayal, Jaisalmer, Pushkar, Sambhar, Gadvaro, Umarkot, Lanka etc are all depicted keeping their narrative distance in consideration. The places inhabited by people who play villainous roles in the narrative, like Jindrao Khichi and Jait Singh Bhati are placed at the extreme corners of the *phad*.

Various episodes narrated in the *phad* constitute a scene; a series of events interrelated to each other. The scenes are represented on the *phad* in a directional manner in relation to other scenes. The entire scene or a particular representation in the scene could be used in multiple ways to depict different meanings at different points of narration. The scenes or characterizations aim to fill all available space on the *phad*, which might not necessarily adhere to a scale. For instance Pabuji is drawn on a much larger scale than his companions or his brother Buroji. Besides, episodes believed to have happened in a place are all depicted together irrespective of the time that may have elapsed between episodes. Attempt is made to fill all the residual space with minor events and

---

[49] Srivastava, 'The Rathor Rajput Hero of Rajasthan', 592.

[50] Smith, *The Epic of Pabuji*, 15.

characters that may have a supplementing role in the *phad* narration. As John Smith views it, the *phad*, "may not be naturalistic, but it does resemble, nature to the limited extent of abhorring the vacuum".[51] Therefore, it would be more useful to look for meanings with the particular context of the narrative of the *phad*. For instance, Umarkot is placed near Lanka, so that the return journey of Pabuji via Umarkot can be depicted. It is so, as the marriage to the princess Phulwanti of Umarkot forms the most important climactic moment in the story. Therefore, the *phad* has its own narrative geography that relates to the manner in which episodes are recited.

Phad vachan takes place at night, and begins with the usual invocations to the gods. Entire epic is never narrated at a time; only certain episodes, which are deemed necessary for the structure of the epic, are recited. The main narration is done by the Bhil *bhopa* with the aid of a *ravanhattho*, anklet bells and an assistant. When *phad vachan* takes place the assistant, who usually is the wife of the bhopa, illuminates each episode with a lamp. Thus, at the point when an event is being narrated it becomes the focal point of the *phad*, shadowing all other representations. During the narration of the *phad*, the audience also participates by adding *hunkaras* at relevant points.

I have argued from the beginning of this work that communities like Rajputs. Charans, Bhils and Raikas, that figure in this epic, are historically linked with each other, not just in terms of occupational structures, but also through processes of emergence of one community from another. Each of these communities plays an important role in the epic. In playing these roles, the characters who represent each community, appear ambiguously similar to each other in the manner that they are depicted and the values and ethics they extol.

Let me begin by exploring the depiction of Rajputs in the narrative. Janet Kamphorst points out that Pabuji's role in the entire epic is that of a cattle rustler and an itinerant warrior, even though he asserts that he is a Rajput and should not be challenged.[52] All the battles fought by Pabuji, feature cattle prominently, whether it is with Khichi, Mirza Khan or Dodo Soomro. When Pabuji acquires the mare Kesar Kalami, his sister-in-law Dod Gaheli, expresses the fear that now he would undertake cattle raids or *dhads*. Were cattle raids by Rajputs a regular feature? Were Rajputs merely the protectors of cattle of Charans and other communities or were they cattle owners and

---

[51] Smith, *The Epic of Pabuji*, 33.

[52] Nainsi, *Khyat*, II, 67, Pabuji points out to his sister-in-law, Dod Gehali, that she should not be taunting a Rajput. Also Janet Kamphorst, *In Praise of Death*, 192.

traders themselves? For instance, why did Jindrao Khichi covet the cattle of the Charans? Or was a change in Rajput identity taking place around the time when the epic was being composed and recited? It is clear that *dhads* were no embarrassment to Rajputs in this period. Pabuji's sister-in-law objects to his acquisition of the mare fearing that now he would undertake cattle raids. What then is the association between the acquisition of horses and cattle raids? Did the ascent of Rajputs as horse riding warriors, while allowing them to augment their cattle wealth also placed them above other groups engaged in similar contests?

Let us examine these questions in the context of the role of Bhils in the epic of Pabuji, which presents the Bhils in a contradictory light. Let us examine these contradictions. Pabuji's epic is played out in a village where even to this date Bhils outnumber any other community. Janet Kamphorst examines several versions of the epic and points out that, on the one hand Bhils are depicted as fellow fighters in the battles undertaken by Pabuji, on the other they are also portrayed as camel eating and opium guzzling tribe despised by other Rajputs. On the one hand, they are shown to make an impressive army, on the other, they are also called *paradhi* fighters. While Rajputs are shown to fight with swords, clubs and maces, the Bhils are depicted as archers.[53] Interestingly, Chando and Dhembo, the Bhil brothers are shown to own and ride horses, unlike Harmal Raika, who is shown to have walked to Lanka. Unlike Rajputs like Buroji, it is the Bhil bridal parties that unhesitatingly join the battle against Jindrao. Therefore, when Deval attempts to create barriers between the lines of blood of martyred Rajputs, Raikas and Bhils, Pabuji forbids Deval and asks her to let the bloodlines unite. This perhaps points to the inherent struggle for identity in this period. While it is clear that the Bhil archers act as subordinate to Pabuji, it can also be pointed out that Bhils and Rajputs are not very unlike and both aspire martial identities. The Pabuji epic underlines two primary reasons for death in battle aspired by Rajputs. One while avenging the death of an ancestor and the second, while protecting cattle. The first is claimed not just by Pabuji's nephew Rupnath, but also by the Bhils in relation to Ano Vaghela, who had killed their father. Regarding the second, the Bhil warriors are also involved in battles over cattle along with Pabuji. Only when a Rajput died nobly while seeking revenge or defending cattle, could he be deified. So, this raises the question of what then distinguishes the Bhils from the Rajputs?

This contradiction in Bhil identity continues to operate even today. Bhil like Rajput appears to have been a generic category that has been used to

---

[53] Kamphorst, *In Praise of Death*, 207.

include a number of groups living in present day Gujarat, Madhya Pradesh, Maharashtra and Rajasthan. Bhils are variously addressed as Thori, Ahari, Paradhi, Sanwal, Nayak etc in these regions though they themselves consider some of these categories as derogatory.[54] In modern phraseology they have often been referred to as Adivasis or Girijan, terms that glaringly allude to a non-Aryan past as opposed to that of an Aryan Rajput one. In the ethnographic reports of the nineteenth century, Bhils have been characterized as lawless bandits and were notified as a criminal tribe. Like Rajputs, Bhils also claim a martial past, derived from their being expert archers. In the late nineteenth century, Bhil martitality was channelized through their induction into Bhil Corps in Mewar and Khandesh.

The medieval history of Rajasthan is replete with evidences of communities like Bhils, Mers, Meos etc. being contestants for control of land and animal wealth along with Rajputs. It is apparent from the Mewar example that Bhil authority was usurped by Rajputs but they continued to challenge Rajput authority through raids. *Dhads* appear to have been as much part of the Bhil claim to power as that of the Rajputs. Unsurprisingly, the main accusation against most of these communities in the nineteenth century is of cattle theft and Bhils among other tribal communities were branded bandits and unruly tribesmen in both, Rajput courtly and British colonial sources. However, the control extended on roads, trade routes and hilly areas by these communities shows that they continuously contested assertions of sovereignty as put forward by their self-proclaimed Rajput, Mughal and British overlords.[55] Therefore, it is not surprising that Bhils allude to a Rajput past, a claim that is validated in the Pabuji epic.

In the Pabuji epic, it is evident that both Bhils and Rajputs were a part of the "medieval pastoral nomadic world of cattle rustlers".[56] Bhils and Mers had been dominated by Sisodiyas and Rathors respectively. Yet, the struggle for Rajput dominance had not fully subjugated these groups and they still played an important role in the mutual struggles of the Rajput groups and were sought out as allies. The disdain expressed towards them points to the fact that they were being steadily and constantly marginalized.

It is important to point out that it is Bhils who appropriated and reformulated the Charan narrative of Pabuji and developed it into an alternate tradition

---

[54] Kamphorst, *In Praise of Death*, 72.
[55] Skaria, 'Being jangli: The politics of wildness', 193–215.
[56] Kamphorst, *In Praise of Death*, 217.

that only they could recite. It is Bhil Bhopas who perform the *mata* recitals patronized by the Rajputs at the temple in Kolu. The *phad vachan*, patronized by the Raikas is also carried out by the Bhil Bhopas, who despite being a low non-priestly caste have become the priests of a Rajput folk deity. It is the process of *phad vachan* that lays the ground for a Bhil counter narrative in which in significant portions it is the Bhil Dhembo, who becomes the hero of the epic. The exploits of Dembho are reminiscent of Hanuman, both in impulsiveness and simplicity of his behavior as well as in the strength that Dembho displays. He is shown to have gargantuan appetite like Hanuman and is willing to follow every command of Pabuji unhesitatingly. He consumes opium, as do all Rajputs, and is revealed to singlehandedly wipe out the army of the Khichi before being stopped by Pabuji from killing Khichi who is his sister's husband. In the sacrificial tradition of the Rajputs, he is shown to feed his entrails to hungry vultures and yet continue to fight. Though Dembho does not conform with the image of the headless Rajput *junjhar*, yet he forms the Bhil focal point in this counter narrative.

Another question to be raised here about Rajput-Bhil relationship is that of Pabuji's mother, the nymph that Pabuji's father is supposed to have married. Dhandhal Rathor comes across this nymph in the forest and tricks her into marrying him. Nymphs or *apsaras* are recurrent figures in Indian mythological traditions, supposedly celestial dancing girls who were sent to disrupt the meditations of sages if they appeared to threaten the cosmological order. In the context of Pabuji epic, where each character is identified with a lineage, the presence of this nymph who has no lineage to speak of, raises certain questions. Janet Kamphorst explores the possibility of this nymph being a tribal woman, perhaps of Bhil or Mer origin.[57] As discussed elsewhere in this work, marriage was seen as a legitimate way of augmenting resources, whether mobile or territorial, in the Rajput world. In such a case a marriage between a wandering Rajput warrior and a Bhil woman would not be very out of place. In fact the Grasias claim to have originated from Rajput-Bhil unions. However, what is interesting is the shrouding of the identity of the Bhil woman in this narrative. She disappears fairly early in the narrative only to appear as a mare, a necessary requirement for a Rajput. However, by demarcating a very clear boundary between herself and her husband, which may not be crossed without her permission she distinguishes her 'wild' world from Dhamdhal Rathor's 'village'. Her appearance as a tigress also alludes to 'untamed ferociousness'

---

[57] Ibid.

of the forest. Therefore, it becomes interesting to question the relationship of Bhils with Pabuji. Can Pabuji instead of being a pure bred Rajput be seen as someone born of mixed Bhil-Rajput origins? Interestingly, in several parts of Rajasthan and Madhya Pradesh, Bhils are addressed as *mamas* or maternal uncles. What if Pabuji were not merely seen as a Rajput *Bhomiyaji*, but as an ancestor by the Bhils? Does it then remain a Rajput narrative or become a Bhil narrative? Bhil Bhopas claim that they recite the *phad* because as Pabuji was ascending the heavens, he left behind a cloth marked by his blood for his Bhil followers.[58] This raises an interesting question about, why a Rajput hero should leave his story behind to be told by his low caste followers. In my understanding it becomes a way of reclaiming Pabuji as one of their own, done much more overtly by Meghwals in the case of Ramadeo of Runecha.[59]

Further, at this juncture the mutual relationships of the Bhil Bhopas and Rajputs with the cattle herding community of Raikas also need to be explored. While it is Bhils that are Pabuji's priests he is actually revered by Raikas. Whenever the Raikas establish a new hamlet or an *oran*, they construct a small shrine of Pabuji. Anthropologist Vinay Kumar Srivastava has raised some interesting questions about the role of Raikas and their relation to the Pabuji myth, particularly about the Raika assertion that camels were brought into Marwar by Pabuji.

Let us examine the epic of Pabuji from the point of view of the Raikas, who along with Gujars are listed as a backward caste in Rajasthan. However, unlike Gujars the Raikas have not been able to consolidate themselves as a strong political group and remain marginalized socially, economically and politically. Raikas claim to have been created by Shiva to take care of the camel, the animal created by Parvati. The first Raika named Pinda was married to a celestial nymph Rai, and had twelve daughters and one son. The twelve daughters married twelve Rajputs, leading to the thirty six subcastes of Raikas, while the son also married a Rajput woman but carried the original caste name of Samar. There exist several other names for the community. They are called Bhut as they stay in the forests and are fearsome to look at, Rebari (*rahbari*), as they live outside the customs, Raika, born of Rai, Othi or camel rider and Dewasi, in whom gods dwell. Pabuji epic raises an interesting question about the role of

---

[58] Elizabeth Wickett, *The Epic of Pabuji ki par in Performance*, World Oral Literature Project, Published by University of Cambridge, 2001, 5.

[59] Dominique-Sila Khan, 'Is God an untouchable? A case of Caste Conflict in Rajasthan, *Comparative Studies of South Asia, Africa and the Middle East*, Spring 1988, 18 (1): 21–29.

Raikas and their relation to Pabuji. While it is Bhils that are Pabuji's priests, he is actually revered by Raikas as he brought camels to the community (and not because he is a cow rescuer). Vinay Srivastava questions, if in that case, were there no camels in Marwar before the time of Pabuji, as the self-image of Raikas suggests that they were appointed camel herders by Shiva?[60]

A prominent character in the epic is Harmal Raika, who is sometimes addressed as Dewasi in the epic. It is Harmal Dewasi who is entrusted with the task of finding the whereabouts of the she-camels. He arrives in Lanka after becoming a Nath ascetic and lives with the herders for six months and wins over their confidence and devotion. When he is returning, they want to gift him a camel but in his role as an ascetic he cannot graze a camel, so he brands the camels in the name of Pabuji and leaves, after which a *dhad* is carried out and these camels are herded away. The traditions suggest that a particular breed of camels, *thok* or *rati bhuri sandni* (reddish-brown she-camel) was brought by Pabuji to Marwar from Sindh. It was originally herded by Sindhi Muslims from a hamlet called Sairo Bagani in village Lankaye in Thar Parkar.[61] The news of the presence of camels in this *dhani* was conveyed by a Charan. Yet there remains the question of what happened to the camels that were brought from Sindh. While Pabuji's epic indicates that they were given to Kelam, Srivastava indicates that half of them were given to Harmal Dewasi, which forms the basis of the assertion that the camels and the responsibility of rearing them was granted to the Raikas by Pabuji.

The emergence of a clear association of Raikas with camels and dissociation of Rajputs would have to be understood in the context of identification of idea of a warrior with horse. In the sandy deserts of the Thar, it was the cushioned feet of the camel that could cover long distances in short durations. In the seventeenth century epic of Dhola Maru, despite the fact that Dhola is a Rajput, the preferred medium of travel is camel, while Pabuji is always depicted with his mare Kesar Kalami. I have pointed out elsewhere that it was in the late fourteenth and early fifteenth century that horses become available on a sustained basis to Rathors, which made it possible for them to appoint guard-posts equipped with horse riders and foot soldiers on the outposts in the desert. While the dominant idea in the Pabuji epic is of cattle protection, the fight among the Rajputs is for the horse, which makes it possible for them to protect as well as raid cattle. On the other hand, it is Raikas who emerge as exclusive camel herders distinguishing them from Rajput warriors.

---

[60] Srivastava, 'The Rathor Rajput Hero of Rajasthan', 610.
[61] Ibid.

The other aspect of the imagery of Rajput warrior was the idea of a warrior as an ascetic. Pabuji is revered as a warrior-ascteic, *Lakshman jatti*, the celibate warrior, the renunciant who is prepared to give up worldly pleasures for the sake of his duty. Harmal Raika becomes a Nath *jogi*, goes through all rites of becoming a *jogi*, including visiting his mother and wife in the guise of a *jogi*. This part of the Pabuji epic has similarities with the tradition of Raja Bharthari and Gopichand which requires the former kings to test their resolve by visiting their palaces as *jogis*.[62] Kolff points out that soldiers continued to be viewed as ascetics who left their worldly attachments behind when they went soldiering.[63] Now, in a world where Bhils, Raikas and Rajputs were all engaged in cattle raiding or *dhads*, the mission undertaken by Harmal is not unusual. His task is commensurate with the expectations from any warrior in this cattle-rustling world. What the Raika thus undertakes to carry out is not a spy mission, which can be treated as a jocular interlude in the epic otherwise filled with pathos of war, but an act of war itself. The warrior ascetic is not merely a mythological presence in Indian history, but the presence of groups of armed ascetics who participated in many a struggle has been attested by several scholars.[64] Harmal Raika thus falls in this tradition where cattle raids, battle and asceticism all come together to make the medieval warrior. Besides, in a world where cattle were regarded as mobile wealth, herding itself would have required martial capabilities. Harmal is thus viewed as a warrior ascetic, an equivalent of Pabuji himself. It is the exhorting of this warrior ascetic tradition that makes the *vachan* of the *phad* of Pabuji reinvent the community identity of Raikas, for they are not just narrating and hearing the exploits of Pabuji but also of their own ancestors. For them the narration of the *phad* is a journey into the self.

Finally, we arrive at the question of the Charan narrative. The relationships between Charans and Rajputs have been explored extensively in the first section of this chapter. While the epic hero Pabuji undertakes the various feats of conquests it is actually the Charani Deval who pushes him towards his predestined fate. Deval who is explained to be a Charani *sagati*, who makes

---

[62] Ann Grodzins Gold, *A Carnival of Parting: The Tales of King Bharthari and King Gopi Chand,* University of California Press, Berkley, 1992.

[63] Kolff, *Naukar, Rajput and Sepoy,* 71–85

[64] David Lorenzen, 'Warrior Ascetics in Indian History', *Journal of the American Oriental Society,* Vol. 98, No. 1 (Jan - Mar, 1978), 61–75. Also see William Pinch for warrior ethics among Gosains in Banaras. *Warrior Ascetics and Indian Empires,* Cambridge University Press, New Delhi, 2006.

the events take the shape that they do. For instance, it is Deval who by the gift of the mare creates the feud in the first place. When Pabuji is proceeding for his wedding she commits the ill omen of stopping him and makes him promise to return if called, to protect her cattle. After he has rescued her cattle, she deliberately delays him by asking him to water the cattle and find the small calf, so that Jindrao Khichi is able get the time to enlist the help of his maternal uncle, Jai Singh Bhati. She is the cause of the final climactic battle between Pabuji and Khichi. Why does Deval do so? It has been suggested by both by Smith and Kamphorst that Deval being the Charani *sagati*, is far more powerful than Pabuji. Deval is a part of the tradition of Charani *sagatis* that are revered by the Rajputs as explained in the previous section. As a *sagati* Deval is the guarantor and protector of the realms not just terrestrial but also celestial. Therefore, it is her duty to preserve the divine pre-ordinance and make it happen. Consequently, putting an end to all Pabuji's prevarications it is Deval who takes him steadily towards his fate as a Rajput warrior ascetic. She accompanies him in the battle-field and controls the events there. She is fearless and does not accept Pabuji's miracles as they are accepted by his Bhil and Raika followers. In the context of Charan-Rajput relations, perhaps it points out that it is the supposed sanctified duty of the Charan to guide a Rajput towards his true dharma and fate. Besides, as horse breeder and trader, Charani Deval appears to control the most important resource required by the Rajput Pabuji. In the episode of Rathor Chunda described by Nainsi, it is a Charan who recognizes him as a Rajput, but also equips him with a horse and weapons. Given the location of Charani goddess temples on routes through the desert, it is clear that Charans exercised significant control over movement through the desert. Repeated references to Rajput armies encountering Charan goddesses and being supplied with water and food on their adventures are indicative of the fact that Charans were indispensable on desert routes. Thus Deval's presence as a dominant factor in Pabuji's adventures is a reflection of the status of Charans in the Rajput world.

However, Charans as a community do not view the Pabuji epic as anything more than a poetic rendition of the life of a Rajput warrior, one of several that Charans composed. On the other hand, the Bhil Bhopa seeks to place himself within the narrative by exploring relationships between Bhils and Pabuji. In fact, Charan and Bhil interface with the Pabuji tradition reveals two opposite processes. While in textual traditions Charans demilitarize themselves, representing themselves either as cattle breeders, traders and priests or through the Charani goddess Deval, in the oral versions, Bhil Bhopas seek a martial past for the Bhils. It can perhaps be explained by the fact that with the sedentarisation of Rajputs, Charans became a part of emerging court culture, receiving *sasan* grants as well as patronage as poets and historians. On the other hand, Bhils and Raikas were pushed to margins of the emerging political culture. Therefore, while the vast Charanic traditions valorize and reiterate a particular kind of Rajput image, it is the oral renditions by communities like Bhils and Kamads that underline the continuities in the pasts of communities like Rajputs, Bhils, Raikas and Charans. Thus, it is actually the Bhils and Raikas who reinterpret the epic through their own understanding of social relations through the act and process of *vachan* of the *phad* of Pabuji, thereby re-appropriating him as one of their own.

It is this constant telling and retelling of the tale of Pabuji that has given rise to a "pluriform Rajasthani tradition".[65] The multifaceted narrative of Pabuji has some sections that relate to a particular community. The devotional hymns dedicated to Deval relate to the Charans. The escapades of Dhembo or the seven Bhil bridal parties relate to the Bhils and the escapades of Harmal Dewasi to the Raikas. It is this multilayered narrative that provides a meaning to all communities that relate to the epic adventures of Pabuji. It also indicates the two way process of emergence of written narratives from oral ones and their subsequent reinterpretation through oral narration. This tradition, while, is seemingly situated in the medieval martial ascetic world of the Rajputs, it has been reframed several times over though the narration by the Bhil Bhopas and by being heard by Raika devotees. It thus, in a sense, constitutes a counter narrative to the prevalent dominant Rajput histories. This process has turned Pabuji into a *lok devata*, who is not merely to be propitiated like a *bhomiaji* but revered in the sense of a deity. Therefore, an attempt at understanding the Pabuji epic in its oral as well as written form leads us to see the changes in the community identity of pastoral and tribal groups. We can also see that oral epics like that of Pabuji (and perhaps also other like that of Devnarayanji, Ramdeoji, Tejaji among others) question the idea of fixed social identities. A notion of ambivalent, flexible, and indeterminate social identities emerges through repeated oral narrations. These epics are able to bring forth both the conflicts and the cohesion in the mutual relations of communities in the processes of transformation of community identities. While the oral narrative

---

[65] Janet Kamphorst, 'The Deification of South Asian War Heroes- Methodological Implications', in *Epic Adventures: Heroic Narratives in Oral Performance Traditions of Four Continents*, Verlag Münster, 2004.

of Pabuji points towards indeterminacy of caste identities in the Thar, on the other hand the oral narrative of Devnarayan, venerated by the Gujars brings out the manner in which a Gujar identity distinct from Rajputs is created through the veneration of Devnarayan and the telling of his story.

Devnarayan: The Cowherd Warrior/God/King

The narrative of Devnarayan is closely associated with emergence of Gujars as a community. Devnarayan, venerated as an avatar of Vishnu, is seen to encompass values of valour as well as compassion in this narrative, and emerges as a hero who leads Gujars towards a victory in the battle against the Sisodiya king of Mewar responsible for the death of his father and uncles, in the Bagravat Mahabharat. He is worshipped in the temples established in his name as well as through the recitation of his story along with the representation on a visual narrative called the *Devnarayan ji ki phad*, quite like the *phad* of Pabuji. The *phad* or the painted scroll, which is one and a half meters wide and eight and a half meters long is treated as a portable temple and has images of Devnarayan and his cousins, Bhunaji, Meduji, Madnoji and Bhangiji, in his court. *Phad vachan* takes place either at the home of a devotee, in front of a temple or at a Gujar community meeting. The Bhopas of Devnarayan are Gujars themselves, who use a double stringed instrument called *bin* or *jantar* to sing the tale as well as to comment on and explain various parts of the narrative. Devnarayan temples are located in Gujar villages where Devnarayan along with his brothers, ancestors as well as attendants is worshipped in the form of iconic bricks.[66]

The epic of Devnarayan traces the life of the ancestors of Devnarayan, as well as the origin of Gujars as a community. The singers of the epic divide it into two halves, the Bagravat Bharat and the Sri Devnarayan Katha. While in the first part the fierce sacrificial warrior acts of the twenty four Bagravat brothers are narrated, the second part extols the divinity of Devnarayan as the incarnation of Vishnu who validates the roles attributed to Gujars as a social group.

> The first part of the epic is located in the Satyuga, when Shiva in a fit of ravenous hunger consumed twenty four *rishis*. Because of this act Shiva ended up defiling a *Yagna* being conducted by Brahma and was told that in order to repent he would have to offer his body to the rishis who would be born

---

[66] Aditya Malik, *Nectar Gaze and Poison Breath: An Analysis and Translation of the Rajasthani Oral Narrative of Devnarayan*, OUP, New York, 2005.

as twenty four sons of the same father in the *Kaliyuga*. In the *Kaliyuga*, the narrative commences in the kingdom of Raja Bisaldev of Ajmer, where a Rajput named Hari Ram killed a lion that had been tormenting the kingdom. While Hari Ram went up to the holy lake of Pushkar to wash himself, with the lion's head on his shoulder, his reflection was seen by a Brahmin woman called Lila Sevri who had taken a vow of never looking at a man. But as a result of this viewing she conceived and Hari Ram and Lila ended up marrying, even though the former was a Kshatriya and the latter a Brahmin. However, the son born to Hari Ram and Lila was a half man and half beast with the head of a lion, thus named Baghsingh or Baghravat. As the king was convinced of his divine nature, Baghravat was adopted by the king but was forced to live in seclusion in a garden because of his fierce appearance, where he was looked after by a Brahmin servant. There being little likelihood of Baghravat marrying because of his appearance, the Brahmin managed to entice twelve women of different castes to enter the garden and circumambulate Baghravat, thus symbolically marrying them to him. Baghravat had two offspring with each of these twelve wives, including a daughter, collectively called Bagravats. The Bagravats were all married to Gujar women, and took up cow herding. Savai Bhoj, whose mother was a Rajput woman, was the most courageous of them all. He encounterd the ascetic Rupnath who blessed the Bagravats with wealth and might for twelve years after which they would be annihilated. Initially the Bargavats turned to piety, but eventually they befriended the Rajput Rana of Mewar, who became their *dharambhai*. Along with Rana they participated in several drinking sessions, to such an extent that alcohol entered the kingdom of Vasuki Nag, who complained to Bhagawan. But Bhagawan could not help because of the devotion of the Bagravats and also because though their annihilation had been decided by the end of twelve years no one was actually willing to carry it out. Bhagawan visited Sadu mata, the wife of Savai Bhoj and explained the dilemma. Sadu mata became aware of the impending doom, but asked for Bhagawan to be born as her son after the Bagravats have been annihilated.

Bhagawan approached Shakti, who agreed to be born as Jaimati, who was adopted by the Rajput king of Bhual. As she grew she insisted to be married to a house which has 24 sons born to the same father. The Bagravats were located but suggest that she marry the Rajput Rana. In a twist of events, Jaimati while apparently marrying the Rana, married Savai Bhoj, by circumambulating his sword. She refused to consummate her marriage to the Rana and eloped with the Bagravats. The Rana gathered armies of 52 forts. Jaimati, who assumed her true form as Bhavani, promised to accompany Bagravats on the condition that they fight one at a time and offer their heads to her, thus assuring that all of them were slain. All the Bagravat women, with the exception of Sadu mata commited sati, and the four surviving infants Bhangiji, Meduji, Madnaji and Bhunaji were raised incognito.

Here the narrative shifts to the birth of Devnarayan, who was brought forth on a lotus flower by Basuki Nag. Devnarayan was raised incognito by his mother in her natal land Malwa, but was eventually informed of his identity by a bard, Chochu Bhat. He decided to return to his ancestral land, and seek revenge for the death of his father and uncles. On his way he married three women, the daughters of the kings of demons and underworld as well as the daughter of the king of Ujjain. He also reunited with his cousins, one of whom had become a Nath ascetic, and another adopted by the Rana himself. Before the final encounter, Devnarayan released his herd of 9,80,000 cows in the fields of the Rana, who though beheaded by the bow-string of Devnarayan was revived. The title of Rana was retracted and he was called Sisodiya (sis + diya, referring to the grant of head). The Rana was ordered to establish the city of Udaipur, which according to the narrative took its name from one of Devnarayan's name, Ud.

Devnarayan, after seeking revenge decided to depart to heavens but stayed long enough to grant two children called Bila and Bili to his queen Pipalde. Bila was stubborn and refused to recognize the divinity of Devnarayan. After several unhappy turns including contracting leprosy and then being cured by Devnarayan, he realized the divine power of his father. He agreed to become the first priest of his temple. Thus having established a place of worship, a lineage of priests, and a community of devotees, Devnarayan finally returned on his celestial chariot to heaven.[67]

There appear to be many similarities between the Pabuji and the Devnarayan epics, prima facie, particularly with the idea of sacrifice in the cause of cow protection, as well as the use of visual narrative or the *phad*. However, the Devnarayan epic, unlike the Pabuji epic, has no confusions about location or ownership. It is clearly identified as an epic tradition belonging to Gujar community of worshippers, with priests and singers who are also Gujars. The epic and the traditions of worship surrounding it have in recent years become the foci of Gujar community identity, which while on the one hand views itself, both as a cattle herding as well as a land owning community. The Devnarayan epic provides a context to the Gujars' roles both as cowherds as well as warriors.

---

[67] Aditya Malik (ed.), *Śrī Devnārāyaṇ Kathā : Marwaḍ kā Pāramparik Gāyan*, DK Printworlds, New Delhi, 2003, 45–82.

In contrast to the Pabuji epic, where the Rajput hero fights and sacrifices his life for a cattle-rearing community, the heroes of the Devnarayan epic are from a cattle rearing community, but who die a 'Rajput' death. Also, again while Pabuji epic conceals mixed caste unions and parentage under the garb of celestial origins, the Devnarayan epic on the other hand constantly foregrounds mixed origins through marriages between men and women of different caste groups. While the Pabuji epic refers to a lost Rajput past of Bhil and Raikas, the Devnarayan epic is about the rejection of a Rajput past. Finally, while the Pabuji epic is about sacrifice, the Devnarayan epic is about acquiring power both terrestrial as well as divine. As Kolff suggests, the Devnarayan epic offers context to a "clash between pastoralism on the one hand and the claims of genealogical status and territorial rule on the other".[68]

The Devnarayan epic becomes the approach to Gujar origin myths, which like so many origin myths of itinerant groups are associated with Shiva. Vishnu takes the form of Mohini (a form of Parvati), in order to annihilate Bhasmasur who has received a boon from Shiva of being able to immolate anyone he places his hand on. After causing self-immolation of Bhasmasur, Visnu appears to Shiva in the form of Mohini, making Shiva ejaculate. Narada receives the semen in his palm and it is eventually poured in the ear of Anjana who became the mother of Hanuman. In this process, Narada happens to wipe his hand on some grass, which, consumed by a cow, impregnates her and the resultant offspring is the Gujar, who is made the protector of all cows by *parabrahma parmeswara*.[69] The other origin myth of the Gujars traces their origin to Syambhu Manu and Satrupa (also identified as Sawai Bhoj and Sadu Mata), who are granted the boon of becoming the parents of Vishnu twice, once in Satayuga, as Rama and second time in Kaliyuga as Devnarayan.[70] The claim to originate from Manu, the celestial father and then the association with the Suryavamsi clan of Rama, are not unusual and are found in almost all oral epic traditions where the hero is found to be an incarnation of a divine figure. The repetition of these myths as part of the Devnarayan epic tradition seeks to underline the position of the Gujars in social hierarchy.

While Gujars have been able to amass political and economic power and in some parts of northern India are seen as a landed community, their social position remains a marginal one. The cultural image of the Gujar is of an

---

[68] Kolff, *Naukar, Rajput and Sepoy*, 84.

[69] Malik, *Nectar Gaze*, 104–107

[70] Ibid., 95–102.

ignorant herder though the historical claims of Gujar past also associate them
with Gurjara-Pratiharas, with long migrations through Thar. However, as the
Devnarayan epic reveals, any Rajput link that the Gujars may claim, comes from
multi-caste marriages that are contracted in the course of the epic rather than
any other claim to descent from the older *kshatriya* clan. The original ancestor
of the Gujars is a Rajput, who marries a Brahmin woman. Their offspring,
Baghravat is married to twelve women of different castes, including Rajputs.
But when it comes to marrying the twenty four sons, no other community is
prepared to give daughters to such a mixed caste family. The King of Ajmer,
Bisaldev, who appears to be a patron of Hari Ram, calls Gujars and shares
hukkah with them, and asks for daughters in marriage. This results in all further
offspring becoming Gujars.

The Bagravats are portrayed in Rajput vocabulary, even though they appear
to have massive herds at all times. When contrasted with their *pagri badal*
brother (adoptive brother through the ritual of exchanging turbans) Rajput
Rana, who is old and feeble, the Bagravat brothers are young, brave and
impetuous. They are also depicted as generous and pious, building temples and
water bodies, a kingly function in pre-modern India. It is their association with
the Rana that leads them to indulge in alcohol and cause alarm in the heaven,
ensuring their annihilation. The possibility of the Rana marrying Jaimati only
arises because the Bagravats refuse as they are not Rajputs. However, this also
becomes the cause of enmity between the Rana and the Bagravats leading
to the annihilation of the Bagravats. By seeking revenge for the death of his
ancestors, what Devnarayan also appears to be seeking is a Rajput death.

However, in this epic, the death that we encounter is of the Rajput Rana.
At this juncture the narrative takes a turn unforeseen in any other similar
narratives of the region. The Devnarayan narrative part from containing their
own origin myths, also describes a Gujar origin myth of the Sisodiya Rajputs.
When the Rana is beheaded, Bhunaji who has been adopted and brought up
by the Rana, wishes him to be revived. Devnarayan returns his head (sis +
diya), thus causing the Rana and his offspring to be called Sisodiya. In the
Rajput narratives, Sisodiya are called so because of their settlement in the
Village Sisoda when they arrived in the Aravali hills.[71] Not only is the Rana's
massive army defeated, but unlike the Rajput narratives of heroism, he is
revived and made to take a name given by a lowly non-Rajput group. Given
that Rajput genealogical traditions place the Sisodiyas of Mewar on the top,

---

[71] Nainsi, *Khyat*, I, 1.

the Gujar perspective of Sisodiyas can be seen as a challenge thrown by this powerful cattle herding group in another frontier region that is the Aravali hills. Another interesting episode that we encounter in this epic is the battle between Rajputs and Gujars who are guarding Devnarayan and his queen Pipalde. In this battle since all the fighters behead each other, when they are revived by Devnarayan, the heads and torsos get mixed up. The resultant Sindhan Rajputs thus are again a community of mixed origins like the Gujars themselves. Thus, given the emphasis placed by the Sisodiyas of Mewar on purity of blood, the allusions to mixing of castes becomes a way of challenging the hierarchy based on origins.

The epic also repeatedly contrasts the temporal power exercised by the Rajputs with the divine power of Devnarayan. When Devnarayan's Bhat Chochu goes to look for his cousin Bhunaji in the kingdom of Rana, he is set a series of unachievable tasks by Nimade, the brother of Rana. At every stage, it is through Devnarayan's agency, that these tasks are achieved. Thus Devnarayan challenges the temporal authority of Rajputs at every level and emerges victorious. However, after his victory, he does not wish to takeover the temporal realm of the Rajput Rana. What he wishes and achieves, is a spiritual following and an identity for his people.

Thus, what the Devnarayan epic achieves for Gujars is an identity that is not an identity reflected through a Rajput mirror. While for Bhils and Raikas worshipping a Rajput god becomes a way of creating a bridge between past and present social identities, the Devnarayan tradition actually seeks the formulation of a very distinct and rooted self identity. It, along with all such traditions, challenges the dominant frames of history in two ways. The first, by establishing processes of identity formations which seek no confirmation from a dominant paradigm. Secondly, through an imaginative web of text which refers to fantastic origins, disapproved of sexual unions, strange encounters and impetuous deeds, it also challenges the established historical imagination. The tradition does not seek to fix the historicity of Devnarayan or the Rana he defeats. It merely wishes to provide a Gujar narrative to their past, that the community itself imagines. In this sense, given the near absence of groups like Gujars, Bhils and Raikas from the historical narratives of 'Rajputana' what these epic traditions offer are indeed counter-narratives.

A Song and its Singers: Dhola-Maru

The epics of Pabuji and Devnarayan, raised certain questions about social identities in a folk religious context. But this is not the only context within which the oral traditions in the Thar circulate. Another genre that we can examine is one of several love tales that circulate in the region. These quasi-historical tales are fraught with emotions of love, separation, mysticism, memory, battle and death. Unlike the traditions of Pabuji, Tejaji, Devnarayanji and Ramdeoji, these do not operate in the realm of the sacred. They are not narrated in religious gatherings but rather on social occasions by various musician castes, variously called Mirasi, Langa, Manganiyar, Dholi, Dhadhi etc. in the Thar. However, these 'love-ballads' also appear to underline similar processes of identity formation as the heroic epic traditions.

The *duhas* of Dhola-Maru are popular in large parts of northern India including Rajasthan, Punjab, Uttar Pradesh, Madhya Pradesh and Chattisgarh. They are performed in various parts of northern India in the form of *swang*, a genre that is viewed both as "folk literature and literature in more classical sense".[72] The ballad of Dhola-Maru has been called a love lyric, a ballad, a folk opera, a romantic lay, or an epic. None, "is completely inaccurate, for it is sung or told in meter; it has fictive historical overtones; it is romance; it is sometimes performed as a folk opera; it has epical proportions".[73] There is no single version of the Dhola-Maru ballad, with written versions incorporating oral variants popular in the region. The variant popular in the Thar is primarily a ballad of separation, collated in the sixteenth century by a Jain poet of Jaisalmer called Kushalalabh.[74] It is seen as a specimen of ancient poetic traditions in languages that include Dingal, Braj and Apabhramsha. While some of the *duhas* compiled by Kushalalabh were available to him in the sixteenth century, he himself added some *chaupais* to the corpus, a tradition that has continued.[75] Therefore, it is not possible to arrive at an 'original'

---

[72] Susan S Wadley, 'Dhola: A North Indian Folk Genre', *Asian Folklore Studies*, Vole 42, 1983, 3.

[73] Ibid

[74] Of the three available recensions of the *duhas*, the Kushalalabh recension is called the Jodhpur recension. Charlotte Vaudeville, 'Leaves from the Desert: The Dhola Maru ra Duha An Ancient Ballad of Rajasthan' in *Myths, Saints and Legends in Medieval India*, Charlotte Vaudeville, compiled by Vasudha Dalmia, OUP, New Delhi 1996, 276. Narottam Das Swami refers to another recension, which begins with the description of the marriage of Raja Pingal of Pugal with Uma, the daughter of Deora Chief of Jalor, who had already been promised to king Ranadhaval of Gujarat. This version includes a description of battle between the Gujarat and Pugal forces in Jalor. Narottam Das Swami, *Dhola Maru ra Duha*, 8.

[75] Vaudeville, 'Leaves from the Desert', 276, Swami, *Dhola Maru ra Duha*, 6.

Dhola-Maru, with hundreds of verses being added to various episodes of the corpus. The Dhola-Maru songs contain elements of *birha* (separation), *sandesh* (messages to a distant lover) and *akshepa* (recriminations and lamentations of wife or lover), common to the tradition of love ballads, including Umar-Marvi jo kisso, Moomal-Mahendro and Lorik-Chanda.

However, the story of Dhola-Maru is also a part of the large three generational epic tradition popular in Braj and Chattisgarh regions, known as Dhola. The central narrative of the epic of Dhola is drawn from the classical tale of Nal-Damayanti from the Mahabharata. The epic begins with the story of Raja Pratham and his wife Manjha, their childlessness, the austerities undertaken by Manjha for a child and the birth of Nal in exile. The epic traces the travails of Nal, raised without awareness of his princely origins, his marriage to an *apsara* called Motini and his reunion with his father through the rendering of the Nal-Puran in Raja Pratham's court, reminiscent of Lava-Kusa singing the Ramayana to Rama in the Uttarakanda of the Valmiki Ramayana. In the second part, a variant of the classical Nalopkhyana of the Mahabharata epic, Raja Nal marries Damayanti or Dumenti, a marriage that marks the beginning of another period of exile for Raja Nal, referred to as *Raja Nal ka Aukha*. Towards the end of the second part, the plot following various exiles of Raja Nal, travels to the desert of Pugal, where a part of the narrative is located. Raja Nal's son Salha Kumar or Dhola is the main protagonist of the version popular in the Thar. A version that integrates all three narrative traditions is compiled by Susan Wadley in her work, *Raja Nal and the Goddess*.[76]

The evolution of a three generational epic like Dhola would evidently have taken place over several centuries and would have been created in several versions. The links associating these versions are rather tenuous, emanating from the fact that Dhola's father is called Raja Nal and he too faces a dispossession like Nal in the Nal Damayanti epic. While the *duhas* of Dhola-Maru are sung in the *khayal* tradition, classically identified with the Mirasis of the Thar, the epic tradition of Dhola is performed in the *swang* tradition. The performances of the Dhola epic are considered to be semi-religious performances, with dedications to Ganesha and Sharda preceeding the performance. The Dhola-Maru songs are never sung in a religious context, but rather in the social context of *kachehris* patronized by the Rajputs where semi-classical musical traditions of Mirasis, Langas and Manganiyars

---

[76] Susan S. Wadley, *Raja Nal and the Goddess: The North India Epic Dhola in Performance*, Indiana University Press, Bloomington and Indianapolis, 2004.

were nurtured. Yet, the folk performers continue to portray these epics as a continuous narrative. This is not very surprising in the context of Indian folk tradition as there is often a conscious attempt to relate several folk narratives and place them within a larger meta-narrative.[77]

This section attempts to understand the popularization of the Dhola-Maru *duhas* and their relationship with ideas of mobility and kingship in the Thar. The tale, as related by Charlotte Vaudeville, is as follows,

> Salhakumar, also called Dhola, son of king of Narvar and Maruni (or Maru) the daughter of king Pingal from Pugal in Marwar are married in infancy, following a chance meeting of both families in Pushkar. Subsequently King Pingal goes back to Pugal, taking with him the infant Maruni. Years pass, and one day the adolescent Maruni sees her spouse Salha in a dream. She immediately falls prey to *viraha*, the pangs of separation, in which she begins to languish. Maruni's parents send some Dhadhis, itinerant musicians, to Dhola, but the messengers do not come back. A merchant passing through informs Maruni's father that Prince Dhola is now married to Malvani, princess of Malwa, and that she had managed to intercept all the messages coming from Pugal before they reached Dhola.
>
> Dhola on the other hand, wants to go back to Maruni in Marwar but the entreaties and strategies of his second wife, Malvani, hold him back. Eventually, he manages to select a camel which would be able to take him to Pugal in a single day. When Malvani finally yields to slumber, Dhola escapes, mounted on his camel. Malvani wakes up and laments her fate. She attempts one last stratagem to make him turn back, but is unsuccessful.
>
> On the way, Dhola meets a Caran, a bard. The Caran tries to distract him from his aim, saying that Maruni has already grown old. But Dhola is reassured by another bard, who sings of the charms and virtues of Maruni. At night Dhola and his camel reach the outskirts of Pugal and stop by a well. A dream reveals to Maruni that her husband is somewhere nearby. The news of Dhola's arrival brings joy to the whole town, and fulfils Maruni's hopes: her joy knows no bounds. The husband and wife are happily reunited.
>
> Fifteen days later, Dhola departs with Maruni: King Pingal showers Dhola with presents and sends an escort. At the second resting place, Maruni is bitten by a snake. Dhola sends back the escort and remains with the lifeless body of his beloved. A yogi passing by, brings back the young girl to life, using his magic powers.
>
> Dhola and Maruni mounted on their faithful camel, set off again on the road to Narvar. They are chased by the wicked Umar Soomra, a Muslim who

---

[77] Susan Wadley, 'Dhola: A North Indian Folk Genre', 5

lusts after Maruni. Umar makes use of a trick to stop the couple but his strategy is found out in time and the couple escapes-the swiftness of their wonderful camel taking them far ahead of Umar's troopers.

Joyfully, Dhola and Maruni come back to Narvar, and Maruni takes her rightful place in the palace as the first wife and favorite of Dhola. The jealous Malvani passes barbed comments on the parched barren lands of Marwar. Dhola supports Maruni and eulogises the land of Marwar. The legend ends with Maruni's victory and Dhola's praise of the women of Marwar.[78]

In the first instance, compared with the epics of Pabuji and Devnarayan, the tale of Dhola Maru is rather simplistic, devoid of multiple layers of meaning as seen in the Pabuji epic. Compared to Rajput heroes, Dhola is a pale character. He does not fight any battles, for protection of cattle or forts, nor does he revenge the slaying of his ancestors. He is simply caught between the machinations of women. He does not even ride a horse, but a camel and it is the speed and sagacity of his camel that makes the journey to Pugal and back possible. It is through the indications of the *dumni* and Maruni that he is able to see Umar Soomra for what he is, and here too instead of fighting he just escapes, that too without removing the hobble of the camel. Even Maruni and Malvani do not represent a moral force like Deval, Sodhi Princess, Jaimati or Sadu Mata, they are love stricken women fighting over a non-warrior hero like Dhola. Even in the context of performances, the travails of lovelorn Maruni provide little excitement or drama as compared to the sacrifice of the Sodhi who lets her newly married husband go to the battle field, or the machinations of Jaimati. As a matter of fact, it is Umar the anti-hero, who forms a far more interesting figure in the folklore of the Thar and to whom numerous references are found in Gujarati, Kutchi, Punjabi and Sindhi folklore.[79]

---

[78] Vaudeville, 'Leaves from the Desert', 277–278. Also, Swami, *Dhola Maru ra Duha*.

[79] Umar Soomro, who is also known as Hamir Soomro in Kutch and Kathiawar, is really not the name of a single chief but rather that of a whole tradition itself. He is not merely a part of the political history but also that of the mythological landscape of the Thar Desert. It is not known how many Umars actually existed and the rule of what area did the Soomras actually command. But it is known that the lore of Umar is sung all over the Thar Desert. Rather than being narrations of battles, these tales emerge out of popular imagination and allude to the most popular motifs of folk tales, the element of love and fantasy. In one instance, Umar Soomra is believed to have been besotted with Sangviyan, a Goddess of Charan origins, who was considered to be the presiding deity of the Bhatis. Umar proposed marriage to Sangviyan and evoked her ire. The goddess turned the flow of a tributary of Sutlej that use to pass through Dhat towards Multan. Since then the saying goes that the water went to Multan. Narayan

However, when seen as a part of the larger Dhola epic, several complexities of
the Dhola-Maru ballad become apparent. While as a love ballad, Dhola Maru
duhas provide no social references to place it in a historical context, Dhola epic,
on the other hand refers to the struggles involved in the rise of Jat kingships in a
region dominated by Rajputs in the seventeenth century.[80] As a *swang* tradition
patronized by Jat peasants, the Dhola appears to mock Rajputs throughout. In
the entire epic, though clearly referred to as a Kshatriya, Raja Nal's Kshatriya
status is not only taken away from him as he is forced to assume several other
caste identities like that of a Baniya, Nat and Ghanchi, in the text, but also
because of lack of characteristic Kshatriya valor in his behavior. Not only does
he constantly lose battles, he relies on magical support from his first wife,
the *apsara* Motini, and the goddess. He succeeds only when he assumes caste
positions other than Kshatriya. According to Susan Wadley, the Braj version
of the epic represents a metaphoric statement on Jat rule, as Nal is forced to
prove his identity in front of kings of fifty two forts, which he is able to do
only when aided by women or goddesses.[81] R P Rana refers to a Jat culture of
resistance to the Kachwaha hegemony of Amber rulers in the seventeenth and
the eighteenth centuries.[82] This not only took the shape of political resistance
that led to the rise of Jat kingships like Bharatpur and Dholpur, but also a
cultural resistance whereby Jats resisted a Vaishnavite hegemony patronized
by Kachhwahas as Braj-Mathura was identified as Vaishnavite region of the
cowherd god Krishna. The political and cultural resistance by Jats drew upon
an older history of being among groups that exercised control over this region.
Also, if Dhola's other name Salhkumar is taken into consideration, it is fairly
close to the Bhatti Salbahan who conquered Punjab, from whom a number of
Rajput as well as Jat (both Hindus and Muslims) communities claim origins.[83]
A *kavitt* about the 13th century Bhati King Salbahan of Jaisalmer, in *Nainsi
ri Khyat,* accounts for cows, buffaloes and camels as his wealth.[84] Salbahan
thus entered the lore of the Thar, as a mighty conqueror, to whom exploits of

---

Singh Bhati's Commentary on Munhata Nainsi's *Vigat*, III, 88. Umar is also the hero/
antihero in Shah Abdul Latif Bhitai's *Umar Marvi jo Kisso*, where Umar the Jam of
Sindh fell in ill-fated love with Marvi, a maiden of the harsh desert of Dhat. Kishni
Khemani, *Bharat ke Geye Sindhi Premakhyan*, Ajmer, 2003.

[80] Wadley, *Raja Nal and the Goddess*, 180–183.

[81] Ibid.

[82] Rana, *Rebels to Rulers*, 122–135.

[83] Vaudeville, 'Leaves from the desert', 287.

[84] Nainsi, *Khyat*, Vol II, 37–38, 'Kavitt Bhati Salvahan ra'.

several warriors and cattle rustlers like him may have been attributed, who could have been Rajputs, Jats, Mers, Bhils, Charans or Rabaris. It is difficult not to notice the absence of the other usual Rajput identity markers like a sword or a horse, or even a genealogy in the case of Salhkumar.

Given the fact that both Rajputs and Jats appear to originate from the mobile cattle rearing and rustling groups, it is not surprising that these groups find references in each other's narratives, while attempting to establish a separate identity at the same time. The dominance of Rajput perspective in the historiography of the region, not only obliterated references to Jat kingship or Jat resistance to Rajput kingship, but also increasingly posited Jats as sturdy, hard working but simple minded peasant community, as opposed to the martial Rajputs. On the other hand the traditions emerging around the Jat hero Tejaji, portray him on horseback with a drawn sword. Believed to have been born around the tenth century in Khidnal, Tejaji laid his life while fighting Minas who had stolen the cows of Jats.[85] Like the Rajput Chauhan folk deity, Gogade, also known as *zahir pir*, another folk deity venerated by the Jats, Tejaji is also known to protect cattle against snake bites. By giving up his life to protect cattle and by allowing a snake to bite him on his tongue, Tejaji achieved something that neither Raja Nal, nor Dhola or even Devnarayan achieve, that is a 'Rajput' death. I would like to view the idea of 'Rajput' death in the context of several groups contesting for dominance in a cattle rearing, rustling and trading world. In the manner in which communities evolve several ways of self-referencing over their long histories, they also evolve several ways of 'othering'. In this process of othering, a noble, self-sacrificial death played an important role, with each community striving to acquire such role models.[86] In my understanding a noble, sacrificial death, usually protecting some other group in a 'Rajput' warrior like fashion, is an important step in the social mobility of a community. These identities continue to be shaped over centuries, thus acquiring very indeterminate and often conflicting features. However, this indeterminacy is actually indicative of the fact that these identities were ambivalent to begin with. Here we can also begin to understand why the traditions of Pabuji, Tejaji, Gogade, Devnarayan, Ramdeoji, Harbhujii etc. acquire an ethical moral paradigm, while Dhola-Maru or Umar-Marvi remain

---

[85] Sukhbir Singh Dalal, *Jat Veer: Kshatriya Vanshtopanna Jatveero ki Kathayein*, Delhi, 1991.

[86] At this point, Stuart Blackburn's 'Deification by Death' model appears quite pertinent, though I would like to think of these deaths as necessary in world of competing identities. Blackburn, 'Patterns of Development of Indian Oral Epics' in *Oral Epics in India*.

love ballads at best. The heroic epics of the Thar follow a definite trajectory in which, the hero is deified after his death and later mythologised. In these epics, ethics and morality become the defining forces. Thus, while "the romantic heroes tend to become unconcerned with morality; the deified heroes represent the moral standards of the region".[87] These moral standards, usually associated with martiality and sacrifice, are important because the identity of the community is shaped around them.

Another aspect of this question of identity that can be explored through Dhola is the relationship of Jats with other peripatetic pastoral groups. Charlotte Vaudeville in her study of Braj region points out that "despite the Vaisnava abhorrence of the bloody rites associated with Devi worship, the pastoral castes, especially, the Jats and Gujars who form the bulk of the autochthonous population of Braj-Bhumi remained attached to the cult of their local goddesses".[88] Jat goddess worship traditions manifest themselves in the epic of Dhola, whether in the *sumeri*, or through the presence of goddesses Durga and Kali throughout the text. Extending Vaudeville's observation further, it would be worthwhile to see these local goddess traditions in relation to the traditions of Charani goddesses, particularly keeping in mind the similar routes of migration supposed to have been undertaken by Jats and Charans from Baluchistan to the Thar. It would not be very surprising to find Charanic traditions finding space in the social, cultural and religious world of communities with shared histories of migrations as well as processes of social mobility. Interestingly, at one point Salhkumar's mother refers to his other wife Malvani as *vajaran*, translated by Vaudeville as a street-walker.[89] But, *vajaran* could well mean *banjaran*, or *vagharan*, both referring to low caste mobile pastoral groups.[90] Besides, in another episode, when Salhkumar seeks Malvani's consent to go to Pugal under false pretenses of an exile and adventure, he wishes to go and serve the King of Idar, get cheap horses from Multan, broad mouthed camels from Kutch, pearls from the ocean and fabric from Gujarat.[91] Kolff studying the traditions of soldering in pre-Mughal north India argues that the "peregrinations of merchants and soldiers belong to the same group of pastoral

---

[87] Wadley, 'Dhola', 10.

[88] Vaudeville, 'Braj Lost and Found' in *Myths, Saints and Legends*, 65.

[89] Ibid.

[90] Swami, *Dhola Maru*, 135. 'Tab boli Champavati, Salhkunvarri maat/ Re Vaajaaran Chohri, Kain kheladai ghaati. (Then, Queen Champavati, Salhakumars's mother cried; 'Now, you slut, what is this little game') Translation by Charlotte Vaudeville, 'Leaves of the Desert', 316.

[91] Swami, *Dhola Maru ra Duha*, 117–120.

activity".[92] In fact exploring the structure of military labour market in northern India, Kolff argues that "until as late as the nineteenth century, there was no lack of men opting for a life spent as errant soldier, migrant labourer, or pack animal trader".[93] Does Salhkumar then belong to this peripatetic community with multiple identities? At this stage to me, it is not important whether Raja Nal or Salhkumar can be identified as Rajputs or Jats, but rather that they belong to the wide community of peripatetics, which includes warrior-ascetic heroes like Pabuji or Tejaji. The ambulatory circuits of these heroes overlap, as do their pursuits. It is on these peripatetic circuits that complex community identities, which refer to both the 'self' and the 'other' emerge.

A dominant emotion that a ballad like Dhola-Maru highlights is that of exile, and thus of separation, *birha*. Both Marvani and Malvani suffer from pangs of separation from Salhkumar. In her analysis of the *Barahmasa* genre of folk songs Charlotte Vaudeville suggests that, in a, "pastoralist, mercantile and soldiering society,...the absent hero is generally represented as a merchant held up by his trading activities in a foreign land, or a *ulagana*, a mercenary soldier in the service of a distant lord".[94] Vaudeville interprets the Apabhramsha word *ulagana* to be a mercenary soldier, vowed to wandering life, which she argues was a feature of western India. Norman Zeigler has identified these wanderings as *vikhau*, periods of exiles and wanderings, "during which the kingdom was unprotected, and there was a resultant mixing of castes and confusion of order. The Rajputs themselves were of necessity forced to take up other and often less divine occupations. They had become dependent and refugees".[95] In the epic of Dhola, *Raja Nal ka aukha*, or his hardships are represented through his period of exile from Narwar. These periods of absences are represented in the folklore through the idea of *birha*, which seems to be an ever present phenomenon in the Thar.[96] While *birha* has largely been understood in the

---

[92] Kolff, *Naukar, Rajput and Sepoy*, 77.

[93] Ibid.

[94] Charlotte Vaudeville, *Barahmasa in Indian Litertures*, Motilal Banarasidass, Delhi, 1986, 33–34. Jhaverchand Meghani points out that in Gujarat Charans composed and sung elegiac *barahmasas* that rendered dirges for "a dead benefactor, a brave warrior, or a close friend". Meghani, 'Elegiac "Chhand" and "Duha"', 46.

[95] Zeigler, *Action, Power and Service*, 112.

[96] Swami, *Dhola Maru ra Duha*, 109.
Hum kumlani kant bin, jal vihuni bel
Binjarari bhai jiyuun, gaya dhunkati melh. (I have wilted like a climber without water. My lover has left me like the kiln of a Banjara)

context of separation of human lovers, its deep philosophical meanings have been explored by Bhakti poets as well as Sufis like Jayasi, Khusro, Kabir, Mirabai, Surdas, Shah Latif among others.[97] In this poetry, *birha* represents the longing for union with the Supreme Being outside as well as within the self. In this philosophy, *birha* represents a torment, but one that is holy and leads to the union with god, who is considered to be the lover. In the Sufi poetry, the torment of separation itself represents a state of ecstasy, the search itself a destination. In a sense, these songs of separation too represent the ultimate paradox of peripatetic life of this region. While submerged in the element of longing, these also point out that nomadic way of life could not be given up, not only because of economic compulsions but because it was a way of life. In this way of life, separation itself was a journey and a destination.

The singing of these songs of separation forms an important part of the social exchange in The Thar. These ballads are the part of the repertoire of minstrel communities like the Langas, Manganiyars, Mirasis, Dhadhis, Dholis etc., who themselves were peripatetic communities.[98] Komal Kothari interprets the word Manganiyar, literally meaning as one who begs, in a social sense. He claims that it rather refers to patron-client relationship between the minstrel communities and their patrons, whose social life could not be considered complete without the presence of a genealogist and a minstrel. Both had important roles to play in the social exchange undertaken by the patron or

---

Nazir Jairazbhoy in his ethnographic work on musician communities of western Rajasthan attests to the continuous presence of songs dealing with separation in Rajasthan, owing to migratory practices in the region. 'Music in Western Rajasthan: Stability and Change' in *Yearbook of the International Folk Music Council*, Vol 9, 1977, 50–66, 52.

[97] An example being Jayasi's poetry. Jayasi writes in the Nagmati *Barahmasa*, '...my heart too swings to and fro/ tossed up and down by Virah's harsh blows!/ Mysterious, unknowable, impassable is the road/ and I wander round and round like a Bhambhiri fly', Vaudeville, *Barahmasa*, 65. Also Surdas who writes, 'How long has it been since Hari left? It seems world's four eons have passed/Surdas says, forgotten in Braj/We're like fish who exist without water', (Uncanny Resemblance) *The Memory of Love: Surdas sings to Krishna* Tr. and Introduction by John Stratton Hawley, OUP, 2009, 128. Kabir too plays with the element of *birha*, 'The Kunjha cranes cry plaintively in the sky/ thunder and ponds are refilled/ But she whom her lord has deserted/endures untold torments', Vaudeville, *Barahmasa*, 43.

[98] Komal Kothari, 'Musicians of the People: The Manganiyars of The Thar' in *Idea of Rajasthan* Vol II, Manohar Publishers and American Institute of Indian Studies, Delhi, 2001, 205–237.

jajman, who could not hope to enter marital relations with members of his own caste if he did not command their services. The minstrels did not only sing in events like births and marriages, they were also entitled to sing laments at death. The singing of *subhraj* makes the most prominent social introduction of a client. If a dispute arose between a patron and his minstrel, the latter could publically disown the patron taking a series of steps. A disowned patron could not hope to remain a respected member of his own community. [99]

The Manganiyars and Langhas are Muslim musicians and are quite different from the Bhopas of Pabuji, as they do not claim to be bards but musicians in a real sense. Using instruments like *rabab, kamayacha, pyaledar sarangi, chautaro, sirimandal* etc., they not only sing songs of birth, marriages and death, but are also entitled to sing in the *kacheris* of the patrons. [100] It is in these assemblies that they sing ballads like Dhola-Maru, Umar-Marvi, Moomal-Rano and Sassi-Punnu. Manganiyars sing classical compositions like *mota git (bada khayal)* and *chota git (chota khayal)*. Some of their ragas have originated in the Thar and are not found in north Indian classical tradition. [101] The names of ragas can sometimes also refer to a person character from the love ballads like raga Maru, Sorath, Rano and Sasvi (Sassi). Manganiyars also sing religious songs like *bhajans, harjas, heli* etc, apart from the love ballads. The most constant emotion in these songs is that of *birha* or separation as in a dominantly mobile culture, separation or *birha* is a constant presence. However, what these singers also sing are the *qalaams* of sufis of Sindh like Shah Abdul Latif Bhitai and Sachal Sarmast. Popularly known as *sindhi qalaams*, these songs themselves are reflective of movement of traditions over time and space. Most of these musician communities travel over long spans to attend to their patrons' needs. Not only do they accumulate varying traditions as parts of their repertoire, they themselves are a part of these plaintive songs of separation, as they too are mobile. Such traditions form an essential part of the constantly mobile lore of the Thar, which has the inherent quality of mobility within, as the culture where it exists is innately mobile. The presence of such traditions has been attested in other mobile communities around the world, particularly the gypsies, who use the "Indian *bhairavi* musical scale, as well as a type of 'mouth music' known as *bol*, which consists of rhythmic syllables that imitate the sound of drum strokes". [102]

---

[99] Ibid.

[100] Jairazbhoy, 'Music in Western Rajasthan', 51.

[101] Ibid.

[102] Isabel Fonseca, *Bury me standing: The Gypsies And Their Journey*, Vintage, 2006, 106.

Finally, I come to the question of pre-dominance and continued popularity of these narratives and musical traditions in the Thar. In my understanding this question needs to be dealt with at two levels. The first being, the mobility of these communities and thus absence of markers like forts, as in case of Rajput histories, necessitates reliance on aural devices like songs and musical instruments and visual narratives like the *phad*. In a mobile culture, memory becomes the most important part of the process of transmission of information over time and space and its interpretation. In a predominantly non-literate society, bards played a critical role in the accumulation and dissemination of information with legendary memory. For Rajputs, this role was played by Charans before their narratives were textualised. For example the *bat* in its oral form registered the shifts in Rajput identity. The textualisation of *bats* also affected the oral historical culture associated with it, and thus Bhats and Charans who were the practitioners of these traditions.

Secondly, and more importantly, frontiers like the Thar not only give rise to multiple political possibilities, but also to multiple historical imaginations. The Thar, as I have argued through this book, as a frontier was home to multiple, fluid and ambivalent identities. Komal Kothari comes to the conclusion that the richest musical traditions come from the cattle breeding regions.[103] My understanding of community identities in the Thar indicates that most communities in the Thar including the Rajputs have been involved with cattle breeding, rearing and trading at some point of time in history, which led to fluid and ambivalent identities. With the emergence of territorial polity, identities also concretized into endogamous caste groups. For dominant groups like Rajputs, written traditions provided the possibility of forgetting and obliterating memories of mixed past and indefinite identities. This was most apparent in the genealogies that emerged from Rajput courts. Since the written historical traditions of the Thar were shaped according to the agro-centric sedentist imagination, the oral traditions carried the narratives of the mobile groups. For the mobile groups, it was in the oral traditions that the possibility of remembering what was being obliterated, existed. Just like the acts of forgetting and obliterating were deliberate, so were the acts of remembering. The oral epic traditions of the Thar not only preserve the forgotten histories of marginalized communities, but also the perspectives of marginalized communities on what is considered is to be the history of the Thar. These traditions are thus counter-narratives to the dominant historical

---

[103] Bharucha, *Rajasthan: An Oral History*, 91.

traditions of the Thar. Moreover, unlike written historical traditions, which become fixed, the oral narratives remain fluid like the identities themselves, the retellings exploring and reformulating community identities. Engaging with oral narratives in the Thar allows the possibility of stepping beyond 'factual' 'verifiable' history and delving into other 'ways of history'.

Conclusions

Nomadic Narratives in the Frontier

The Thar Desert, for the longest period in its history, has fostered cultures linked by networks of mobility. The arid Thar, located between the Indo-Gangetic plains and the Indus valley, connects the north Indian plains with the riverine systems of the Indus. There exists a long history of mobility through this region, with armies, merchants, pastoralists, ascetics and bards having constantly criss crossed the desert. Mobility across frontiers promoted connectedness of economies and cultures, as is visible in the oral and written traditions of the region that are constantly woven around the motif of travel, and display similarity of themes, characters and patterns. The Thar Desert was thus a geographical, ecological, political, social and cultural frontier where innumerable invasions and migrations provided space to multiple identities as well as multiple historical imaginations.

Such conceptualization thus permits imagining the Thar as a frontier, which allowed groups to move horizontally and vertically, and in process acquire identities that were far ambivalent than what they appear to be today. Between eleventh and thirteenth centuries, the Thar provided space to displaced groups, warriors in search of employment or adventure and local tribal-pastoral groups with ability to replace older clans. From the fourteenth century onwards the Thar witnessed development of a wide range of sedentary political formations that increasingly centred themselves around the idea of 'Rajput'. The emergence of 'Rajput' states, in the Thar region, that is the Bhati state of Jaisalmer and the Rathor states of Jodhpur and Bikaner, resulted from gradual transformation of mobile pastoral and tribal groups into landed sedentary agrarian ones. The process of settlement involved both control over mobile resources through raids, battles and trade, as well as channelizing of these resources into agrarian expansion. Kinship structures as well as marital and martial alliances were instrumental in this transformation. This was not merely political, but also a social and economic transformation, whereby the basis of state formation shifted from mobile pastoral wealth and pasture lands to agrarian, cultivable lands. Such a shift in arid and semi arid landscapes meant that the states were required to constantly negotiate and balance the need for a settled sedentary agrarian base amidst a mobile population dependent on circulation of resources. In this manner, exploring the Thar desert provides an opportunity to study a region where sedentary state formations emerged amidst predominantly mobile population. Exploring the Thar also allows investigating processes of identity formation in frontier zones, where Rajput identity often appears enmeshed with nomadic, pastoral, tribal and Muslim identities.

By the time nineteenth century census reports and tribe-caste compendia were drawn, Rajputs refer to themselves as a closed endogamous group. The process of closing of ranks also involved a process of 'othering' by which other communities were attributed all 'non-Rajput' values. In the colonial ethnographic accounts rather than referring to Rajputs as having emerged from other communities, Bhils, Mers, Minas, Gujars, Jats, Raikas, all lay a claim to a Rajput past from where they claim to have 'fallen'. Historical processes, however, suggest just the opposite. Before Rajputs came to be seen as landed aristocratic caste group, Rajput appears to have been an umbrella term used to refer to a number of groups with similar aspirations and capability. In fact as Ziegler suggests, there was no contradiction in being a Rajput and a Muslim at the same time. However, the emergence of Rajput genealogical orthodoxy by the sixteenth century closed the ranks of Rajput 'caste' group, and excluded several groups with similar claims. Nevertheless the idea of Rajput as a martial group remained dominant and a number of old and new martial groups like Bhils, Mers, and later Sikhs and Marathas continued to claim Rajput 'status'. The difference between Rajput 'caste' and Rajput 'status' can also be seen through their locations. While the aristocratic Rajput 'caste' was located within the boundaries of evolving sedentary state formations, the marginal aspirational 'spurious' Rajputs were located on the frontiers of polities, as well as those of emerging empires, whether Mughal or British.

History of 'Rajputana' has largely been approached through a dominant Mughal agro-centric perspective on Rajput state formation. However, a closer look at medieval Rajput polity reveals intense struggle for control over circulation of resources in the arid Thar region. Rajput states like Jodhpur, Bikaner and Jaisalmer developed around fortifications in the fourteenth century and attempted to evolve administrative structures that centered on agricultural revenue appropriation, even though a significantly large proportion of population remained engaged in occupations that required mobility across the Thar, particularly in pastoralism and trade. Unlike the popular perception, even Rajputs remained engaged with nomadic pastoralism, animal husbandry and cattle trade till much later than is assumed. Munhata Nainsi in his seventeenth century chronicles, *Munhata Nainsi ri Khyat* and *Marwar ra Paraganan ri Vigat* refers to a number of disputes between Rajputs that involved cattle raids. Also, a close reading of the lore regarding Rajput folk deities like Pabuji, Mallinath, Gogaji and Ramdeo, who are viewed as protectors of cattle herding communities actually indicates the intense struggle for control over cattle and pasturelands that Rajputs were engaged in. Rajputs extended patronage to Brahmins and bardic communities like Bhats and Charans who composed detailed genealogies linking Rajput clans to older Kshatriya lineages as well as celestial sources, which not only legitimized their claims to aristocracy but also distanced them from their tribal pastoral origins. By late sixteenth century, a major part of the Thar had been incorporated into the Mughal empire. The aristocratic self definitions that both the Mughals as well as the Rajputs evolved were complimentary. Notions of Rajput valour and royalty were established not merely through Charan panegyrics and genealogies, but also through Mughal recognition of these values as 'Rajput'. The need for a clear genealogical order emerged as claims to aristocracy as well as to the seats of power had to be substantiated both to internal as well as external forces, like the Mughals and the British. This established a genealogical orthodoxy, increasingly sustained through caste endogamy, which determined the location of pure Rajputhood. In fact, Mughals were accommodated among the ranks of superior upper caste Rajputs through marital alliances, while older marital allies, like Bhils, Mers, and Pathans were increasingly excluded. Mughal-Rajput marriages established a kind of parity between the two, and legitimised both Mughals and Rajputs as royal lineages. This created a confusion in the understanding of location of true 'Rajputhood' which on the one hand was represented by the valiant, noble, chivalrous 'Mughal'/ 'Anglo-Saxon' Rajput, and on the other the *barwuttea* of the frontier. The *barwuttea* or the outlaw, which at various times included groups like Bhils, Mers, Sodhas, Bidawats, Khokhars, Marathas, Sikhs, Satnamis, too laid a claim to true Rajputhood on the basis of its resistance to authority whether the ruling Rajput clans, Mughals or British.

An engagement with this confusion challenges the dominant understanding of the rule of Rajput lineages like Bhatis and Rathors, and by extension of indirect rule of Mughals and British, as extending uniformly over the Thar region. Outside of the areas controlled by main and the cadet branches of Bhatis and Rathors, the authority of these Rajput lineages in the Thar region remained a highly negotiated one. Jarechas, Sammas, Soomras, Sodhas, Daodpotras also controlled large parts of the Thar region and its nearabouts. Besides, the fringe areas of the Thar Desert were constantly 'despoiled' by communities like Bidawats, Sehraes, Sodhas, Mers, Bhils, and Meenas etc, all of who claimed to be Rajputs or have a Rajput ancestor. So, while a Mughal pattern was negotiated and implemented in parts of Jodhpur and Bikaner states, the fringe areas of these kingdoms were often claimed by groups that bore closer resemblance to what ruling Rajput lineages would have been like a couple of centuries earlier. It is in these frontier borderlands that the authority of the states was sometimes conveyed via the recalcitrant Rajput thikanadars through taxes like *bolawo* and *rukhwali ri bhachch*, which were protection taxes levied by the *thikanadars* on trade routes. Here it also becomes important to point out that the disintegration of the Mughal Empire in the eighteenth century provided an opportunity to the Rajput rulers in the Thar to renegotiate issues of authority in core and outlying areas. The expressions of contests over authority were particularly visible on trade routes and markets that were zealously guarded by the Rajput states to ensure a smooth exchange and transit of commodities. When Maratha and Pindari incursions threatened the Rajput states that were forced to pay heavy indemnities, it was commercial revenue that was often used to offset the demands. Nevertheless, the exactions made by thikanadars, as well as their involvement in recurrent instances of theft and robbery on routes passing through their territories were evidences that the control exercised by the Rajput states was not uncontested.

Indirect rule was imposed in Rajput states of the Thar region following the treaties that were negotiated between the Rajput states and the British between 1812 and 1818. British interventions in Rajputana, as the Rajput states were increasingly addressed, were motivated by two primary reasons. The passage to the frontier region of Sindh was through the states of Jodhpur, Bikaner and Jaisalmer. Apart from the strategic value of this passage it was also an important trade corridor, used for example, to channelize opium produced in Malwa, through the emporia town of Pali, into Sindh. Another important reason was to gain control over various sources of salt production spread all over this region that competed with British salt from Bengal, which was finally achieved by the 1870s. Indirect rule in Rajput states of the Thar region implied their incorporation into Western Rajputana States Residency and Bikanir Agency and the appointment of Political Agents and Residents in Rajput courts.

British officials primarily attempted to understand and then reorganise the vast space of the Thar into a comprehensible and ordered space. The fluidity of boundaries in the Thar as well as the operation of multiple levels of authority on travel networks appeared baffling to the British administrators who attempted to order what they understood as vacant untamed space. Their encounters with a space that they considered hostile engendered a need to create networks and systems that could be controlled. This meant extending control over outlying areas frequented by mobile groups and then repopulating them with agricultural communities. Expansion of cultivation, extension of artificial irrigation through canals, construction of roads and railway lines, reorganisation of commercial traffic, human and cattle census, fodder farms, forest conservation schemes were all introduced in this period in order to render a 'hostile wasteland' useful. The construction of railway lines through Rajputana was aimed at regulating traffic in commodities, in particular salt, through this region. Practices like 'Through Traffic', which allowed merchants to deal in salt without engaging with local salt merchants or intermediaries, were introduced along the railway lines.

However, the new networks of roads and railways often ignored the older networks, leading to the inevitable decline of older commercial towns and networks. With the construction of railway lines, it was also imagined that instead of herds moving towards pastures, railways would transport fodder, particularly in order to mitigate the severity of famines. Thus, new circulatory networks were expected to regulate the mobile space and make it ordered and comprehensible. These interventions in the adminstrative structures and practices irrevesibly altered networks of circulation as well as older sytems of control that operated on these networks. A fallout of these systems of control was the increased surveillance of mobile groups like Banjaras, Bhils, Minas, Mers, Sansis, Kalbelias, Charans and Bhats, which could also be understood in the context of nineteenth century ideology of suspicion towards mobile communities. This resulted in criminalising what were considered as legitimate mobilities in the frontier regions of the Thar, for examples that of Banjaras trading in salt, or that of Bhats and Charans negotiating alliances. The late nineteenth century emerges as the time when the circulatory networks were decisively reordered.

Rajput princely states, weighed by internal and external struggles, as well as inevitable encounter with modernity found themselves either incapable or unwilling to intercede in the manner in which circulatory networks were being altered. The transfer of administrative responsibilities from Oswal *mutsaddis* to Brahmin *musahibs* from outside Rajputana, also eroded the control that the Rajput princes exercised over administrative matters. Yet, the power of the princes was not completely negated as they remained the sources of legitimacy in their respective states. While the Mughal empire had legitimised the notion of 'Rajput', the British further consolidated it. James Tod's *Annales and Antiquities of Rajast'han*, placed the Rajput princes of the Thar in the category of Anglo-Saxon Knights of Europe. In doing so, however, it was the power of the cadet branches and *thikanadars* that was eroded. Thus years of indirect rule repositioned the internal dynamics of the Rajput ruling community, centralizing it more than it had ever been in the past. Yet, it did not come to reside absolutely in the ruling Rajput kings, who increasingly became titular heads of an emerging bureaucracy shaped by British indirect rule.

A consequence of Mughal and British engagement with Rajputana was the 'systematization' of the history of the region, which was conflated with the history of the ruling Rajput clans. Thus histories of 'other' Rajput and non-Rajput groups were relegated to the margins. However, a number of oral narratives popular among these groups provide interesting alternate perspectives into processes of state formation and emergence of social identities the in frontiers of the Thar. Most of these narratives relate adventures of Rajput heroes, venerated as protectors of cattle and cattle herding communities. The folk lore around Rajput folk deities like Mallinath, Gogade Chauhan, Harbhuji, Pabuji, Ramdeoji as well as the Jat hero Tejaji and the Gujar deity Devnarayan displays their close association with cattle rearing communities or ones associated with by products like hides and wool. The operative pattern of most of these epic traditions insists that these heroes laid their lives protecting cattle and cattle herders. A deeper engagement with these narratives unravels multiple narratives within, which bring forth the complexity of the relationships of nomadic pastoral and tribal groups with Rajputs. The various versions of these narratives highlight the shift in Rajput image, from that of a cattle herder, rustler and trader to that of protector of cattle herders. These narratives can well be read as Bhil, Raika or Gujar narratives, where by the fluidity and indeterminacy of Rajput identity is bared. Rather than as a protector of cattle herding groups, a Rajput hero like Pabuji can well be venerated as an ancestor of the Bhils. In this sense these narratives can well be understood as counter-narratives, a term used quite effectively by Shail Mayaram in case of Meo narratives, where martiality in narratives becomes a way of re-examining identities. These narratives reveal multiple perspectives on the processes through which Rajput as well as other identities were consolidated.

While written narratives constantly insist on exclusivity of Rajputs, the oral narratives underline their closer associations with mobility and mobile communities, underlining the counter claim that Rajputs were once one of these groups. The writing of Rajput histories by their Charan bards, British administrators as well as through documents like census reports and tribe-caste compendia could not prevent the circulation of these oral narratives on the circulatory networks. Even as indigenous historical narratives like *Khyats* and *vats* were discredited in the nineteenth century by chronological histories of British administrator-historians, the oral narratives continued to have a circulation of their own. It is the oral epic traditions of the Thar, along with its peripatetic musical traditions, that underline the idea that just like its nomadic narratives, its history too has to be primarily one of its nomadic travellers.

Bibliography

Primary Sources

Unpublished documents

Rajasthan State Archives, Bikaner

Sanad Parwana Bahis, No 1–62 VS 1821/1764 CE to VS 1867/1810 CE.

Jodhpur ri Sayer Bahi, VS 1855/1798 CE.

Kagad ri Bahi, VS 1840/1783 CE to VS 1856/1799 CE.

Zagat Bahis, VS 1817/1760 to VS 1857/1800 CE.

Jaisalmer Mehkama Khas, 1898–1910.

Mehkama Khas Jodhpur - (Administration, Boundary, Survey. Forests, Custom, Salt, PWD, Miscellaneous) 1878–1917.

Mehkama Khas Bikaner, Regency Council - (Home Department, Revenue, Political, Home,) 1878–1910.

Mehkama Khas State Council Bikaner - (Political, Revenue, PWD, Ghaggar Canal) 1896–1910.

Rajasthan State Archives Branch Office, Jodhpur

Thane Siwana ra Jama Kharach ro Navo, VS 1861-62/1804-1805 CE.

Sirkar Jalor re Sayer ri Thane ri Bahi, VS 1861-62/1804-1805 CE.

National Archives of India

Foreign Deptt (Political, Secret, Internal and General Consultations) 1830–1917.

Home Political, 1836–1920.

Western Rajputana States Agency/Residency, 1805–1942.

Rajasthani Chronicles

Bhati, Narayan Singh (Ed.), *Munhata Nainsi Ri Likhi Marwar Ra Paraganan Ri Vigat*, (I–III), Rajasthan Oriental Research Institute, Jodhpur, 1966–74.

———— (Ed.), *Jaisalmer ri Khyat* by Mehta Ajit Singh, *Parampara*, 1981, 57–58.

———— (Ed.), *Maharaja Takhatsingh ri Khyat*, Rajasthan Oriental Research Institute, Jodhpur, 1991.

———— (Ed.), *Maharaja Mansingh ri Khyat*, Rajasthan Oriental Research Institute, Jodhpur, 1997.

Bhati, Hukam Singh, (Ed.), *Bikaner ri Khyat* by Dayaldas Sindhayach, Rajasthani Shodh Sansthan and Maharaja Man Singh Pustak Prakash, Jodhpur, 2005.

Chand, Likhmi, *Tawarikh Jaisalmer*, Rajasthani Granthagar, Jodhpur, (1899) 1999.

Sakaria, Badri Prasad, (Ed.), *Munhata Nainsi ri Khyat*, (I–IV), Rajasthan Oriental Research Institute, Jodhpur, (1962) 1984–93.

Singh, Brajesh Kumar, (Ed.), *Maharaja Vijaysinghji ri Khyat*, Rajasthan Oriental Research Institute, Jodhpur, 1997.

Printed Reports

Census of Marwar State, Report by Phiroze R. Kothiwala, Census Superintendent Marwar State, Part I, Marwar State Press, Jodhpur, (1911) 1916.

Census Report Bikaner State, Compiled by Rai Bahadur Babu Kamta Prasad B A, Home Minister of Council and Superintendent Census operations, Bikaner State, 1911.

Dane, Richard M, 'The Manufacture of Salt in India', *Journal of the Royal Society of Arts*, Vol. 72, No. 3729 (May 9), 1924, 402–418.

Fagan, P J, *Settlement Report of Bikanir*, 1893.

Grierson, Sir George, *Linguistic Survey of India*, Vol. IX, Part II, 1908.

Hewson, F T, *Report on Marwar Customs*, 1884.

Livestock Census of India, 2012.

Loch, W, *Report on Marwar Customs*, 1892.

Manual for the Guidance of Native States in Rajputana and Central India for the Control and Reclamation of Criminal Tribes, 1896.

Manual of Northern India Salt Revenue Department, Vol I and II, Allahabad, 1905.

Marwar Administration Reports, 1891–1910.

National Research Centre on Camel, Bikaner, Vision, 2030, published July 2011.

Prasad, Sukhdeo, *Marwar Famine-relief Report*, 1899.

Rajasthan Development Report, 2003, Planning Commission, GOI.

Report on Relief Operations Undertaken in the Native States of Marwar, Jaisalmer, Bikaner and Kishangarh during the scarcity of 1891–92.

Report on Salt Royalty Dispute between Jaipur and Jodhpur, 15th December, 1883.

Report on Settlement of Criminal Tribes in Marwar, 1890–91.

Report on the Administration of Northern India Salt Revenue Department, 1883–84.

Singh, Munshi Hardayal, *Report Mardumshumari Raj Marwar,* 1891 (III), Jagdish Singh Gehlot Shodh Sansthan, Jodhpur, (1896) 1997.

———, *Majmui Halaat wa Intijaam Raj Marwar Babat San 1883–84, Mutabik Samvat* 1940.

Tessitori, LP, 'A Scheme for the Bardic and Historical Survey of Rajputana', *JPASB,* (New Series), Vol X, No. 10, 1914.

———, A Progress Report on the Preliminary work done during the year 1915 in connection with the Proposed Bardic and Historical Survey of Rajputana, *JPASB, (New Series)* Vol. XII, No: 3, 1916.

———, A Progress Report on work done during the year 1916 in connection with the Bardic and Historical Survey of Rajputana, *JPASB, (New Series)* Vol. XIII, No: 4, 1917.

———, A Progress Report on the work done during the year 1917 in connection with the Bardic and Historical Survey of Rajputana, *JPASB, (New Series)* Vol. XV, No: 41, 1919.

Gazetteers

Erskine, Major K D, Imperial Gazetteer of India, Provincial Series: Rajputana, Calcutta, 1908.

Erskine, Major K D, *Rajputana Gazetteers: The Western Rajputana States Agency and Bikaner Agency,* Allahabad, (1909) 1919.

Powlett, P W, *Gazetteer of the Bikaner State,* 1874, Bikaner: Government Press, 1932.

Walters, C K M, *Gazetteers of Marwar, Mallani and Jeysalmer,* Calcutta, 1877.

Memoirs, Travelogues and Ethnographic Works

Adams, A, *The Western Rajputana States: A Medico Topographical and General Account of Marwar, Sirohi and Jaisalmer,* Junior Army and Navy Stores, London, 1889.

Boileau, A H E, *Narrative of a Tour Through Raiwara Embracing the Princely States of Jaisalmer, Jodhpur and Bikaner in 1836,* Baptist Mission Press, Calcutta, 1837.

Crooke, W, *Tribes and Castes of the North Western Provinces and Oudh,* 4 Vols, Calcutta, 1894–96.

Cunningham, A, *The Ancient Geography of India,* Trubner and Co., London, 1871.

Enthoven, R E, *Tribes and Castes of the Bombay Presidency,* (4 Vols), Cosmo, New Delhi, (1921–23), 1975.

Festing, Gabrielle, *From the Land of Princes*, with a preface by Sir G Birdwood, London, Smith Elder and Co., London, 1904.

Forbes, Alexander Kinloch, *Hindoo Annals of the Province of Goozerat in Western India*, Edited with Historical Notes and Appendices by H G Robinson in two volumes, Low Price Publications, Delhi, (1921) 1997.

Hendley, Col. Thomas Holbein, *General Medical History of Rajputana*, Indian Medical Service, Calcutta, 1900.

Hobson - Jobson, *A Glossary of Colloquial Anglo-Indian Words and Phrases and of Kindred Terms Etymological, Historical, Geographical and Discursive*, Munshiram Manoharlal, Delhi, (1903) 1968.

Rose, H A, *Glossary of the Tribes and Castes of the Punjab and the North West Frontier Province*, 3 Vols, Civil and Military Gazette Press, Lahore, 1911–19.

Russell, R V, *The Tribes Castes of Central Provinces of India*, (4 vols.) Macmillan and Co. Ltd., London, 1916.

Tod, James, *Annals and Antiquities of Rajast'han or, the Central and Western Rajpoot States of India*, 2 vols, (Ed.) Douglas Sladen, Rupa and Co, New Delhi, (1829) 1997.

Tod, James, *Origin, Progress and Present State of Pindarees*, Nagpur Govt. Press, Nagpur, 1920.

Tod, James, *Travels in Western India Embracing a Visit to the Sacred Mounts of Jains and the Most Celebrated Shrines of Hindu Faith Between Rajputana and Indus with an Account of the Ancient City of Nehrwalla*, Munshiram Manoharlal, Delhi, (1839) 1997.

Secondary Sources

Agrawal, A, 'Mobility and Control among Nomadic Shepherds: The Case of the Raikas', *Human Ecology*, Vol. 22, No. 2 (Jun), 1994, 131–144.

———, *Greener pastures: Politics, Markets, and Community among a Migrant Pastoral People*, Duke University Press, Durham, North Carolina, USA and London, 1999.

———, *The Grass is Greener on the Other Side: A Study of the Raikas, Migrant Pastoralists of Rajasthan*, IIED, London, 1992.

———, 'I don't need it but you can't have it: Analysing Institutional Conflicts between Farmers and Pastoralists', Pastoral Development Network, Overseas Development Institute, London, 1994, 36–55.

Agrawal, Govind, *Churu Mandal ka Shodhpurna Itihas*, Lok Sanskriti Shodh Sansthan, Nagar-Shri, Churu, 1974.

Alam, Muzaffar and Sanjay Subrahmanyam (Eds.), *The Mughal State, 1526–1750*, OUP, Delhi, 1998.

Allchin, B and Allchin, F R, *The Rise of Civilization in India and Pakistan*, Cambridge, 1982.

Amin, Shahid, 'On retelling the Muslim Conquest of North India' in *History and the Present* (Eds.), Partha Chatterjee and Anjan Ghosh, Permanent Black, Delhi, 2002.

Appadurai, Arjun (Ed.), *The Social Life of Things: Commodities in Cultural Perspective*, Cambridge, 1986.

Aquil, Raziuddin and Partha Chatterjee (Eds.), *History in the Vernacular*, Permanent Black, Ranikhet, 2008.

Arnold, David and Ramachandra Guha (Eds.), *Nature, Culture and Imperialism: Essays on Environmental History of South Asia*, OUP, New Delhi, 1995.

Asopa, J N, *Origin of the Rajputs*, Bharatiya Vidyapith, Delhi, 1976.

Babb, Lawrence A, 'Rejecting Violence: Sacrifice and the Social Identity of Trading Communities', *Contributions to Indian Sociology*, 32; 1998, 387–407.

Babb, Lawrence, *Alchemies of Violence: Myths of Identity and Life of Trade in Western India*, Sage, New Delhi, 2004.

Babb, Lawrence, Varsha Joshi and Michael Meister (Eds.), *Multiple Histories: Culture and Society in the Study of Rajasthan*, Rawat Publishers, Jaipur, 2002.

Bajekal, Madhavi, 'The State and Rural Grain Market in Eighteenth Century Eastern Rajasthan', *IESHR*, 25(4), 1988, 443–473.

Banks, Marcus, 'Why Move? Regional and Long Distance Migrations of Gujarati Jains?' in *Migration: The Asian Experience, (Eds.)* Judith Brown and Rosemary Foot, Macmillan, London, 1994, 131–148.

Bannerjee, A C, *The Rajput States and British Paramountacy*, Delhi, 1980.

Barfield, Thomas J, *The Perilous Frontier- Nomadic Empires and China*, Basil Blackwell, Cambridge, Massachusetts, 1989.

Barhat, Shiv Dutt Dan, *Jodhpur Rajya ka Itihas*, 1753–1800, Jaipur, 1991.

Barrow, Ian J, *Making History, Drawing Territory: British Mapping in India c. 1756–1905*, OUP, New Delhi, 2003.

Barth, Frederik, *Nomads of South Persia, The Basseri Tribe of Khamseh Confederacy*, London, 1961.

Basu, Raj Sekhar, 'Rights over Wastelands' and New Narratives of the Paraiyan Past' (1860–1900), *SIH*, 24, 2, 2008, 265–293.

Bayly, C A, *Rulers, Townsmen and Bazaars: North Indian Society in the Age of British Expansion*, Cambridge, 1983.

————, 'State and Economy in India over Seven Hundred Years' *The Economic History Review*, New Series, Vol. 38, No. 4 (Nov), 1985, 583–596.

————, *Indian Society and the Making of the British Empire*, Indian Edition, Cambridge University Press, 1990.

————, 'Knowing the Country: Empire and Information in India', *MAS*, Vol. 27, No. 1, Special Issue: How Social, Political and Cultural Information Is Collected, Defined, Used and Analyzed (Feb) 1993, 3–43.

Beck, Brenda, *The Three Twins: The telling of a South Indian Folk Epic*, Indiana University Press, 1982.

Berland, Joseph, 'Servicing the ordinary folk: Peripatetic peoples and their niche in south Asia', in *Nomadism in South Asia,* (Eds.) Aparna Rao and Michael Casimir, OUP, 2003, 2008,105–124.

Bhadani, B L, 'Well Irrigation in Marwar in the 17th Century' *Shodh Patrika*, XL, 1, Jan–March, 1989, 54–70.

Bhadani, B L, 'Pastoral sector in the Economy of 17th Century Marwar', Paper presented at Second International Seminar on Rajasthan, Udaipur, 1991, 34–35.

Bhadani, B L, *Peasants, Artisans and Entrepreneurs: Economy of Marwar in the Seventeenth Century*, Rawat Publications, Jaipur and New Delhi, 1999.

Bhargava, V S, *Marwar and the Mughal Emperors, 1526–1748*, Munshiram Manoharlal, Delhi, 1966.

Bhargava, V S, *Rajasthan ke Itihas ka Sarvekshan*, College Book Depot, Jaipur, 1971.

Bharucha, Rustom, *Rajasthan: An Oral History: Conversations with Komal Kothari,* Penguin Books, New Delhi, 2003.

Bhati, Hari Singh, *Pugal Ka Itihas,* Bikaner, 1989.

Bhatnagar, Rasmi Dube, 'A Poetics of Resistance: Investigating the Rhetoric of Bardic Historians of Rajasthan' in *Muslims, Dalits and the Fabrications of History, Subaltern Studies XII*, (Eds.) Shail Mayaram, M S S Pandian and Ajay Skaria, Permanent Black, New Delhi, 2005, 224–279.

Bhattacharya, Neeladri, Pastoralists in a Colonial World', in *Nature, Culture and Imperialism: Essays on Environmental History of South Asia* (Eds.), David Arnold and Ramachandra Guha, OUP, Delhi, 1995, 49–85.

————,' 'Introduction', *SIH*; 14; 1998, 165–171.

————, 'Predicaments of Mobility: Peddlars and Itinerants in Nineteenth Century Northwestern India', in *Society and Circulation Mobile People and Itinerant Cultures in South Asia* 1750–1950 (Eds.), Claude Markovits, et al Permanent Black, Delhi, 2000, 163–214.

Bhattacharya, Sukumar, *The Rajput States and The East India Company*, Munshiram Manoharlal, New Delhi, 1972.

Bhukya, Bhangya, *Subjugated Nomads: The Lambadas under the Rule of Nizam*, Orient Blackswan, Delhi, 2010.

Blackburn, Stuart H and A K Ramanujan (Eds.), *Another Harmony: New Essays on Folklore in India*, University of California Press, Berkeley, 1986.

Blackburn, Stuart H, Peter J Claus, Joyce B Flueckiger and Susan S Wadley (Eds.), *Oral Epics in India*, University of California Press, Berkeley, 1989.

Blake, Stephen T, 'The Patrimonial Bureaucratic Empire of the Mughals', *JAS*, 39, 1979, 77–94.

Brara, R, 'Are grazing Lands "wastelands"? Some Evidence from Rajasthan', *EPW*, February 22, 1992, 411–418.

Brown, Mark, *Penal Power and Colonial Rule*, Routledge, 2014.

Busch, Allison, *Poetry of Kings: The Classical Hindi Literature of Mughal India*, OUP, NY, 2012.

Chandra, Satish, *Parties and Politics at Mughal Court*, Peoples Publishing House, New Delhi, 1979.

Chandra, Satish, Raghubir Singh and G D Sharma (Eds.), *Marwar under Jaswant Singh 1638–1768: Jodhpur Hukumat Ri Bahi*, Book Treasure, Jodhpur, 1993.

Chatterjee, Partha and Anjan Ghosh (Eds.), *History and the Present*, Permanent Black, Delhi, 2002.

Chattopadhyaya, B D, *Aspects of Rural Society and Rural Settlements in Early Medieval India*, K P Bagchi and Co., Calcutta, 1990.

Chattopadhyaya, B D, *The Making of Early Medieval India*, OUP, Delhi, 1998, 2012.

Choyal, Shiv Singh, 'Marwar Ke Dhai Ghar', *Maru Bharati*, vol.2, July, 1961.

Cohn, Bernard, 'Networks and Centres in the Integration of Indian Civilisation' in *An Anthropologist among Historians and Other Essays* (Eds.), B. Cohn and Ranajit Guha OUP, New Delhi, 1987, 78–88.

Cohn, Bernard, 'Regions Subjective and Objective' in *An Anthropologist among Historians and Other Essays* (Eds.), Bernard Cohn and Ranajit Guha, OUP, New Delhi, 1987, 100–136.

Cohn, Bernard 'The Census and Objectification in South Asia', in *An Anthropologist among Historians and Other Essays* (Eds.), B. Cohn and Ranajit Guha, OUP, New Delhi, 1987, 224–254.

Copland, Ian, *The British Raj and the Indian Princes: Paramountcy in Western India: 1857–1930*, Orient Longman, Bombay, 1982.

Dalal, Sukhbir Singh, *Jat Veer: Kshatriya Vanshtopanna Jatveero ki Kathayein*, Delhi, 1991.

Dandekar, Ajay 'Landscapes in Conflict: Flocks, Hero-stones, and Cult in early medieval Maharashtra', *SIH*; 7, 1991, 301–324.

Dangwal, DD 'State, Forests and Graziers in the hills of Uttar Pradesh: Impact of Colonial Forestry on Peasants, Gujars and Bhotiyas' *IESHR*; 34; 405–435, 1997.

Datta, Rajat, 'Commercialisation, Tribute, and the Transition from late Mughal to Early Colonial in India' *MHJ*; 6; 2003, 259–291.

Dawood, N I (Ed.), *Ibn Khaldun, An Introduction to History: Muqaddimah*, Tr Franz Rosenthal, Routledge and Kegan Paul, London, 1967.

Deshpande, Prachi, *Creative Pasts: Historical Memory and Identity in Western India 1700–1960*, Permanent Black, Ranikhet, 2007.

Devra G S L, 'Desertification and Problem of Delimitation of Rajputana Desert during the Medieval Period' in *Human Ecology*, Special Issue, no.7, 1999, 97–107.

_____,'A Rethinking on the Politics of Commercial Society in PreBritish India: From Mutsaddi to Marwari', Occasional Papers, NMML, New Delhi, 1987.

_____,'Nature and Incidence of Rokad Rakam (Non-agricultural Taxes) in the Land Revenue System of the Bikaner State (1650–1700 AD)', *Proceedings of Indian History Congress*, Vol. 37, 1976, 190–95.

_____, *Rajasthan ki Prashasnik Vyavastha*, Bikaner, 1991.

Devy, G N, *A Nomad Called Thief: Reflections on Adivasi Silence*, Orient Longman, New Delhi, 2006.

Digby, Simon, *Sufis and Soldiers in Aurangzeb's Deccan; Malfuzat-I Naqshbandiya*, Delhi, 2001.

Digby, Simon, *Warhorse and Elephant in Delhi Sultanate: A Study in Military Supplies*, OUP, Oxford, 1971.

Eaton, Richard, *The Rise of Islam and the Bengal Frontier, 1204–1760*, University of California Press, Berkeley, 1993.

Edney, Matthew H, *Mapping an Empire: The Geographical Construction of British India 1765–1843*, University of Chicago Press, 1999.

Embree, A T, 'Frontiers into Boundaries: From the Traditional to the Modern State', in *Realm and Region in Traditional India* (Ed.), R. G. Fox, Durham, 1977, 255–81.

Farooqui, Amar, *Smuggling as Subversion: Colonialism, Indian Merchants, and the Politics of Opium, 1790–1843*, Lexington Books, 2005.

Finnegan, R, *Oral Poetry: Its Nature, Significance and Social Context*, Cambridge University Press, Cambridge, 1977.

Fisher, Michael, *Indirect Rule in India: Residents and the Residency System*, OUP, Delhi, 1992.

Fox R G, *Kin, Clan, Raja and Rule: State-Hinterland Relations in Preindustrial India*, University of California Press, 1971.

Freitag, Jason, *Serving Empire Serving Nation: James Tod and the Rajputs of Rajasthan*, Brill, Leiden, 2009.

Fonseca, Isabel, *Bury me standing: The Gypsies And Their Journey*, Vintage, 2006.

Gehlot, J S, *History of Rajputana*, Jodhpur, 1937, Reprint. Rajasthan Sahitya Mandir, Jodhpur, 1980.

Gellner, Ernest, 'War and Violence', in *Anthropology and Politics: Revolutions in the Sacred Grove*, Oxford, 1995, 160–179.

Gilmartin, David, 'Cattle, Crime and Colonialism: Property as negotiation in north India', *IESHR*; 40; 2003, 33–56.

Gold, Ann G, *A Carnival of Parting: The Tales of King Bharthari and King Gopichand as Sung and Told by Madhu Natisar Nath of Ghatiyali, Rajasthan*, University of California Press, Berkeley, 1992.

Gold, Ann Grodzins and Bhoju Ram Gujar, 'Of Gods, Trees and Boundaries: Divine Conservation in Rajasthan', *Asian Folklore Studies*, Vol. 48, No. 2, 1989, 211–229.

Gold, Ann Grodzins and Bhojuram Gujar, *In Times of Trees and Sorrows: Nature, Power and Memory in Rajasthan*, Duke University Press, Durham, 2002.

Gold, Ann Grodzins, *Fruitful Journeys: The Ways of Rajasthani Pilgrims*, University of California Press, Berkeley, 1988.

Gold, Daniel and Ann Grodzins Gold, 'The Fate of the Householder Nath', *History of Religions*, Vol. 24, No. 2 (Nov.) 1984, 113–132.

Gommans, Jos, 'The Eurasian Frontier after the First Millenium A D: Reflections along the Fringe of Time and Space', *MHJ*, 1, 1998, 125–143.

Gommans, Jos, 'The Silent Frontier of South Asia, C. A.D. 1100–1800', *JWS*, Vol. 9,1, Spring, 1998, 1–23.

Gommans, Jos, *Mughal Warfare, Indian Frontiers and High Road to Empire*, 1500–1700, Routledge, London, 2002.

Gordon, Stewart, *The Marathas: 1600–1800*, The New Cambridge History of India II. 4, Cambridge University Press, 1993.

Guha, Ranajit, *Dominance without Hegemony: History and Power in Colonial India*, Harvard University Press, Cambridge, 1997.

Gupta, B L, *Trade and Commerce in Rajasthan in the Eighteenth Century*, Jaipur, 1991.

Habib, Irfan, *Agrarian Systems of Mughal India, 1556–1707*, OUP, Delhi, 1963, 1999.

Habib, Irfan, *An Atlas of the Mughal Empire*, OUP, Delhi, 1992.

Halbawachs, Maurice, *On Collective Memory* (Ed. and Tr.), Lewis Coser, University of Chicago Press, Chicago, 1992.

Hardiman, David, 'Usury, Dearth and Famine in Western India', *Past & Present*, No. 152 (Aug), 1996, 113–156.

Harlan, Lindsay, 'On Headless Heroes: Pabuji from the Inside out' in *Multiple Histories: Culture and Society in the Study of Rajasthan* (Eds.), Lawrence Babb et al, Rawat Publishers, Jaipur, 2002.

Harlan, Lindsay, *Religion and Rajput Women: The Ethic of Protection in Contemporary Narratives*, Munshiram Manoharlal, Delhi, 1994.

Hawley, John Stratton, *The Memory of Love: Surdas sings to Krishna*, Tr. and Introduction by John Stratton Hawley, OUP, 2009.

Henige, David, *The Chronology of Oral Tradition: Quest for a Chimera*, Clarendan Press, Oxford, 1974.

Hiltebeitel, A, *Draupadi among Rajputs, Muslims and Dalits: Rethinking India's Oral and Classical Epics*, Oxford University Press, New Delhi, 2001.

Hobsbawm, Eric and Terence Ranger, *The Invention of Tradition*, Cambridge University Press, 1983.

Hooja, Rakesh and Rajendra Hooja (Eds.), *Desert, Drought and Development*, Jaipur, 1999.

Hooja, Rima, *A History of Rajasthan*, Rupa and Co., New Delhi, 2006.

Ibrahim, Farhana, *Settlers, saints and Sovereigns: An Ethnography of State Formation in Western India*, Routledge, New Delhi, 2008.

Inden, R. "Ritual, Authority, and Cyclic Time in Hindu Kingship". In J.F. Richards (Ed.), *Kingship and Authority in South Asia*, Oxford University Press, Delhi 1998, 41–91.

Jain M S, *Concise History of Modern Rajasthan*, Vishwa Prakashan, Delhi, 1993.

Jain V K, *Trade and Traders in Western India AD 1000–1300*, Munshiram Manoharlal, Delhi, 1990.

Jain, K C, *Ancient Cities and Towns of Rajasthan: A study of Culture and Civilization*, Motilal Banarasidas, Delhi, 1972.

Jairazbhoy, Nazir, 'Music in Western Rajasthan: Stability and Change' in *Yearbook of the International Folk Music Council*, Vol 9, 1977, 50–66.

Jaisal, Hari Ballabh Maheshwari, *Jaisalmer Rajya ka Madhyakaleen Itihas*, Rajasthani Granthagar, 1997, Jodhpur.

Jodha N S, 'Common Property Resources and Rural Poor in Dry Regions of India' *EPW*, Vol. 21, No. 27 (Jul. 5), 1986, 1169–1181.

Kachchwaha, O P, *Famines in Rajasthan, (1900–1947 AD)*, Hindi Sahitya Mandir, Jodhpur, 1985.

Kamphorst, Janet, 'The Deification of South Asian Epic Heroes-Methodological implications' in *Epic Adventures: Heroic Narratives in the Oral Performance Traditions of Four Continents* (Eds.), Jan Janson and H. J. Maiereds, LIT Verlag, Muenster, 2004, 89–97.

Kamphorst, Janet, *In Praise of Death, History and Poetry in Medieval Marwar* (South Asia), Leiden University press, Leiden, 2008.

Kapur, Nandini Sinha, *State Formation in Rajasthan: Mewar during the Seventh-Fifteenth Centuries*, Manohar Publishers, Delhi, 2002.

———, 'The Bhils in the Historic Setting' in *Mobile and Marginalized Peoples: Perspectives from the Past* (Eds.), Rudolf C Heredia & Shereen Ratnagar, Manohar, New Delhi, 2003.

Kasturi, Malavika, *Embattled Identities: Rajput Lineages and the Colonial State in Nineteenth Century North India*, OUP, New Delhi, 2002.

Kavoori, Purnendu, *Pastoralism in Expansion: The Transhuming Herders of Western Rajasthan*, New Delhi, OUP, 1999.

Kela, Shashank, *A Rogue and a Peasant Slave: Adivasi Resistance 1800–2000*, Navayana, New Delhi, 2012.

Kelly, V, 'Ramdeo Pir and the Kamadiya Panth' in *Folk, Faith & Feudalism* (Eds), N.K. Singhi and R. Joshi, Rawat Publications, Jaipur, 1995.

Kerr, Ian J, *Building the Railways of the Raj: 1850–1900*, OUP, Delhi, 1997.

Khan, Dominique-Sila, 'Is God an untouchable? A case of Caste Conflict in Rajasthan', *Comparative Studies of South Asia, Africa and the Middle East*, Spring 1988, 18 (1): 21–29.

———, *Conversions and Shifting Identities: Ramdev Pir and the Ismailies in Rajasthan*, Manohar, New Delhi, 1997, 2003.

Khazanov, Anatoly M, *Nomads and the Outside World* (Tr. Julia Crookenden), University of Wisconsin Press, Madison, 1984.

———, 'Nomads in the History of Sedentary World' in *Nomads in the Sedentary World* (Eds.), Anatoly Khazanov and Andre Wink, Curzon Press, Surrey, 2001.

Khemani, Kishni, *Bharat ke Geye Sindhi Premakhyan*, Ajmer, 2003.

Khoury, Philip S and Joseph Kostiner (Eds.), *Tribes and State Formation in the Middle East*, I B Tauris and Co., London, 1990.

Kohler-Rollefson, Ilse, 'The Raika Dromedary breeders of Rajasthan: A Pastoral System in Crisis', *Nomadic Peoples*, 30: 1992, 74–83.

———, *Camel Karma: Twenty Years Among India's Camel Nomads*, Tranquebar Press, Chennai, 2014.

Kolff, D H A, 'Sanyasi Trader-Soldiers' and Comment by Stewart Gordon, *IESHR*, Vol VIII, No. 1, March, 1971, 213–220.

———, *Naukar, Rajput and Sepoy: The Ethnohistory of Military Labour Market in Hindustan 1450–1850*, Cambridge University Press, Cambridge, 1990.

Kothari, K., 'Performers, Gods, and Heroes in the Oral Epics of Rajasthan' in *Oral Epics in India*, (Eds.), S H Blackburn, P J Claus, J B Flueckiger and S S Wadley, California University Press, Berkeley, 1989, 103–17.

———, 'Musicians of the People: The Manganiyars of Western Rajasthan' in *Idea of Rajsthan: Explorations in Regional Identity*, Vol II, (Eds.), Karine Schomer et al., Manohar Publishers and American Institute of Indian Studies, Delhi, 2001, 205–237.

Kradin, Nikolay N, 'Nomadism, Evolution and World-Systems: Pastoral Societies in Theories of Historical Development', *JWSR*, viii, 1ii, fall, 2002, 368–388.

Kulke, Hermann, (Ed.) *The State in India: 1000–1750*, OUP, Delhi, 1995.

Kumar, Mayank, 'Situating the Environment: Settlement, Irrigation and Agriculture in Pre-colonial Rajasthan', *SIH* 24; 2008, 211–233.

Kurlansky, Mark, *Salt: A World History*, Vintage Books, London, 2003.

Lalas, Sitaram,'Madhyakalin Rajasthani Gadya Sahitya', in *Parampara* (Ed.) N S Bhati, Rajasthani Shodh Sansthan, Chopasani, 15–16, 1963, 237–253.

Lattimore, Owen, *Studies in Frontier History: Collected Papers*, OUP, 1962.

Leshnik LS and GD Sontheimer, *Pastoralists and Nomads in South Asia*, Otto Harrassowitz, Weisbaden, 1975.

Lodrick, Deryck O, 'Rajasthan as a Region: Myth or Reality' in *Idea of Rajasthan*, vol I, (Eds.), Karine Schomer et al., Manohar Publishers and American Institute of Indian Studies, New Delhi, 2001.

———, 'Symbol and Sustenance: Cattle in South Asian Culture' *Dialectical Anthropology*, 29/1, 2005, 61–84.

Lorenzen, David, 'Warrior Ascetics in Indian History' *Journal of the American Oriental Society*, 98, 1978, 61–75.

Ludden, David, 'History Outside Civilization and the Mobility of South Asia', *South Asia*, Vol. XVII, no. 1, 1994, 1–23.

Malik, Aditya, *Sri Devnarayan Katha*, South Asia Institute, Heidelberg University, Delhi Branch, DK Printerworld, Delhi, 2003.

———, *Nectar Gaze and Poison Breath: An Analysis and Translation of Rajasthani Oral Narrative of Dev Narayan*, OUP, New York 2005.

Maloo, Kamala, *The History of Famines in Rajputana, (1858–1900)*, Himanshu Prakashan, Udaipur, 1987.

Markovits, Claude, Jacques Pouchepadass and Sanjay Subrahmanyam (Eds.), *Society and Circulation: Mobile People and Itinerant Cultures in South Asia 1750–1950*, Permanent Black, Delhi, 2000.

Markovits, Claude, 'The Political Economy of Opium Smuggling in Early Nineteenth Century India: Leakage or Resistance?' in *Expanding Frontiers in South Asian and World History, Essays in Honour of John F Richards* (Eds.), Richard M Eaton, Munis D Faruqui, David Gilmartin and Sunil Kumar, Cambridge University Press, Delhi, 2013.

Mayaram, Shail, *Resisting Regimes: Myth, Memory and the shaping of a Muslim Identity*, OUP, Delhi, 1997.

————, 'Mughal State Formation: The Mewati Counter-perspective', *IESHR*, 1997, 34 (2), 169–197.

————, *Against History, Against State: Counterperspectives from the Margins*, Columbia University Press, New York, 2003.

————, 'Kings versus Bandits: Anti-Colonialism in a Bandit Narrative' in *JRAS*, Third Series, Vol. 13, No. 3 (Nov, 2003), 315–338.

Meghani, Jhaverchand, 'Elegiac "Chhand" and "Duha" in Charani Lore', *Asian Folklore Studies.* Vol 59, No. 1, 2000, 41–58.

Menon, Jaya, 'Mobility and craft production', in *Mobile and Marginal Peoples: Perspectives from the Past* (Eds.), Rudolf Heredia and Shereen Ratnagar, Manohar, New Delhi, 2003, 89–120.

Metcalf, Thomas R, *Ideologies of the Raj*, Indian Edition, Foundation Books, New Delhi, 1998, 2005.

Moxham, Roy, *The Great Hedge of India*, Harper Collins of India, New Delhi, 2001.

Mughal, M. Rafique, *Ancient Cholistan: Archeology and Architecture*, Ferozesons, Lahore, 1999.

Mukhia, Harbans, *The Mughals of India*, Blackwell Publishing, 2004, 2008.

Nahta, Agarchand, 'Rajasthani Baton evam Khyaton ki Parampara', in *Parampara*, (Ed.) N S Bhati, Rajasthani Shodh Sansthan, Chopasani, 11, 1961, 114–124.

Nigam, Sanjay, 'Disciplining and Policing the 'criminals by birth', (Part 1) The making of a colonial stereotype-The criminal tribes and castes of North India, *IESHR*, 27, 2, 1990, 131–164.

Nora, Pierre, *Realms of Memory: Rethinking the French Past*, 3 Vols, Columbia University Press, New York, 1996.

Ojha, G H, *Jodhpur Rajya Ka Itihas*, Part I and II, Vedic Yantralaya, Ajmer, 1938, 1941.

Ojha, G H (Ed.), *Muhnot Nainsi ki Khyat,* (Tr. Ram Narayan Dugar), Rajasthani Granthagar, Jodhpur, (1934), 2010.

Ong, Walter, *Orality and Literacy: The Technologizing of the Word*, Routledge, 2002.

Pannikar, K N, *British Diplomacy in North India: A Study of the Delhi Residency*, 1803–1857, Delhi, 1968.

Peabody, Norbert,'Tod's Rajasthan and the Boundaries of Imperial Rule in Nineteenth Century India', *MAS* 30, no 1, 1996, 185–220.

Peabody, Norbert, *Hindu Kingship and Polity in Precolonial India*, Cambridge University Press, 2006.

Phillips, David T, *Peoples on the Move, Introducing the Nomads of the World*, Piquant, 2001.

Pinch, William, *Warrior Ascetics and Indian Empires*, Cambridge University Press, New Delhi, 2006.

Radhakrishna, Meena, *Dishonoured by History: Criminal Tribes and British Colonial Policy*, Orient Longman, Delhi, 2001.

Ramusack, Barbara N, *The Indian Princes and their States*, Cambridge University Press, Cambridge, 2004.

Rana, R P, *Rebels to Rulers: The Rise of Jat Power in Medieval India*, Manohar, New Delhi, 2006.

Rangarajan, Mahesh, *Fencing the Forest: Conservation and Ecological Change in India's Central Provinces 1860–1914*, OUP, Delhi, 1996.

Rao, V N, David Shulman, Sanjay Subrahmanyam, *Textures of Time: Writing History in South India 1600–1800*, Permanent Black, Delhi, 2001.

Ratnagar, Shereen 'Pastoralism as an issue in Historical Research, *SIH* 7, 1991,181–193.

Ray, Rajatkanta, 'Colonel James Tod, Munhota Nainsi and the Rajputs', *IHR*, Vol. XXV, No2, June, 1999, 100–111.

Reu, V N, *Glories of Marwar and the Glorious Rathors*, Archeological Department, Jodhpur, 1943.

Richards, J F, *The Mughal Empire*, Cambridge, Indian Edition, Foundation Books, New Delhi,1993, 2000,

Robbins, Paul, 'Nomadization in Rajasthan, India: Migration, Institutions, and Economy' *Human Ecology*, Vol. 26, No. 1 (Mar) 1998, 87–112.

Rosin, Thomas R, 'The Tradition of Groundwater irrigation in North western India', *Human Ecology*, 21(1), 1993.

Roy, A B and S R Jakhar, 'Late Quaternary Drainage and Disorganization and Migration and Extinction of Vedic Saraswati' in *Current Science*, Vol 81, no 9, November, 2001.

Roy, Tirthankar, 'Changes in Wool Production and Use in Colonial India', Gokhle Institute, 2002.

Rudolph, Susanne H. and Lloyd Rudolph, *Essays on Rajputana: Reflections on History, Culture, Administration*, Concept Publishing Company, New Delhi, 1984.

Sahai, Nandita Prasad, *Politics of Patronage: The State, Society and Artisans in Early Modern Rajasthan*, OUP, New Delhi, 2006.

Sahel, K L, 'Rajasthani Sahitya mein Ghoro ka Sthan', *Shodh Patrika*, Vol 14, No 1, 1964, 65–69.

Sahel, K L, 'Rajasthan ki Varsha Sambandhi Kahavatein', *Maru Bharati*, Vol IV, 1956, 8–18.

Sarda, H B, *Ajmer: Historical and Descriptive*, Fine Art Printing Press, Ajmer, 1941.

Satya, Laxman D, *Ecology, Colonialism and Cattle: Central India in the Nineteenth Century*, OUP, New Delhi, 2004.

Scott, James C, *Seeing Like a State: How Certain Schemes to Improve Human Condition Have Failed*, New Haven and London, 1998.

Scott, James, *The Art of Not being Governed: An Anarchist History of Upland South East Asia*, Yale University Press, New Haven, 2009.

Sethia, Madhu, *Rajput Polity: Warriors, Peasants And Merchants*, 1700–1800, Rawat Publishers, Jaipur, 2003.

Shah, P R, *Raj Marwar During Paramountcy: A study in Problems and Policies up to 1923*, Sharda Publishing House, Jodhpur, 1982.

Shankar, Girija, *Marwari Vyapari*, Krishna Jansevi, Bikaner, 1988.

Sharma, Dasaratha, *Rajasthan through the Ages*, Rajasthan State Archives, Bikaner, 1966.

————, *Lectures on Rajput History and Culture*, Motilal Banarassidas, Delhi, 1970.

————, *Early Chauhan Dynasties*, (Second Revised Edition), New Delhi, 1975.

Sharma, G D, 'The Marwaris: The Economic Foundations of an Indian Capitalist Class', in *Business Communities of India: A Historical Perspective* (Ed.), Dwijen Tripathi, Delhi, 1984, 187–207.

————, *Rajput Polity: A Study of Politics and Administration of Marwar State, 1638–1749*, Manohar Publishers, Delhi, 1977.

Sharma, G N, *Social Life in Medieval Rajasthan*, Shivlal Agrawal and Sons, Agra, 1968.

Sharma, Vijay Paul, Ilse Kohler-Rollefson, John Morton, *Pastoralism in India: A Scoping Study*, Centre for Management in Agriculture, Indian Institute of Management, Ahmedabad and League for Pastoral Peoples.

Sheikh, Samira, 'Alliance, Genealogy and Political Power: The Chudasamas of Junagarh and Sultans of Gujarat' *MHJ*, 11, 1, 2008, 29–61.

————, *Forging a Region: Sultans, Traders and pilgrims in Gujarat, 1200–1500*, OUP, N. Delhi, 2010.

Singh, Chetan, 'Conformity and Conflict: Tribes and the Agrarian Systems', in *The Mughal State 1526–1750*, (Eds.) Muzaffar Alam and Sanjay Subrahmanyam, OUP, Delhi, 1998.

————, 'Forests, pastoralists and Agrarian society', in *Nature, Culture, and Imperialism: Essays on Environmental History of South Asia* (Eds.), David Arnold and Ramchandra Guha, OUP, New Delhi, 1995. .

Singh, Dilbagh, 'The Role of Mahajans in the Rural Economy in Eastern Rajasthan during the 18 Century', *Social Scientist*, Vol. 2: 22, 1974, 20–31.

————, *State, Landlords and Peasants: Rajasthan in the Eighteenth Century*, Manohar Publishers, Delhi, 1990.

Singh, Karni, *Relations of House of Bikaner with the Central Powers*, Munshiram Manoharlal, Delhi, 1974.

Singha, Radhika, 'Settle, mobilize, verify: Identification practices in colonial India', *SIH*, 16, 2, 2000, 151–198.

Sinha, Nitin, 'Mobility, control and criminality in early colonial India, 1760s–1850s' *IESHR*; 45; 2008, 1–33.

Skaria, Ajay, 'Being jangli: The politics of wildness', *SIH*; 14; 1998, 193–215.

————, *Hybrid Histories: Forests, Frontiers and Wildness in Western India*, OUP, Delhi, 1999.

Smith, J D, 'An Introduction to the Language of the Historical Documents from Rajasthan', *MAS*, 9: 1975, 433–64.

————, 'The Singer or the Song? A reassessment of Lord's "oral theory"' *Man*, 12, 1977, 141–53.

————, 'Scapegoats of the gods: the ideology of the Indian epics' in *Oral Epics in India* (Eds.), S H Blackburn, et al., California University Press, Berkeley, 1989, 176–93.

————, 'Worlds apart: Orality, Literacy, and the Rajasthani Folk-Mahabharata', *Oral Traditions* 5/1, 1990, 3–19.

————, *The Epic of Pabuji, A Study, Transcription and Translation*, Katha Books, New Delhi, 1991.

————, 'Winged words revisited: diction and meaning in Indian epic', *Bulletin of the School of Oriental and African Studies*, 62/ 2, 1999, 267–305.

Sneath, David, *The Headless State: Aristocratic Orders, Kinship Society and Misrepresentations of Nomadic Inner Asia*, Columbia University Press, New York, 2007.

Snodgrass, J G, 'The centre cannot hold: Tales of hierarchy and poetic composition from modern Rajasthan', *Journal of the Royal Anthropological Institute*, 10/2, 2004. 261–285.

Somani, R V, *History of Jaisalmer*, Rawat Publications, Jaipur, 1990.

Sontheimer, Gunther-Dietz, *Pastoral Deities in Western India* (Tr. Anne Feldhaus), OUP, Delhi, 1993.

————, 'The Dhangar: A Nomadic Community in a Developing Agricultural Environment' in *Nomadism in South Asia* (Eds.), Aparna Rao and Michael Casimir, OUP, New Delhi, 2003, 2008, 364–398.

Sopher, David, 'Indian Pastoral Castes and Livestock Ecologies' in *Pastoralists and Nomads in South Asia*, (Eds.), L S Leshnik and G D Sontheimer, Otto Harrasowittz, Weibaden, 1975.

Sreenivasan, Ramya, 'Honoring the Family: Narratives and Politics of Kinship in Pre-colonial Rajasthan' in *Unfamiliar Relations: Family and History in South Asia* (Ed.), Indrani Chatterjee, Permanent Black, Delhi, 2004, 46–73.

————, 'The 'Marriage' of 'Hindu' and 'Turak' Medieval Rajput Histories of Jalor', *MHJ*, 7, (1), 2004, 87–108.

————, *The Many Lives of a Rajput Queen: Heroic Pasts in India c.1500–1900*, Permanent Black, Delhi, 2007.

————, 'Faith and Allegiance in Mughal Era', in *Religious Interactions in Mughal India* (Eds.), Vasudha Dalmia and Munis Faruqui, OUP, New Delhi, 2014, 157–191.

————, 'Warrior Tales in Hinterland Courts in North India, *c* 1370–1550' in *After Timur Left: Culture and Circulation in Fifteenth Century North India* (Eds.), Francesca Orsini and Samira Sheikh, OUP, New Delhi, 2014, 111–130.

Srivastava, Vinay Kumar, 'The Rathor Rajput Hero of Rajasthan: Some Reflections on John Smith's The Epic of Pabuji', *MAS*, Vol. 28, No. 3, July, 1994, 589–614.

————, *Religious Renunciation of a Pastoral People*, OUP, 1997, Delhi.

Stein, Burton, 'The Segmentary State: Interim Reflections', in *The State in India :1000–1750* (Ed.), Hermann Kulke, OUP, Delhi, 1995.

Stern, Henri, 'Power in Traditional India: Territory, Caste and Kinship in Rajasthan', in R G Fox (Ed.), *Realm and Region in Traditional India*, New Delhi, 1977.

Subrahmanyam, Sanjay, *Merchants, Markets and the State in Early Modern India*, OUP, Delhi, 1990.

————, 'Portfolio Capitalists and Political Economy of Early Medieval India', *IESHR*, 23, 1988, 358–77.

Swami, Narottam Das (Ed.), *Dhola Maru ra Duha*, Rajasthani Granthagar, Jodhpur, 1995.

Taft, Frances, 'Honor and Alliance: Reconsidering Mughal-Rajput Marriages' *The Idea of Rajasthan, Explorations in Regional Identity*, Vol II, (Eds.) K. Schomer et al, Manohar and American Institute of American Studies, New Delhi, 2001, 217–242.

————, 'Royal Marriages in Rajasthan', *Contributions to Indian Sociology*, 7: 1973, 64–80.

Talbot, Cynthia, 'Becoming Turk the Rajput Way: Conversion and Identity in an Indian Warrior Narrative', *MAS,* 43, no. 01 (2009): 211–243.

————, 'Contesting Knowledges in Colonial India: The Question of Prithviraj Raso's Historicity' in *Colonial and Modern Constructions of the Past: Essays in the Honour of Thomas R Trautmann*, Yoda Press, New Delhi, 2011, 171–212.

————, 'Justifying Defeat : A Rajput Perspective on the Age of Akbar,' *JESHO*, 55 (2012): 329–68.

Tambs-Lyche, Harald, *Power, Profit, and Poetry: Traditional Society in Kathiawar, Western India*, Manohar, Delhi, 1997.

Tambs-Lyche, Harald, 'Marriage and Affinity among Virgin Goddesses', in *The Feminine Sacred in South Asia. Le sacré au feminine en Asie du Sud.* (Ed.), Harald Tambs-Lyche, Manohar, Delhi, 2004, 63–87.

Thapar, R, 'The Image of the Barbarian in Early India'. *Cultural Pasts: Essays in Early Indian History*, OUP, 2000, 235–271.

————, 'The Historian and the Epic' in *Cultural Pasts: Essays in Early Indian History*, OUP, 2000, 613–630.

————, 'Genealogical Patterns as Perceptions of the Past' in *Cultural Pasts: Essays in Early Indian History*, OUP, 2000, 709–754.

————, *The Past Before Us: Historical Traditions of Early North India*, Permanent Black, 2013.

Thompson, Gordon R, 'The Carans of Gujarat: Caste Identity, Music, and Cultural Change' *Ethnomusicology*, Vol. 35, No. 3 (Autumn) 1991, 381–391.

Thompson, Gordon, 'The Barots of Gujarati-Speaking Western India: Musicianship and Caste Identity', *Asian Music*, Vol. 24, No. 1 (Autumn, 1992 - Winter, 1993), 1–17.

Timberg, T, 'The Study of a Great Marwari Firm: 1860–1914', *IESHR*, Vol VIII, No 1, March 1971, 264–283.

Timberg, T, *The Marwaris: From Traders to Industrialists*, Vikas Publishers, New Delhi, 1978.

Vansina, J, *Oral Tradition: A Study in Historical Methodology*, tr. H M Wright, Transaction Publishers, New Brunswick, NJ, 1965, 2006.

Vansina, J, *Oral Tradition as History*, James Currey Ltd, Oxford, First edition: 1985, 1997.

Vaudeville, Charlotte, *Barahmasa in Indian Literatures*, Motilal Banarasidass, Delhi, 1986.

———, 'Leaves from the Desert: The Dhola Maru ra Duha An Ancient Ballad of Rajasthan' in *Myths, Saints and Legends in Medieval India*, compiled by Vasudha Dalmia, OUP, New Delhi, 1996, 277–278.

Vyas, R P, *Role of Nobility in Marwar*, Jain Bros, New Delhi, 1969.

Wadley, Susan S, 'Dhola A North Indian Folk Genre', *Asian Folklore Studies*, Vol 42 No. 1, 1983, 3–25.

Wadley, Susan S, "A Bhakti Rendition of Nala-Damayantī: Todarmal's 'Nectar of Nal's Life'" in *International Journal of Hindu Studies*, Vol. 3, No. 1 (Apr), 1999, 27–56.

———, *Raja Nal and the Goddess: The North Indian Epic of Dhola in Performance*, Indiana University Press, Bloomington, 2004.

Westphal-Hellbusch, S, 'Changes in the meaning of ethnic names as exemplified by the Jat, Rabari, Bharvad and Charan in North-western India, in *Pastoralists and nomads in South Asia* (Eds.), L.S. Leshnik, G.D. Sontheimer, Otto Harrassowitz, Wiesbaden, 1975, 117–138.

Wickett, Elizabeth, *The Epic of Pabuji ki par in Performance*, World Oral Literature Project, Published by University of Cambridge, 2001.

Wink, Andre, *Al-Hind. The Making of the Indo Islamic World, The Slave Kings and the Islamic Conquest 11th to 13th centuries*, Vol II, OUP, NY, 1999.

———, 'India and the Turko-Mongol Frontier' in *Nomads in the Sedentary World* (Eds.), Anatoly Khazanov and Andre Wink, Surrey, 211–233, 2001.

Zaidi, S Inayat, 'Akbar and the Rajput Principalities: Integration into Empire', in *Akbar and his India* (Ed.), Irfan Habib, OUP, Delhi, 1997.

Zaidi, Sunita, 'The Mughal State and tribes in seventeenth century Sind', *IESHR*, 26, 343–362, 1989.

Zeigler Norman, 'Marvari Historical Chronicles', *IESHR*, January–March, Vol XIII, No 1, 219–250, 1976.

———, 'Some Notes on Rajput Loyalties during the Mughal Period 'in *The Mughal State* (Eds.), Alam and Subrahmanyam, OUP, 2006, 168–213.

————, *Action, Power and Service in Rajasthani Culture: A Social History of the Rajputs of the Middle Period Rajasthan*, Unpublished PhD Thesis, University of Chicago, Illinois, 1973.

————, 'Evolution of the Rathor State of Marvar: Horses, Structural Change and Warfare', in Karine Schomer et al (Eds.), *The Idea of Rajasthan: Explorations in Regional Identity*, Vol. II, Manohar Publishers and American Institute of Indian Studies, New Delhi, 1994.

Appendix-I

Jodhpur King List

Rao Siha/Sihoji	1212–1273
Rao Asthan	1273–1292
Rao Dhuhadji	1292–1308
Rao Raipal	1308–1313
Rao Kanha	1313–1323
Rao Jalhansi	1323–1328
Rao Chhadoji	1328–1344
Rao Tidoji	1344–1357
Rao Tribhuvansi	
Rao Mallinath	
Rao Salkha	1357–1378
Rao Viramde	1378–1383
Rao Chundo/Chundaji	1384–1423
Rao Kanha	1423–1424
Rao Sata	1424–1427
Rao Rinmal	1427–1438
Rao Jodhaji	1453–1489 (Spent intervening years recovering Mandor)
Rao Satal	1489–1492
Rao Sujo	1492–1515
Rao Gango	1515–1532
Rao Malde	1532–1562

Rao Chandrasen	1562–1581 (In exile as Jodhpur was taken over by Akbar)
Motaraja Udai Singh	1583–1595
Raja Sur Singh	1595–1619
Raja Gaj Singh I	1619–1638
Maharaja Jaswant Singh I	1638–1678
Maharaja Ajit Singh	1707–1724
Maharaja Abhay Singh	1724–1749
Maharaja Ram Singh	1749–1751
Maharaja Bakhat Singh	1751–1752
Maharaja Beejay Singh	1752–1793
Maharaja Bheem Singh	1793–1803
Maharaja Man Singh	1803–1843
Maharaja Takhat Singh	1843–1873
Maharaja Jaswant Singh II	1843–1895
Maharaja Sardar Singh	1895–1911
Maharaja Sumer Singh	1911–1918

Source: Vishveswar Nath Reu, *Marwar ka Itihas*, Maharaja Man Singh Pustak Prakash, Jodhpur, 1940, 1999, Vol II, 678–681.

Appendix-II

Bikaner King List

(Starts with Rao Bika who was Rao Jodha's son)

Rao Bikaji	1465–1504
Rao Nara	1504–1505
Rao Lunkaran	1505–1526
Rao Jaitsi	1526–1542
Rao Kalyan Mal	1542–1573
Raja Rai Singh	1573–1612
Raja Dalpat Singh	1612–1614
Raja Surat Singh	1614–1631
Raja Karan Singh	1631–1667
Maharaja Anup Singh	1667–1698
Maharaja Sarup Singh	1698–1700
Maharaja Sujan Singh	1700–1736
Maharaja Zorawar Singh	1736–1746
Maharaja Gaj Singh	1746–1787
Maharaja Raj Singh II	1787
Maharaja Surat Singh	1787–1828
Maharaja Ratan Singh	1828–1851
Maharaja Sardar Singh	1851–1872
Maharaja Dungar Singh	1872–1881
Maharaja Ganga Singh	1887–1943

Source: Vishweswar Nath Reu, *Marwar ka Itihas*, Maharaja Man Singh Pustak Prakash, Jodhpur, 1940, 1999, Vol II, 682–683.

Appendix-III

Jaisalmer King List

Bhati*	
Bachrao*	
Vijayrao*	
Majamrao*	
Mangalrao*	
Kehar I*	
Tanu*	
Vijayrao Chudalo*	
Devraj*	
Mundh*	
Vachu*	
Dushajh*	
Rawal Jaisal*	
Rawal Salbahan	1190–1200
Rawal Bijal	1200
Rawal Kalhan	1200–1219
Rawal Chachigadev I	1219–1251
Rawal Karan	1215–1283
Rawal Lakhansen	1300–1304
Rawal Punyapal	1304–05
Rawal Jaitsi I	1305–1313
Rawal Mulraj I	1313–1315

Rawal Duda	1319–1331
Rawal Gharsi	1343–1361 (Jaisalmer occupied by Khalji forces)
Rawal Kehar II	1361–1391
Rawal Lakshman	1396–1436
Rawal Vairsi	1436–1446
Rawal Chachigadev II	1447–1470
Rawal Devidas/Devkaran	1470–1506
Maharawal Jaitsingh II	1506–1527
Rawal Karamsi	1527–1528
Rawal Lunkaran	1528–1550
Rawal Maldev	1550–1561
Rawal Harraj	1561–1577
Rawal Bhim Singh	1577–1613
Rawal Kalyan Das	1613–1627
Rawal Manohar Das	1627–1650
Rawal Ramchandra	1650
Rawal Sabal Singh	1650–1659
Rawal Amar Singh	1659–1701
Rawal Jaswant Singh	1701–1707
Rawal Budh Singh	1707–1721
Rawal Akhay Singh	1722–1761
Maharawal Mulraj II	1761–1820
Maharawal Gaj Singh	1820–1846
Maharawal Ranjit Singh	1846–1864
Maharawal Bairisal	1864–1891
Maharawal Shalivahan	1891–1914

Source: Harivallabh Maheshwari Jaisal, *Jaisalmer Rajya ka Itihas*, Gwalior, 1999.

*Dates uncertain. *Source*: *Munhata Nainsi ri Khyat*, Vol II Ed. Badri Prasad Sakriya, 9–10.

Index